D1753252

FIRE FOLLOWING EARTHQUAKE

EDITED BY
Charles Scawthorn
John M. Eidinger
Anshel J. Schiff

Technical Council on Lifeline Earthquake Engineering
Monograph No. 26
January 2005

Published by the American Society of Civil Engineers

Library of Congress Cataloging-in-Publication Data

Fire following earthquake / Charles Scawthorn, John M. Eidinger, and Anshel J. Schiff, editors.
   p. cm. -- (Technical Council on Lifeline Earthquake Engineering monograph ; no. 26)
  Includes bibliographical references and index.
  ISBN 0-7844-0739-8
   1. Lifeline earthquake engineering. 2. Fires. I. Scawthorn, Charles. II. Eidinger, John M. III. Schiff, Anshel J. IV. Monograph (American Society of Civil Engineers. Technical Council on Lifeline Earthquake Engineering) ; no. 26.

TA654.6.F57 2004
624.1'762--dc22
                                                                                                   2004049717

The material presented in this publication has been prepared in accordance with generally recognized engineering principles and practices, and is for general information only. This information should not be used without first securing competent advice with respect to its suitability for any general or specific application.

The contents of this publication are not intended to be and should not be construed to be a standard of the American Society of Civil Engineers (ASCE) and are not intended for use as a reference in purchase of specifications, contracts, regulations, statutes, or any other legal document.

No reference made in this publication to any specific method, product, process, or service constitutes or implies an endorsement, recommendation, or warranty thereof by ASCE.

ASCE makes no representation or warranty of any kind, whether express or implied, concerning the accuracy, completeness, suitability, or utility of any information, apparatus, product, or process discussed in this publication, and assumes no liability therefore.

Anyone utilizing this information assumes all liability arising from such use, including but not limited to infringement of any patent or patents.

ASCE and American Society of Civil Engineers—Registered in U.S. Patent and Trademark Office.

Photocopies: Authorization to photocopy material for internal or personal use under circumstances not falling within the fair use provisions of the Copyright Act is granted by ASCE to libraries and other users registered with the Copyright Clearance Center (CCC) Transactional Reporting Service, provided that the base fee of $25.00 per chapter plus $.50 per page is paid directly to CCC, 222 Rosewood Drive, Danvers, MA 01923. The identification for this book is 0-7844-0739-8/05/$25.00 + $.50 per page. Requests for special permission or bulk copying should be addressed to Permissions & Copyright Department, ASCE.

Copyright © 2005 by the American Society of Civil Engineers.
All Rights Reserved.
Manufactured in the United States of America.

Preface

Earthquakes are a continuing threat to mankind. The usual earthquake hazards of ground shaking, liquefaction, landslide, surface faulting and tsunami all combine to lead to destruction to life and property. Many building codes and other types of standards and guidelines already exist to deal with these hazards.

But much less well understood is the risk due to Fire Following Earthquake (FFE).

While building codes require seismic bracing of sprinklers and, in some cases, alternative water supplies, there are no guidelines, standards or codes that specifically address FFE. Since the 1980s there has been a growing interest in this issue amongst a relatively small number of researchers, practicing engineers and fire department officials. Recognizing the need to expand the knowledge base of the FFE hazard to a wider audience, TCLEE gathered a group of leading experts in the field to prepare this monograph. These experts have attempted to bring together in this monograph, a thoughtful and thorough examination of the many aspects of the FFE hazard.

While this monograph provides an excellent source of information about qualitative and quantitative methods to deal with the FFE hazard, this monograph should not be considered a Guideline, Standard or Code. There is still much to learn about the best ways to address the risk due to fire following earthquake. Perhaps at this point in time, it would be best for each community to develop a "defense-in-depth" approach to deal with this hazard. Not all communities need to adopt all the mitigation strategies. This Monograph should provide suitable source material for a rational approach to help build better disaster-resistant communities.

Charles Scawthorn, F. ASCE
John M. Eidinger, M. ASCE
Anshel J. Schiff, M. ASCE
Editors

Cover Picture

This aerial photograph of Balboa Boulevard in Los Angeles taken shortly after the 1994 Northridge earthquake illustrates the interaction of earthquakes, lifelines and fire. Due to a large landslide on nearly flat ground, water, sewer, and gas pipelines under Balboa Boulevard failed. The result was torrents of water from broken pipes flowing down the street. Gas escaping under pressure from the gas main was ignited by a passing pickup truck. The radiant heat from the resulting gas flare caused ignitions and fires that destroyed several homes adjacent to the street. Two aerial optical-fiber cables supported on poles were also destroyed. The optical-fiber cables contained major trunks connecting central offices, the failure of which disrupted telephone communications. Responding firefighters found no pressure in the hydrants, due to the broken water mains, and resorted to drafting water from backyard swimming pools.

TCLEE

The purpose of the Technical Council on Lifeline Earthquake Engineering (TCLEE) is to advance the state-of-the-art and practice of lifeline earthquake engineering. Members of TCLEE participate in the development of guidelines, pre-standards and standards for the seismic design and construction of lifelines. They encourage lifeline organizations, industries and associated manufacturers, associations and professionals to consider earthquakes and their impacts in the planning, design, emergency planning and operations of lifelines. They also serve as a primary resource for establishing broad consensus on lifeline issues; identify and prioritize research needs related to lifeline planning, design, construction and operation; and support and/or conduct programs for education and technology transfer on lifeline seismic issues.

TCLEE includes seven technical committees: Earthquake Investigations, Electrical Power and Communications; Gas and Liquid Fuels; Port; Seismic Risk; Transportation and Water and Wastewater. Membership to technical committees is open to practicing engineers, academics, manufacturers, public policy planners and other interested people. While many of TCLEE members are also members of ASCE, the wide variety of disciplines involved with lifelines allows that the committees are equally open to mechanical, electrical and other types of engineers, scientists, academics and planners.

Given the wide range of issues dealing with Fire Following Earthquake, the people who contributed to this Monograph include members from many of TCLEE's committees.

TCLEE Monograph Series

These publications may be purchased from ASCE, telephone 1-800-548-ASCE (2723), World Wide Web http://www.asce.org .

Kiremidjian, Ann, Editor, Recent Lifeline Seismic Risk Studies, TCLEE Monograph No. 1, 1990.

Taylor, Craig, Editor, Seismic Loss Estimates for a Hypothetical Water System, TCLEE Monograph No. 2, 1991.

Schiff, Anshel J., Editor, Guide to Post Earthquake Investigations of Lifelines, TCLEE Monograph No. 3, 1991.

Cassaro, Michael, Editor, Lifeline Earthquake Engineering, Proceedings of the 3rd U.S. Conference, TCLEE Monograph No. 4, August, 1991.

Ballantyne, Donald, Editor, Lifeline Earthquake Engineering in the Central and Eastern U.S., TCLEE Monograph No. 5, September, 1992.

O'Rourke, Michael, Editor, Lifeline Earthquake Engineering, Proceedings of the 4th U.S. Conference, TCLEE Monograph No. 6, August, 1995.

Schiff, Anshel J. and Buckle, Ian, Editors, Critical Issues and State of the Art on Lifeline Earthquake Engineering, TCLEE Monograph No. 7, October, 1995.

Schiff, Anshel J., Editor, Northridge Earthquake: Lifeline Performance and Post-Earthquake Response, TCLEE Monograph No. 8, August 1995.

McDonough, Peter W., Seismic Design for Natural Gas Distributors, TCLEE Monograph No. 9, August, 1995.

Tang, Alex and Schiff, Anshel, Editors, Methods of Achieving Improved Seismic Performance of Communication Systems, TCLEE Monograph No. 10, September 1996.

Schiff, Anshel, Editor, Guide to Post-Earthquake Investigation of Lifelines, TCLEE Monograph No. 11, July, 1997.

Werner, Stuart D., Editor, Seismic Guide to Ports, TCLEE Monograph No. 12, March, 1998.

Taylor, C, Mittler, E., Lund, L., Overcoming Barriers: Lifeline Seismic Improvement Programs, TCLEE Monograph No. 13, September 1998.

Schiff, A.J., Editor, Hyogoken-Nanbu (Kobe) Earthquake of January 17, 1995 – Lifeline Performance, TCLEE Monograph No. 14, September, 1998.

Eidinger, John M., and Avila, Ernesto A., Editors, Guidelines for the Seismic Upgrade of Water Transmission Facilities, TCLEE Monograph No. 15, January, 1999.

Elliott, William M. and McDonough, Peter, Editors, Optimizing Post Earthquake lifeline System Reliability, Proceedings of the 5th National Conference on Lifeline Earthquake Engineering, TCLEE Monograph No. 16, September, 1999.

Tang, Alex K, Editor, Izmit (Kocaeli) Earthquake of August 17, 1999 Including Duzce Earthquake of November 12, 1999 – Lifeline Performance, TCLEE Monograph No. 17, March, 2000.

Schiff, Anshel J., and Tang, Alex K. Editors, Chi-Chi Taiwan Earthquake of September 21, 1999 – Lifeline Performance, TCLEE Monograph No. 18, 2000.

Eidinger, John M., Editor, Gujarat (Kutch) India M 7.7 Earthquake of January 26, 2001, and Napa M 5.2 Earthquake of September 3, 2000, TCLEE Monograph No. 19. June 2001.

McDonough, Peter W., Editor, The Nisqually Washington Earthquake of February 2001, Lifeline Performance, TCLEE Monograph No. 20, February 2002.

Taylor, Craig E. and Van Marke, Erik H., Editors, Acceptable Risk Process – Lifelines and Natural Hazards, TCLEE Monograph No. 21, March 2002.

Heubach, William F., Editor, Seismic Screening Checklists for Water and Wastewater Facilities, TCLEE Monograph No. 22, 2003.

Edwards, Curtis L., Editor, Atico Peru M_w 8.4 Earthquake of June 23, 2001, TCLEE Monograph No. 23, October 2002.

Lund, L.V., Editor. Lifeline Performance of El Salvado Earthquake of January 13 and February 13, 2001, TCLEE Monograph No. 24, 2003

Beavers, J.E., Editor, Advancing Mitigation Technologies and Disaster Response for Lifeline System: Proceedings of the sixth U.S. Conference and Workshop on Lifeline Earthquake Engineering, TCLEE Monograph No. 25, August 2003.

TCLEE Publications

Duke, C. Martin, Editor, The Current State of Knowledge of Lifeline Earthquake Engineering, 1977.

Dowd, Munson, Editor, Annotated Bibliography on Lifeline Earthquake Engineering, 1980.

Smith, D. J. Jr., Editor, Lifeline Earthquake Engineering: The Current State of Knowledge, 1981.

Hall, William, Editor, Advisory Notes on Lifeline Earthquake Engineering, 1983.

Nyman, Douglas, Editor, Guidelines for the Seismic Design of Oil and Gas Pipeline Systems, TCLEE Committee on Gas and Liquid Fuel Lifelines, 1984.

Coopers, James, Editor, Lifeline Earthquake Engineering: Performance, Design and Construction, 1984.

Eguchi, Ron, Crouse, C. B., Lifeline Seismic Risk Analysis - Case Studies, 1986.

Cassaro, Michael and Martinez-Romero, E., Editors, The Mexico Earthquakes - 1985 Factors Involved and Lessons Learned, 1986.

Wang, Leon R. L. and Whitman, Robert, Seismic Evaluation of Lifeline Systems - Case Studies, 1986.

Cassaro, Michael and Cooper, James, Editors, Seismic Design and Construction of Complex Civil Engineering Systems, 1988.

Werner, Stuart D. and Dickenson, Stephen E., Editors, Hyogoken-Nanbu (Kobe) Earthquake of January 17, 1995: A Post-Earthquake Reconnaissance of Port Facilities, TCLEE Committee on Ports and Harbors Lifelines, 1996.

ASCE Manual

Schiff, Anshel J., Editor, Guide to Improved Earthquake Performance of Electric Power Systems, ASCE Manual 96.

Table of Contents

ABSTRACT II
PREFACE III
TCLEE IV
TCLEE MONOGRAPH SERIES V
TCLEE PUBLICATIONS VI
LIST OF TABLES XIV
LIST OF FIGURES XVI

1 INTRODUCTION 1
   1.1 THE PROBLEM OF FIRE FOLLOWING EARTHQUAKES 1
   1.2 OUTLINE OF THIS MONOGRAPH 3
   1.3 ACKNOWLEDGEMENTS 5
   1.4 REFERENCES 6

2 HISTORICAL FIRES FOLLOWING EARTHQUAKES 8
   2.1 INTRODUCTION 8
   2.2 THE 1906 SAN FRANCISCO, CALIFORNIA EARTHQUAKE AND FIRES 10
      2.2.1 Seismological and Overall Damage Aspects *10*
      2.2.2 Ignitions and Fire *11*
      2.2.3 Water System Performance *16*
   2.3 THE 1923 TOKYO EARTHQUAKE AND FIRES 19
      2.3.1 Seismological and Overall Damage Aspects: *20*
      2.3.2 Ignitions and Fire: *21*
      2.3.3 Damage to Water Supply Components: *23*
      2.3.4 Impacts of Water Supply Damage on Firefighting *23*
      2.3.5 Summary of the 1923 Kanto earthquake *24*
   2.4 THE 1989 LOMA PRIETA, CALIFORNIA EARTHQUAKE AND FIRES 24
      2.4.1 Seismological and Overall Damage Aspects *25*
      2.4.2 Water Supply and Fire Protection in the City of San Francisco *25*
      2.4.3 Ignitions *28*
      2.4.4 Marina Fire and use of the PWSS *29*
      2.4.5 Summary *31*
   2.5 THE 1994 NORTHRIDGE, CALIFORNIA EARTHQUAKE AND FIRES 31
      2.5.1 Seismological and Overall Damage Aspects *31*
      2.5.2 Fire Service Resources *31*
      2.5.3 Water System Performance *32*
      2.5.4 Ignitions and Fires *33*
      2.5.5 North Balboa Blvd. Fire *35*
   2.6 THE 1995 HANSHIN (KOBE), JAPAN EARTHQUAKE AND FIRES 39
      2.6.1 Seismological and Overall Damage Aspects *39*
      2.6.2 Fire Service Resources *39*
      2.6.3 Ignitions and Fires *39*
      2.6.4 Water System Performance *41*
      2.6.5 Comparative Analysis: Hanshin and Northridge events *42*
   2.7 THE 1999 MARMARA, TURKEY, EARTHQUAKE AND FIRES 43
      2.7.1 Seismological and Overall Damage Aspects *43*
      2.7.2 Tupras petroleum refinery fire *43*
   2.8 THE 2000 NAPA EARTHQUAKE 46
   2.9 SUMMARY OBSERVATIONS 47
   2.10 CREDITS 47
   2.11 REFERENCES 47

3 NON EARTHQUAKE FIRES AND CONFLAGRATIONS 51

- 3.1 INTRODUCTION 51
- 3.2 HISTORIC CONFLAGRATIONS 56
  - 3.2.1 *64 Rome* 56
  - 3.2.2 *1666 London* 56
  - 3.2.3 *1871 Chicago* 58
- 3.3 EARLY 20TH C. CONFLAGRATIONS 61
  - 3.3.1 *1904 Baltimore* 61
  - 3.3.2 *1923 Berkeley* 61
- 3.4 WW2 EXPERIENCE 66
  - 3.4.1 *Hamburg* 66
  - 3.4.2 *Dresden* 66
  - 3.4.3 *Japan* 66
  - 3.4.4 *Analysis* 68
- 3.5 1991 EAST BAY HILLS 68
  - 3.5.1 *Overview* 68
  - 3.5.2 *Fire Spread* 69
  - 3.5.3 *Water System Performance in the Oakland Hills Firestorm* 75
  - 3.5.4 *Fire Flow Guidelines for Water Systems in High Fire Risk Areas* 81
  - 3.5.5 *Alternative Strategies for High Fire Risk Hillside Areas* 82
- 3.6 1993 SOUTHERN CALIFORNIA 85
- 3.7 HIGHRISE FIRES 86
  - 3.7.1 *1988 First Interstate Bank Building* 86
  - 3.7.2 *1991 Meridian Plaza* 88
  - 3.7.3 *1993 World Trade Center* 88
  - 3.7.4 *2001 World Trade Center* 89
  - 3.7.5 *Highrise fire following earthquake* 90
- 3.8 INDUSTRIAL EXPERIENCE 93
- 3.9 SUMMARY OBSERVATIONS 95
- 3.10 CREDITS 97
- 3.11 REFERENCES 97

4 ANALYSIS AND MODELING 99

- 4.1 INTRODUCTION 99
- 4.2 IGNITION 104
  - 4.2.1 *Ignition Rate* 104
  - 4.2.2 *Locations and Causes of Ignitions* 109
  - 4.2.3 *Further Examination of Causes of Ignitions* 112
- 4.3 FIRE REPORT AND RESPONSE 115
- 4.4 FIRE GROWTH AND SPREAD 115
  - 4.4.1 *The Hamada Model* 115
  - 4.4.2 *The TOSHO Model* 121
- 4.5 FIRE RESPONSE AND SUPPRESSION 124
- 4.6 FINAL BURNT AREA 127
  - 4.6.1 *HAZUS* 127
  - 4.6.2 *URAMP^{SC}* 130
  - 4.6.3 *SERA* 135
  - 4.6.4 *RiskLink* 140
- 4.7 INSURANCE ASPECTS OF FIRE FOLLOWING EARTHQUAKE 144
- 4.8 EXAMPLE: VANCOUVER, B.C. 144
- 4.9 CREDITS 150
- 4.10 REFERENCES 150

5 FIRE DEPARTMENT OPERATIONS FOLLOWING EARTHQUAKE 154

	5.1	INTRODUCTION	154
	5.2	NATURE OF THE PROBLEM	154
	5.3	DEPARTMENTAL COMMAND CONSIDERATIONS	156
	5.4	CONFLAGRATION TACTICS	160
	5.5	PLANNING	164
	5.6	REFERENCES AND BIBLIOGRAPHY	168

6 POTABLE WATER SYSTEMS 169

- 6.1 POTABLE WATER SYSTEM FACILITIES AND FUNCTIONS 170
  - *6.1.1 Water Sources* *171*
  - *6.1.2 Potable Water Transmission* *171*
  - *6.1.3 Potable Water Treatment* *172*
  - *6.1.4 Water Storage* *172*
  - *6.1.5 Water Distribution* *172*
  - *6.1.6 Emergency Operation Centers, Administration, Operation and Maintenance Facilities* *173*
- 6.2 EARTHQUAKE PERFORMANCE OF POTABLE WATER SYSTEMS 173
  - *6.2.1 Source Facility Performance* *173*
  - *6.2.2 Transmission Facility Performance* *174*
  - *6.2.3 Water Treatment Facility Performance* *177*
  - *6.2.4 Water Distribution Facility Performance* *178*
  - *6.2.5 Water Storage Facility Performance* *189*
  - *6.2.6 Administration, Operations and Maintenance Facility Performance* *192*
- 6.3 DIRECT IMPACTS OF POTABLE WATER SYSTEM EARTHQUAKE PERFORMANCE 192
- 6.4 SECONDARY IMPACTS OF POTABLE WATER SYSTEM EARTHQUAKE PERFORMANCE 194
- 6.5 POTABLE WATER SYSTEM EARTHQUAKE MITIGATION STRATEGIES 194
- 6.6 SUMMARY 197
- 6.7 CREDITS 197
- 6.8 REFERENCES 197

7 GAS SYSTEMS AND FIRE FOLLOWING EARTHQUAKE 199

- 7.1 INTRODUCTION 199
- 7.2 UNDERSTANDING THE NATURAL GAS DISTRIBUTION SYSTEM 199
  - *7.2.1 Natural Gas Basics* *199*
  - *7.2.2 The Natural Gas Delivery System* *200*
- 7.3 EARTHQUAKE PERFORMANCE OF NATURAL GAS SYSTEMS 202
- 7.4 CAUSES OF POST-EARTHQUAKE IGNITION OF NATURAL GAS 203
- 7.5 OPTIONS TO REDUCE GAS FIRES FOLLOWING EARTHQUAKES 205
  - *7.5.1 Customer Actions to Improve Natural Gas Safety in Earthquakes* *205*
  - *7.5.2 Utility Actions to Improve Natural Gas Safety in Earthquakes* *206*
- 7.6 COMMUNITY PREPAREDNESS AND RESPONSE PLANNING 207
- 7.7 KEY ISSUES 209
- 7.8 REFERENCES 209

8 ELECTRIC POWER AND FIRE FOLLOWING EARTHQUAKE 215

- 8.1 GENERAL DESCRIPTION OF POWER SYSTEMS 215
  - *8.1.1 Power Generation* *215*
  - *8.1.2 Power Transmission System* *215*
  - *8.1.3 Power Distribution System* *216*
  - *8.1.4 Ancillary Facilities and Functions* *216*
  - *8.1.5 Emergency and Backup Power Systems* *218*
- 8.2 EARTHQUAKE PERFORMANCE OF POWER SYSTEMS 218
  - *8.2.1 Overview* *218*
  - *8.2.2 Power Transformer Sudden Pressure Relay* *219*
  - *8.2.3 Distribution System Damage* *219*
  - *8.2.4 Service Center Communications* *219*

	8.2.5	*Utility PBX Congestion*	220
	8.2.6	*Emergency Power*	220
	8.2.7	*Reliable Source of Fuel for Emergency Vehicles*	221
8.3		DIRECT IMPACTS OF POWER SYSTEM PERFORMANCE ON FIRE FOLLOWING EARTHQUAKE	221
8.4		SECONDARY IMPACTS OF POWER SYSTEM DISRUPTION ON FIRE FOLLOWING EARTHQUAKE	221
8.5		KEY ISSUES RELATED TO FIRE FOLLOWING EARTHQUAKE	222
8.6		MITIGATION STRATEGIES	222
	8.6.1	*Seismic Shutdown of Selected Feeders*	222
	8.6.2	*Coordinated Restoration of Electrical Service*	222
	8.6.3	*Enhance Communications to Service Centers*	222
	8.6.4	*Fuel for Emergency Vehicles*	223
	8.6.5	*Seismic Qualification of Substation Equipment*	223
	8.6.6	*Seismic Upgrade for Key Power System Structures*	223
	8.6.7	*Department of Energy Emergency Fuel Supply*	223
8.7		REFERENCES	223

9 COMMUNICATION SYSTEMS AND FIRE FOLLOWING EARTHQUAKE ... 224

9.1		GENERAL DESCRIPTION OF COMMUNICATION SYSTEMS	224
	9.1.1	*Public Switch Network*	224
	9.1.2	*Wireless Network*	226
	9.1.3	*Private Networks*	227
	9.1.4	*Mobile Radio Systems*	227
	9.1.5	*Fire-Service Communications*	228
	9.1.6	*Other Means of Communication*	229
9.2		EARTHQUAKE PERFORMANCE AND VULNERABILITY	230
	9.2.1	*Overview*	230
	9.2.2	*Earthquake Performance of Wireline Systems*	230
	9.2.3	*Earthquake Performance of Wireless Systems*	231
	9.2.4	*Earthquake Performance of Mobile Radio Systems*	232
	9.2.5	*Public Service Answering Points*	232
9.3		DIRECT IMPACTS OF COMMUNICATION SYSTEM PERFORMANCE ON FIRE FOLLOWING EARTHQUAKE	232
	9.3.1	*Inability to Access Emergency Services Through 911 System*	232
	9.3.2	*Dispatch Function Disrupted by Damage to Radio Communication Systems*	232
9.4		SECONDARY IMPACTS OF COMMUNICATION SYSTEM DISRUPTION ON FIRE FOLLOWING EARTHQUAKE	232
9.5		SUMMARY OF KEY ISSUES RELATED TO FIRE FOLLOWING EARTHQUAKE	233
9.6		MITIGATION STRATEGIES	233
	9.6.1	*Communication System Congestion*	233
	9.6.2	*Batteries for Handsets*	234
	9.6.3	*Add Call Box Outside of Firehouses*	234
	9.6.4	*Seismic Installation of Base-Station Radios*	235
	9.6.5	*Congestion on Radio Systems*	235
	9.6.6	*Backup Power for Cell Site*	235
	9.6.7	*Assure that Important Structures will have Adequate Seismic Performance*	235
	9.6.8	*Review and Seismically Upgrade Central Office Building HVAC*	235
	9.6.9	*Cell Site Installation*	235
	9.6.10	*Adding External Emergency Utility Hookups at Central Offices*	235
	9.6.11	*Requesting Service Priority for Critical Facilities*	235

10 ROADWAY SYSTEMS AND FIRE FOLLOWING EARTHQUAKE ... 237

10.1		OVERVIEW	237
10.2		DIRECT IMPACTS OF THIS ROADWAY LIFELINE ON FFE ISSUES	237
10.3		SECONDARY IMPACTS ON OTHER LIFELINES	238
	10.3.1	*Road and Bridge Damage also Damages Water Lines*	238
	10.3.2	*Road and Bridge Damage also Damages Communication Lines*	238

    10.3.3   Road and Bridge Damage Delays or Prevents the Delivery of Services and Supplies 238
    10.3.4   Utility Service Crews Unable to Restore Services 238
10.4   EARTHQUAKE VULNERABILITIES OF ROADWAY SYSTEMS 239
    10.4.1   Past Earthquake Performance with Direct Impacts on FFE 239
    10.4.2   Past Earthquake Performance with Secondary Impacts on other Lifelines 244
10.5   MITIGATION STRATEGIES 245
10.6   KEY ISSUES RELATED TO FIRE FOLLOWING EARTHQUAKE 245
10.7   RESEARCH NEEDS AND EDUCATIONAL MATERIALS ON FFE 246
    10.7.1   Seismic Risk Assessment to Roadway System and its Components, and Mitigation 246
    10.7.2   Use of Technology for Emergency Response Operations 246
10.8   REFERENCES 247

11   METHODS FOR MITIGATING FIRES FOLLOWING EARTHQUAKES 249

11.1   REDUCTION OF DAMAGE 250
11.2   AUTOMATIC SUPPRESSION 257
11.3   CITIZEN SUPPRESSION 259
11.4   COMMUNICATIONS 261
11.5   FIRE DEPARTMENT ASSETS 261
11.6   WATER 263
11.7   PORTABLE WATER SUPPLY SYSTEM (PWSS) 264
11.8   CREDITS 267
11.9   REFERENCES 267

12   HIGH PRESSURE WATER SUPPLY SYSTEMS 268

12.1   INTRODUCTION 268
12.2   SAN FRANCISCO AUXILIARY WATER SUPPLY SYSTEM (AWSS) 268
12.3   VANCOUVER, B.C. DEDICATED FIRE PROTECTION SYSTEM (DFPS) 272
12.4   BERKELEY SALTWATER FIRE SYSTEM (SFS) 275
    12.4.1   System Performance Goals 275
    12.4.2   System Layout 276
    12.4.3   Coverage of the System 277
    12.4.4   Earthquake Hazard 280
    12.4.5   Pipeline Design 283
    12.4.6   Reliability Analysis 285
12.5   KYOTO 289
12.6   CREDITS 293
12.7   REFERENCES 293

13   SEISMIC RETROFIT STRATEGIES FOR WATER SYSTEM OPERATORS 295

13.1   INTRODUCTION 295
13.2   PIPE REPLACEMENT OR PIPE BYPASS? 295
13.3   ABOVE GROUND ULTRA LARGE DIAMETER HOSE PIPE BYPASS 297
13.4   ABOVE GROUND LARGE DIAMETER FLEXIBLE HOSE PIPE BYPASS 302
13.5   HYDRAULICS 302
13.6   CREDITS 303
13.7   REFERENCES 303

14   BENEFITS AND COSTS OF MITIGATION 304

14.1   INTRODUCTION 304
14.2   BENEFIT-COST ANALYSIS 305
14.3   SEISMIC HAZARD CURVES 306
14.4   SEISMIC PERFORMANCE OF EXISTING SYSTEMS 307
    14.4.1   Physical Damages to Facility or Component 307
    14.4.2   Restoration Times and Service Outages 309
    14.4.3   Economic Losses to Customers From Service Outages 310

    14.4.4   Casualties *311*
    14.4.5   Example: Total Damages and Losses Before Upgrade *312*
    14.4.6   Seismic Performance of Upgraded System *314*
14.5   BENEFIT-COST ANALYSIS AND RESULTS 315
    14.5.1   Benefits (Avoided Damages and Losses) *316*
    14.5.2   Net Present Value Calculations *316*
    14.5.3   Benefit-Cost Results *317*
    14.5.4   Interpretation of Benefit-Cost Analysis *318*
14.6   LIMITATIONS OF BENEFIT-COST ANALYSIS AND OTHER DECISION MAKING APPROACHES 319
    14.6.1   Limitations of Benefit-Cost Analysis *319*
    14.6.2   Deterministic Damage and Loss Estimates: Scenario Studies *319*
    14.6.3   Probabilistic Damage and Loss Estimates: Threshold Studies *320*
14.7   EXAMPLE APPLICATION 321
14.8   CONCLUSIONS 324
14.9   REFERENCES 324

GLOSSARY 326

INDEX 330

List of Tables

Table 2-1 U.S. 20th century post-earthquake ignitions 9
Table 2-2 Fire Departments affected by the 1994 Northridge Earthquake 32
Table 2-3 Fire Following the January 17, 1994 Northridge Earthquake 34
Table 2-4 Water usage, Balboa Blvd. Fire 38
Table 2-5 Post-Earthquake fire ignitions, Jan. 17, 1995 Hanshin Earthquake 40
Table 2-6 Hanshin and Northridge Earthquakes: comparative analysis 42
Table 3-1 Selected North American large fires and conflagrations 52
Table 3-2 Selected North American wildland fires 55
Table 3-3. Principal factors contributing to conflagrations in US and Canada 1914-1942 95
Table 3-4 Hazardous materials incidents in past earthquakes 96
Table 4-1 Fires following U.S. earthquakes - 1906 – 1989, used for Eq (3) 106
Table 4-2 Approximate No. of SFED per ignition, vs. MMI 107
Table 4-3 General sources of ignition, LAFD data, Northridge Earthquake 110
Table 4-4 Property use for 77 LAFD Earthquake-related fires 4:31 TO 24:00 hrs, January 17, 1994 110
Table 4-5 Forms of heat ignition, 77 LAFD Earthquake-related fires 4:31 TO 24:00 hrs, January 17, 1994 111
Table 4-6 Material first ignited for 77 LAFD Earthquake-related fires 4:31 To 24:00 hrs, January 17, 1994 112
Table 4-7 Constants for eqns. 4-1 to 4-6 117
Table 4-8 HAZUS fire following earthquake estimates, New York City 129
Table 4-9 HAZUS validation study – fire following earthquake 129
Table 4-10 Ignition rates by occupancy 130
Table 4-11 Building densities, by occupancy 131
Table 4-12 Mean area burned as a function of building density 134
Table 4-13 Mean area burned as a function of building density BRUSH ZONE CONDITIONS 134
Table 4-14 Number of fire engines within service area 136
Table 4-15 Number of fire engines outside of service area available for mutual aid 136
Table 4-16 Value of structures burned, x $1,000,000 137
Table 4-17 Scenario events, fires and losses (Loss in Year 2000 C$millions, % of total value at risk) 149
Table 5-1 Approximate No. of SFED per ignition, vs. MMI 165
Table 6-1 Seismic upgrade programs, various US water utilities 169
Table 6-2 Seismic upgrade programs, various Japanese water utilities 169
Table 6-3 US water system seismic upgrades – forecast 170
Table 6-4 Reported statistics for main pipe and service lateral repairs 185
Table 6-5 Buried pipe vulnerability functions 187
Table 6-6 Ground shaking - constants for fragility curve 188
Table 6-7 Permanent ground deformations - constants for fragility curve 189
Table 6-8 Earthquake characteristics for tank database 190
Table 6-9 Complete tank database 191
Table 6-10 Fragility curves, tanks, as a function of fill level 191

Table 6-11 Fragility curves, welded steel tanks, as a function of fill level and anchorage 192
Table 7-1 Earthquake ignitions in selected earthquakes 210
Table 7-2 Valves and alarm devices that assist in limiting natural gas to customer facilities 211
Table 7-3 Approximate costs for actions to limit natural gas flow after earthquakes 212
Table 7-4 Summary of community actions to improve natural gas safety 213
Table 10-1 Summary of street and freeway closures for the Loma Prieta and Northridge Earthquakes (source: Perkins et al., 1997) 243
Table 12-1 System service areas and coverage ratios 280
Table 12-2 Scenario earthquakes 280
Table 12-3 Ground shaking levels (median levels) 281
Table 12-4 System reliability. baseline case 287
Table 12-5 Alignment alternatives 287
Table 12-6 Alignment reliabilities 288
Table 12-7 Selected historic earthquakes affecting Kyoto, Japan 289
Table 14-1 Expected annual frequency of earthquakes 306
Table 14-2 Example fragility curve for a water treatment plant 308
Table 14-3 Damage estimates before upgrade 309
Table 14-4 Restoration time estimate 310
Table 14-5 Estimated reduction in EBMUD service area gross regional product (GRP) per system day of water service interruption ($000) 313
Table 14-6 (a) Scenario damages and losses: before upgrade 314
Table 14-6 (b). Expected Annual Damages and Losses: Before Upgrade 314
Table 14-7(a) Scenario damages and losses: after upgrade 315
Table 14-7(b). Expected Annual Damages and Losses: After Upgrade 315
Table 14-8 Benefit-cost results 318
Table 14-9 Probabilistic damage and loss estimates 320
Table 14-10 Fire ignitions for San Diego after scenario earthquakes 321
Table 14-11 Structures burned – Rose Canyon M 6.5 scenario earthquake 322
Table 14-12 Costs, benefits and benefit cost ratios for various levels of seismic upgrade 323

List of Figures

Figure 2-1 MMI map 1906 San Francisco earthquake 10
Figure 2-2 San Francisco 1906 Fire – Ignitions at 6:00 am (top) and Spread at 7:15 am and 8:30 am (bottom), Central Business District 12
Figure 2-3 San Francisco 1906 Fire – final extents in solid line; many collapsed masonry buildings in shaded area. 13
Figure 2-4 San Francisco 1906 collapsed building, fire in background 13
Figure 2-5 View down Sacramento St. during San Francisco 1906 fire.....................................14
Figure 2-6 San Francisco destruction by fire: View looking west from Telegraph Hill, showing unburned houses on summit of Russian Hill. St. Francis Roman Catholic Church, with excellent brick walls in foreground. 14
Figure 2-7 Damage to the Transmission Pipelines Serving San Francisco, 1906....................17
Figure 2-8 Failure of Pilarcitos 30 inch pipeline. Pipe telescoped and was thrown sideways on the wooden bridge crossing Small Frawley Canyon. San Andreas fault crosses pipe diagonally. 18
Figure 2-9 (a) map of the 1904 San Francisco water system (b) showing final burnt area 19
Figure 2-10 Kanto and Chubu Districts showing areas principally affected by liquefaction 20
Figure 2-11 Distribution on earthquake intensity in Tokyo..21
Figure 2-12 Outbreak of fires in the city of Tokyo (Kanto earthquake of 1923)......................22
Figure 2-13 Firespread in Central Tokyo, showing direction and hourly progress of flame front..22
Figure 2-14 San Francisco AWSS Plan 25
Figure 2-15 San Francisco PWSS..26
Figure 2-16 MWSS Pipe Breaks, Marina District, 1989 Loma Prieta Earthquake 26
Figure 2-17 1989 Loma Prieta Earthquake, Marina Fire, SFFD deployment.......................30
Figure 2-18 Locations Earthquake-related fires, 1994 Northridge Earthquake 33
Figure 2-19 LAFD Fires, 4:31 to 24:00 hrs, January 17, 1994 34
Figure 2-20 LAFD Incident Response Types, 1308 Incidents 4:31 to 24:00 hrs, January 17, 1994 35
Figure 2-21 North Balboa Blvd. Incident, 1994 Northridge Earthquake (top) Aerial View, (bottom) Surface View 36
Figure 2-22 LAFD Deployment 5:35 am, No. Balboa Blvd., 1994 Northridge Earthquake 38
Figure 2-23 Aerial View Burnt Area, 1995 Kobe Earthquake..41
Figure 2-24 Tupras Refinery damage due to fire following earthquake (a) Burnt and collapsed tank; (b) Top of stack collapsed into unit, severely damaging heater unit. Top of stack is in foreground of photo. 45
Figure 3-1 Great Fire of London 1661 57
Figure 3-2 Ruins of Chicago Fire of 1871 59
Figure 3-3 Baltimore Fire, 1904 59
Figure 3-4 Chicago Fire 1871: (l) Extent of Burnt area; (r) path and time of flamefront arrival..60
Figure 3-5 Berkeley 1923...62

Figure 3-6 Berkeley 1923 – North of U.C. Berkeley Campus 62
Figure 3-7 Map of 1923 Berkeley Hills Water System, Fire Spread and Response..............63
Figure 3-8 Time Line of Structures Burned, 1991 East Bay Hills Fire 69
Figure 3-9 Rate of Fire Spread, 1991 East Bay Hills Fire 70
Figure 3-10 Point of Origin and Progress of 1991 East Bay Hills Fire: contours show origin at 11am, and progressive extent of fire at 11:15, 11:30, 12:00, 1:00pm, 2:00 pm, 3:00 pm, 4:00 pm and 5:00 pm..75
Figure 3-11 1991 East Bay Hills fire - narrow roads in Hills area, and burnt abandoned automobiles indicative of the congestion during the evacuation............84
Figure 3-12 West limit of burnt area, 1991 East Bay Hills fire, showing some houses burnt, others surviving...85
Figure 3-13 First Interstate Bank Building burned-out floors..87
Figure 3-14 Highrise Building Fire Protection System Schematic 93
Figure 4-1 Fire following earthquake process..100
Figure 4-2 Fire department Operations Time Line..101
Figure 4-3 Post-earthquake ignition rate, based on San Fernando and later data – (t) Ignitions per thousand SFED vs. MMI, and (b) Ignitions per million sq. ft. floor area vs. PGA (No. data = 59) 107
Figure 4-4 Ignition Model for Equation (3) 109
Figure 4-5 Fire Ignition rate as a Function of Season and Time of Day.............................113
Figure 4-6 Ignition Rates – Summer and Winter Fuzzy Weights 114
Figure 4-7 Ignition Rates –Time of Day Fuzzy Weights..114
Figure 4-8 Ignition Rates – Summer Evening Fuzzy Weights..114
Figure 4-9 Santa Monica Fire Department Incident Reports January 17, 1994 Northridge Earthquake, 0431-2400...116
Figure 4-10 Typical Fire Spread...118
Figure 4-11 Fire spread estimated with Hamada equations, vs. fire spread observed in U.S. non-earthquake conflagrations 119
Figure 4-12 Probability of Crossing Firebreak..121
Figure 4-13 Flowchart of the HAZUS Earthquake Loss Estimation Methodology 128
Figure 4-14 Overall Approach for Estimating Potential Losses and Benefits Associated with Mitigation Measures 132
Figure 4-15 Example Natural Hazard Disclosure (Fire) Map, Orange County CA 135
Figure 4-16 Fire Following Earthquake Time History Analysis - EBMUD.....................139
Figure 4-17 Ignitions and deployment of fire engines for San Francisco - RiskLink Simulation..141
Figure 4-18 Active and controlled fires for San Francisco - RiskLink Simulation............141
Figure 4-19 Geographical Distribution of Peak Ground Acceleration for the 1906 San Francisco Earthquake (RiskLink Simulation)..142
Figure 4-20 Geographical Distribution of Mean Burnt Area for the 1906 San Francisco Earthquake (RiskLink Simulation)..143
Figure 4-21 Scenario Earthquake Events, Analysis of Fire Following Earthquake, Lower Mainland, British Columbia (Source: ICLR, 2001).................................146
Figure 4-22 Mean number of ignitions, M 9.0 Subduction zone event 149

Figure 5-1 Map of liquefaction zones and landslide-prone areas, Disaster Readiness and Response Plan, City of Seattle. Available online at http://www.ci.seattle.wa.us/emergency_mgt/resources/plans.htm 167
Figure 6-1. Schematic Diagram of a Typical Water System..170
Figure 6-2. Wrinkled Los Angeles Aqueduct. This pipe remained in service 175
Figure 6-3. Wrinkled 48" Balboa Boulevard Pipeline. This pipe leaked........................... 175
Figure 6-4. Buried Pipeline Damage in the San Francisco Bay Area, 1989 Loma Prieta Earthquake 179
Figure 6-5. City of Kobe Customer Water Service Restoration Time................................181
Figure 6-6. Damaged Pipeline, Kobe 1995 181
Figure 6-7. Damaged AC Pipeline, Adapazari, Turkey 1999 182
Figure 6-8. Damaged Steel Transmission Pipe, Chi-Chi Earthquake 1999 183
Figure 6-9. Pipe Fragility Model, For Wave Propagation 186
Figure 6-10. Pipe Fragility Model, For Permanent Ground Deformations........................186
Figure 6-11. Serviceability Index vs. average break rate...193
Figure 7-1 Typical Gas Meter Arrangement 200
Figure 7-2 Natural Gas Delivery System (provided by Pacific Gas & Electric Company)..201
Figure 10-1 Bridges and Tunnels Across San Francisco Bay 239
Figure 10-2 The Copper River Bridge #2 was closed after the 1964 Great Alaska earthquake due to approach settlement and shifted truss spans (NAS, 1973). 240
Figure 10-3 Landslide blocking Route 17 after the Loma Prieta Earthquake.....................241
Figure 10-4 Excavation under Balboa Blvd (after the 1994 Northridge earthquake) resulted in several homes burning down, one of which can be seen in the background (Lund)..244
Figure 11-1 Fire following earthquake process, with mitigation opportunities in bold 250
Figure 11-2 Strapping water heater (a) plan view, showing 'plumber's tape' strap around water heater and attachment to framing using rigid tubing; (b) elevation view; (c) photo of actual water heater – note 'plumber's tape' strap around 'belly' of the water heater 252
Figure 11-3 (a) How to shut off a gas valve, and (b) actual gas service to a home, showing pressure regulator (flat round plate-like object) and valve, with 'gas wrench' wired to pipe, and writing on side of house showing location 252
Figure 11-4 Example of automatic gas shut-off valve (a) typical installation, (b) close-up of valve 253
Figure 11-5 (top) diagram, and (bottom) example of cripple wall bracing using plywood 254
Figure 11-6 Watsonville in 1989 Loma Prieta earthquake, examples of houses fallen off their foundations 255
Figure 11-7 Vancouver, B.C. Central Business District, with overhead electric lines – an obvious fire hazard, especially following an earthquake......................................256
Figure 11-8 Test of intumescent paint 259
Figure 11-9 San Francisco Fire Department's NERT (Neighborhood Emergency Response Team home page (Source: http://www.sfnert.org/) 259

Figure 11-10 (a) Example of single bay unreinforced masonry fire station, built in 1913, and (b) strengthened in 1990s per indicated structural reinforcement scheme. 262
Figure 11-11 PWSS system, consisting of Hose Tender towing HydroSub portable pumping unit 265
Figure 11-12 Oakland FD Portable Water Supply System Hose Tender, showing portable hydrants above rear axle), Gleeson pressure-reducing valves (to right of wheel) and hose ramps strapped underneath chassis 265
Figure 11-13 In-line portable hydrant, with Gleeson pressure-reducing valve attached.....266
Figure 11-14 PWSS hose ramps over LDH, permitting traffic to freely pass over the hose, which is otherwise an obstacle..266
Figure 12-1 Schematic AWSS and Sources of Supply: Twin Peaks Reservoir, Cisterns, Pump Station and Fireboat(s) 270
Figure 12-2 San Francisco Cistern – Elevation (top) and Plan (bottom).........................271
Figure 12-3 Dedicated Fire Protection System (DFPS), City of Vancouver, B.C. 273
Figure 12-4 DFPS Pump Station Schematic 273
Figure 12-5 Vancouver, B.C. DFPS (a) False Creek pump station (foreground, nearing completion in 1995); (b) welded steel pipe under construction, 1995; (c) False Creek station proof test, 1995. 274
Figure 12-6. City of Berkeley Proposed Salt Water Fire System...277
Figure 12-7. City of Berkeley – Surface Geology 282
Figure 12-8 Generalized Map of Kyoto, Japan 290
Figure 12-9 Kyoto. Upgrade option using new transmission and local grid pipeline network 291
Figure 12-10 Kyoto. Upgrade option using new pipeline in subway, with pump stations drawing water from local water supplies 291
Figure 12-11 Kyoto. The proposed system could feed handlines wielded by firefighters over perhaps as much as a 1,000 m radius, without need of fire engines 292
Figure 12-12 Kyoto. Fire coverage area by new redundant water system 292
Figure 13-1. Flex Hose Deployment in EBMUD System...298
Figure 13-2. Flex Hose Deployment in EBMUD System...299
Figure 13-3. Flex Hose Attachment To Fire Hydrant..302
Figure 13-4. Deployment of Flex Hose..301
Figure 13-5. Alternate Deployment of Flex Hose 301
Figure 13-6. Deployment of 5" Diameter Flex Hose...302
Figure 13-7. Use of Ramp Over A 5" Diameter Flex Hose 303
Figure 14-1. San Diego Water System and Nearby Earthquake Faults..............................321
Figure 14-2. Scenario Losses to San Diego – As Is Water System.....................................322
Figure 14-3. Scenario Losses to San Diego – Upgraded Water System.............................323

1 Introduction

1.1 The Problem of Fire Following Earthquakes

Earthquakes cause damage by a variety of damaging agents, including fault rupture, shaking, liquefaction, landslides, fires, release of hazardous materials, tsunami etc. Shaking is present in all earthquakes, by definition, and is the predominant agent of damage in most earthquakes. Occasionally, however, building characteristics and density, meteorological conditions and other factors can combine to create a situation in which fire following earthquake, or post-earthquake conflagration, is the predominant agent of damage. Large fires following an earthquake in an urban region are relatively rare phenomena, but have occasionally been of catastrophic proportions.

It is worthwhile defining the term conflagration, as it is variously used and confused. A conflagration is a large destructive fire – in the fire service, in the urban context, a conflagration usually denotes a large fire that spreads across one or more city streets. Conflagration usually connotes such a fire with a moving front – that is, a *mass fire*, as distinguished from a *firestorm*. The term conflagration is sometimes used for a large fire destroying a complex of buildings, but will not be typically used in that sense in this Monograph. A glossary appears at the end of this Monograph, which further defines these terms.

In both Japan and the United States, fire has been the single most destructive seismic agent of damage in the twentieth century. The fires following the San Francisco 1906 and Tokyo 1923 earthquakes rank as the two largest peace time urban fires in man's history, and were both terribly destructive. While not widely perceived today by the public or even many professionals in the earthquake or fire service fields, fire following earthquake is recognized by professionals specializing in this field as continuing to pose a very substantial threat in both countries.

Although fire following the 1906 earthquake was the overwhelming cause of the damage San Francisco and Santa Rosa, and has continued as a significant cause of damage since, it has received relatively little attention in the US. This is perhaps due to several factors:

- Earthquakes historically have been the professional concern of seismologists and structural engineers, who as a class of professionals are largely uninformed of fire,

- Fire protection engineers and fire service personnel have similarly ignored earthquakes, seeing their goal as the mitigation of chronic fire losses by code implementation and other techniques, rather than as earthquake response,

- Major conflagrations were a common occurrence in the US prior to WW2, so that the 1906 experience was seen as more of a conflagration than an earthquake phenomenon. The subsequent decline in US urban conflagrations, due to improved fire and building codes, and to improved fire service response due primarily to radios, has only increased this sense of "it can't happen here".

Less apparent but perhaps even more important factors contributing to the lack of attention paid to post- earthquake fire has been:

- The lack of a major urban US earthquake since 1906. It is little appreciated that it takes a major earthquake under windy conditions, striking a large urban region, to create the conditions of dozens or hundreds of ignitions overwhelming the fire service, to create a conflagration. San Francisco 1906 and Tokyo 1923 fulfilled these conditions. Earthquakes since 1906 (1933 Long Beach, 1964 Alaska, 1971 San Fernando, 1987 Whittier, 1989 Loma Prieta and 1994 Northridge in the US; 1968 Tokachi-oki, 1978 Miyagiken-oki, 1984 Nihonkai- chubu and 1995 Kobe in Japan) have generally not fulfilled these conditions. Note however, that there were many ignitions in the earthquakes of 1971, 1989, 1994 and 1995, and that there were conflagrations of many acres in Kobe in 1995.

- The general lack of awareness of the existence of an analytical framework within which to model the many factors involved in post-earthquake fire, and to quantify these factors and the outcome: many small fires, or conflagration?

That large fires following earthquakes remain a problem is demonstrated by ignitions following recent earthquakes such as the 1994 Northridge and 1995 Kobe earthquakes, as well as several recent large non-earthquake conflagrations, including the 1991 East Bay Hills and 1993 Southern California wild fires. While long a concern to fire departments and the insurance industry, consideration of the problem has been subject to debate regarding the likelihood and severity of post-earthquake fires in any future events.

Until the early 1990s, perhaps the only group at all concerned with post-earthquake fire has been the insurance industry, who due to 1906 are quite aware of the potential for catastrophic loss due to this phenomena. Steinbrugge (1982) presents probably the best summary of knowledge deriving from this field. Scawthorn et al (1981), based on work by Hamada (1951), Horiuchi (n.d.), Kobayashi (1979) and others in Japan, developed a probabilistic post-earthquake fire ignition and spreading model, which has subsequently been applied at two levels:

- Jurisdictional: a detailed modeling, with ignitions, fire loading, engine location and other parameters modeled grid wise at about the 10 hectare level of resolution, Due to the sizable data collection and computational effort involved, this model has only been applied to one US jurisdiction, the City of San Francisco (Scawthorn, 1984), and

- Regional: a coarser model based on approximations derived from the Jurisdictional model. Applied to the San Francisco and Los Angeles and other regions (AIRAC, 1987; NDC, 1992; ICLR, 2001), this model permitted for the first time-quantified estimates of the aggregate losses due to fire following earthquake. This work has largely served the needs of the insurance industry.

Spurred by the 1989 Loma Prieta earthquake, many water utilities became more aware of earthquake risk and the potential for great economic and fire losses to their communities. The East Bay Municipal Utility District conducted a comprehensive examination of their water system, including detailed modeling of fire ignitions and spread. Eidinger and Lee (1995) extended the prior fire ignition and spread models of Hamada and Scawthorn to include detailed modeling of water system performance.

The fact that fire following earthquake has been little researched or considered in North America is particularly surprising when one realizes that the conflagration in San Francisco after the 1906 earthquake was the single largest urban fire in history to that date. It remains today the single largest earthquake loss in U.S. history, in terms of life and economic loss. The loss over three days of more than 28,000 buildings within an area of 12 km^2 was staggering: $250 million in 1906 dollars, and over 3,000 killed[1]. That fire has since only been exceeded in a peacetime urban fire by the conflagration following the 1923 Tokyo earthquake, in which over 140,000 people were killed and 575,000 buildings destroyed (77% of the buildings destroyed were by fire) (Usami, 1996).

Fires following large earthquakes are a potentially serious problem, due to the multiple simultaneous ignitions which fire departments are called to respond to while, at the same time, their response is impeded due to impaired communications, water supply and transportation. Additionally, fire departments are called to respond to other emergencies caused by the earthquake, such as structural collapses, hazardous materials releases, and emergency medical aid.

Most recently, the January 17, 1994, Northridge (M 6.6) and January 17, 1995, Kobe (M 7.2) earthquakes have again emphasized the importance of the fire following earthquake problem. Still, it must also be recognized that fires are not always a major issue, as for example in the Bhuj, India 2001 (M 7.7) or Atico, Peru 2001 (M 8.4) earthquakes. It is hoped that this monograph will shed some light as to why the built environment in the United States, Japan, Canada and New Zealand is susceptible to fire following earthquake, whereas the built environment in India and Peru is not.

1.2 Outline of this Monograph

This Monograph is divided into 14 chapters:

Chapter 2 examines the historical record of fires following earthquake, including the 1906 San Francisco, 1923 Tokyo, 1989 Loma Prieta, 1994 Northridge, 1995 Kobe, 1999 Turkey and 2000 Napa earthquakes. These earthquakes provide a range of possible FFE outcomes, from the very severe (1906 San Francisco, 1923 Tokyo) to the moderately

---

[1] Exact number of fatalities is unknown – until the 1980's, it was believed approximately 700 had been killed. Research by Gladys Hansen, San Francisco Librarian, indicated that far more people killed had not been accounted for. In painstaking research over many years, she slowly gathered evidence from letters of the time, gathered from all over the world, of many more deaths. Of particular interest was the fact that many minority fatalities, especially in San Francisco's large Chinatown, were known in 1906, but not included in the official count. Her work is on going as of this writing, and the count is still increasing. See Hansen and Condon, 1989.

severe (1995 Kobe) to the modest (1989 Loma Prieta, 2000 Napa). The attributes of each earthquake are examined, and from this we can hopefully extract the "good things to do" to limit the FFE hazard in future earthquakes. The record offers compelling evidence that fire following earthquake remains a very significant problem today, with the potential for catastrophic losses, at least in countries with a large urban wood building inventory such as the US, Canada, Japan and New Zealand.

Chapter 3 presents the historical record of a number of non-earthquake related fire conflagrations. Even though these fires were not-earthquake related, many of their attributes have direct bearing to the post-earthquake environment. In particular, we can examine the effectiveness of undamaged water systems in these fires.

Chapter 4 presents a framework for the analysis of fires following earthquakes, which permits the differentiation of situations where scattered fires may occur, versus that where a conflagration will develop. Methodologies based on this analytical framework are used in the water, fire service, insurance and related lifeline arenas to estimate the potential losses that may occur due to fire following earthquake. These estimates are a key ingredient in the development of effective mitigation programs. Some of the models presented have already been codified into computer codes, including HAZUS (public domain, available from FEMA) and other computer codes in use in the private consulting arena such as URAMP (ABS Consulting), SERA (G&E Engineering Systems Inc.) and RiskLink (RMS). Not all of the models presented are in complete agreement with each other, and the user must be aware that advances in modeling are still to be made in the future.

Chapter 5 presents a summary discussion of fire department operations following an earthquake. Considerations for both the senior management of a fire department, and line officers are presented, and the need for a jurisdiction-specific earthquake emergency plan element is emphasized.

Chapter 6 presents a summary discussion of the role of water systems in post-earthquake fire suppression. Quantitative tools are presented to help understand the amount of damage that the water system might undergo in the earthquake. Clearly, the more damaged the water system, the higher the risk of fire spread.

Chapter 7 addresses the role of natural gas systems in the FFE hazard. The issue of ignitions and gas-fed fuel supply is examined. The pros- and cons- of automatic gas shut-off systems is examined.

Chapter 8 addresses how electric power systems impact the fire following earthquake hazard. Electric power is one of the main sources of fire ignitions. Mitigation strategies are presented to limit the number of fire ignitions due to electric power.

Chapter 9 presents the role of communication systems in the FFE hazard. The breakdown or overload of communication systems hampers the ability to rapidly report fire ignitions to fire departments. The longer the time needed to respond to a fire, the more difficult it is to control it once fire crews arrive on site.

Chapter 10 addresses the possible impacts of a damaged transportation network on the FFE hazard.

Chapter 11 examines the various methods employed, or potentially useful, for reducing the potential losses due to fire following earthquake. While the problem of fire following earthquake is daunting, there are a number of methods available for reducing the potential for catastrophic conflagrations to occur. Some of these methods are good practice for the fire service or lifeline operators under normal non-earthquake conditions, while others are specific to the unique problems of fires following earthquakes.

Chapter 12 examines the design and implementation of special-purpose high-pressure water supply systems as built in San Francisco CA and Vancouver B.C., designed in Berkeley CA, and proposed for Kyoto, Japan.

Chapter 13 examines the use of ultra-large diameter (12-inch diameter) flexible hose for post-earthquake delivery of water for fire fighting purposes.

Chapter 14 presents a quantitative framework that can be used to examine the cost-effectiveness of various mitigation strategies.

The flow of the Monograph is thus from reviewing historical experience, which illustrates and confirms the magnitude and nature of the problem, to engineering analysis of the problem and fire department operations, then to a detailed discussion of the interaction of each major element of the infrastructure, or lifeline, with the problem, and finally with four chapters examining in detail the most essential element of the infrastructure with regard to this problem – water supply. This monograph includes a glossary and index.

1.3 Acknowledgements

This monograph could not have been written without the great dedication of a number of people. The following people were the prime and contributing authors for each chapter in this monograph.

Chapter 1 and Introductory Materials. Charles Scawthorn and John Eidinger

Chapter 2. Charles Scawthorn

Chapter 3. Charles Scawthorn and John Eidinger

Chapter 4. Charles Scawthorn, John Eidinger, Chris Mortgat

Chapter 5. Charles Scawthorn and Chief Frank T. Blackburn

Chapter 6. William Huebach and John Eidinger

Chapter 7. Doug Honegger

Chapter 8. Anshel Schiff

Chapter 9. Anshel Schiff and Alex Tang

Chapter 10. Nesrin Basoz

Chapter 11. Charles Scawthorn

Chapter 12. Charles Scawthorn and John Eidinger

Chapter 13. John Eidinger

Chapter 14. John Eidinger and Ken Goettel

The editors of this report were Charles Scawthorn (Consulting Engineer), John Eidinger (G&E Engineering Systems Inc.) and Anshel Schiff (Precision Measurement Instruments).

The editors and authors would like to thank the contributions of many people and organizations who directly supported this effort, including Chief Don Parker (Vallejo Fire Department), Chief Don Manning (Los Angeles City Fire Department, ret.), Chiefs Emmet Condon (ret.), Mario Treviño, John Lo and many other members of the San Francisco Fire Department, Chief W. Wittmer and Capt. Mark Hoffman (Oakland Fire Dept.), Gary Bard (Berkeley Fire Department, ret.), FF G. Beer (FDNY), Capt. J. Donelan (Coalinga Fire Dept.), Capt. H. Nakachi (Kobe Fire Dept.), Prof. Thomas D. O'Rourke and Mr. Harry Stewart of Cornell University, Dr. Fred Krimgold (Virginia Tech), Prof. Michael O'Rourke (Rensselaer), Drs. M. Khater and F. Waisman and Mr. A. Cowell (EQECAT), Dr. K. Porter (Calif. Inst. Tech.), Dr. Ai Sekizawa (Fire Research Institute, Japan), Profs. M. Kobayashi, K. Toki, Y. Yamada and H. Iemura (Kyoto Univ., Japan), Drs. Chris Mortgat and Wei Min Dong (RMS), Alex Tang (Nortel Networks, ret.), Dr. Ken Goettel (Goettel and Associates Inc.), Joe Young (EBMUD), Riley Chung, and John Graves (National Communications System). Elaine Beattie assisted with production of this tome.

Special gratitude is due to Chief Frank T. Blackburn (San Francisco Fire Department, ret.), who pioneered the development of the PWSS and led the fire service in California to a greatly improved understanding of the problems of water supply and fire following earthquake.

1.4 *References*

AIRAC.1987 Fire Following Earthquake, Estimates of the Conflagration Risk to Insured Property in Greater Los Angeles and San Francisco, prepared for the All-Industry Research and Advisory Council, Oak Brook, IL by C. Scawthorn, Dames & Moore, San Francisco.

Eidinger, J., Goettel, K., and Lee, D., (1995) Fire and Economic Impacts of Earthquakes, in Lifeline Earthquake Engineering, Proceedings of the Fourth US Conference, TCLEE Monograph 6, ASCE, August.

Hamada, M. (1951) On Fire Spreading Velocity in Disasters, Sagami Shobo, Tokyo (in Japanese).

Hansen, G., and E. Condon, 1989, Denial of Disaster: The Untold Story and Photographs of the San Francisco Earthquake and Fire of 1906. San Francisco, CA: Cameron and Company.

Horiuchi, S. (n.d.) Research on Estimation of Fire Disasters in Earthquakes, Disaster Prevention Section, Comprehensive Planning Bureau, Osaka Municipal Government, Osaka (in Japanese).

Kobayashi, M. (1979) A Systems Approach to Urban Disaster Planning, Ph.D. Dissertation, Kyoto University, Kyoto, Japan (in Japanese).

ICLR. 2001 Assessment of Risk due to Fire Following Earthquake Lower Mainland, British Columbia, report prepared for the Institute for Catastrophic Loss Reduction, Toronto, by C. Scawthorn, and F. Waisman, EQE International, Oakland, CA.

NDC (1992) Fire Following Earthquake - Conflagration Potential in the Greater Los Angeles, San Francisco, Seattle and Memphis Areas, prepared for the Natural Disaster Coalition by C. Scawthorn, and M. Khater, EQE International, San Francisco, CA.

Scawthorn, C., Yamada, Y. and Iemura, H. (1981) A Model for Urban Post-earthquake Fire Hazard, DISASTERS, The International Journal of Disaster Studies and Practice, Foxcombe Publ., London, v. 5, n. 2.

Steinbrugge, Karl V., Earthquakes, Volcanoes and Tsunamis – An Anatomy of Hazards, Skandia America Group, 280 Park Avenue, New York NY 10017, 1982.

2 Historical Fires Following Earthquakes

2.1 Introduction

This Chapter discusses selected U.S. and foreign earthquakes, with emphasis on the fires following these events. The purpose of this Chapter is several-fold, including:

- To document the record of fires following earthquake, to demonstrate that it is a recurring problem. Not all persons, even in the fire service, understand that following a large earthquake, numerous ignitions will occur simultaneously, in an environment of confusion, damaged water supply, impaired communications and multiple demands on the fire service resources greatly exceeding its capacity.

- To provide data and narratives for specific events, in order that engineers, fire service personnel and emergency planners can understand the nature of what they will be confronted with.

- To emphasize that in selected cases, in large cities with large wood building inventories such as in the US, Canada, Japan, New Zealand and elsewhere, that the multiple simultaneous ignitions that occur following a large earthquake have the potential to grow into a mass conflagration, of catastrophic proportions. Table 2-1 lists all US 20th Century events with post-earthquake ignitions. In the next sections we discuss selected events, generally providing an overview of the event, and data on the ignitions, performance of the water system, and overall result.

Table 2-1 U.S. 20th Century Post-earthquake Ignitions

M	D	Year	Earthquake	M	City or Area Affected	Ignitions	MMI
Apr	18	1906	San Francisco	8.3	Berkeley	1	VIII-IX
Apr	18	1906	San Francisco	8.3	Oakland	2	VII-IX
Apr	18	1906	San Francisco	8.3	San Francisco	52	VII-X
Apr	18	1906	San Francisco	8.3	San Jose	1	VIII
Apr	18	1906	San Francisco	8.3	Santa Clara	1	VIII-IX
Apr	18	1906	San Francisco	8.3	San Mateo Co.	1	VIII
Apr	18	1906	San Francisco	8.3	Santa Rosa	1	X
Mar	11	1933	Long Beach	6.3	Los Angeles	3	VI-VII
Mar	11	1933	Long Beach	6.3	Long Beach	19	IX
Mar	11	1933	Long Beach	6.3	Norwalk	1	VII-VIII
Jul	21	1952	Kern County	7.7	Bakersfield	1	VIII
Mar	22	1957	San Francisco	5.3	San Francisco	1	VII
Mar	28	1964	Alaska	8.3	Anchorage	7	X
Oct	1	1969	Santa Rosa	5.7	Santa Rosa	2	VIII
Feb	9	1971	San Fernando	6.7	Burbank	7	VII
Feb	9	1971	San Fernando	6.7	Glendale	9	VI-VII
Feb	9	1971	San Fernando	6.7	Los Angeles	128	VI-VII
Feb	9	1971	San Fernando	6.7	Pasadena	2	VII
Feb	9	1971	San Fernando	6.7	San Fernando	3	IX
Oct	15	1979	El Centro	6.4	El Centro	1	VII
May	2	1983	Coalinga	6.5	Coalinga	4	VIII
Apr	24	1984	Morgan Hill	6.2	Morgan Hill	4	VII
Apr	24	1984	Morgan Hill	6.2	San Jose	5	VIII
Jun	8	1986	N. Palm Springs	5.9	N. Palm Springs	2	VI-VII
Oct	1	1987	Whittier	6	Whittier	38	VI
Oct	17	1989	Loma Prieta	7.1	Daly City	3	VI
Oct	17	1989	Loma Prieta	7.1	Berkeley	1	VI
Oct	17	1989	Loma Prieta	7.1	Marin Co.	2	VI
Oct	17	1989	Loma Prieta	7.1	Mountain View	1	VII
Oct	17	1989	Loma Prieta	7.1	San Francisco	26	VII
Oct	17	1989	Loma Prieta	7.1	Santa Cruz	1	VIII
Oct	17	1989	Loma Prieta	7.1	Santa Cruz Co.	24	VII-VIII
Jan	17	1994	Northridge	6.8	Los Angeles	77	VI-IX
Jan	17	1994	Northridge	6.8	Santa Monica	15	VIII
Sep	3	2000	Napa	5.0	Napa	1	VI

2.2 The 1906 San Francisco, California Earthquake and Fires

The April 18, 1906 earthquake was the most devastating earthquake in US history, and the largest urban fire in history to that time. The fire that resulted is still the largest urban fire in US history, and has only been exceeded by the fires that followed the 1923 Tokyo earthquake, and the incendiary attacks on Japan and Germany in World War 2.

2.2.1 Seismological and Overall Damage Aspects

This earthquake caused the longest rupture observed in the contiguous United States – from San Juan Bautista to Point Arena, where it passes out to sea, with additional displacement observed farther north at Shelter Cove in Humboldt County – indicating a potential total length of rupture of 430 kilometers. Fault displacements were predominantly right lateral strike-slip, with the largest horizontal displacement - 6.4 meters - occurring near Point Reyes Station in Marin County. The surface of the ground was torn and heaved into furrow-like ridges. Roads crossing the fault were impassable, and pipelines were broken. The region of destructive intensity extended over a distance of 600 kilometers with strong shaking over a wide area, Figure 2-1.

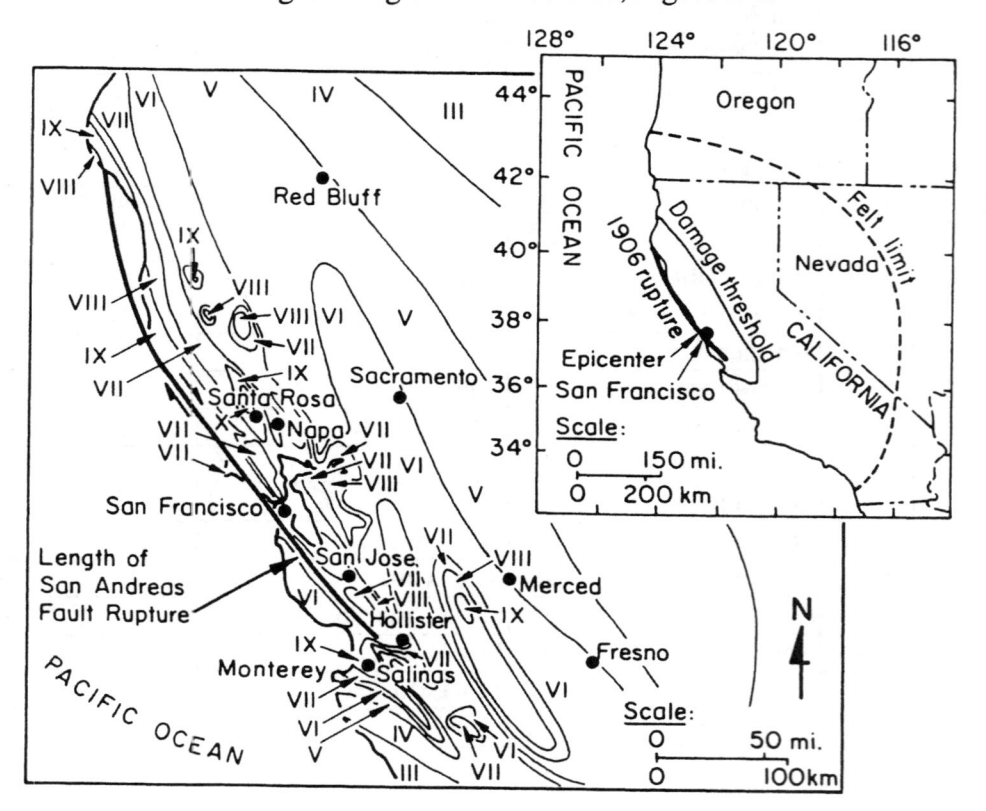

Figure 2-1 MMI map 1906 San Francisco earthquake

On or near the San Andreas Fault, some buildings were destroyed but other buildings, close to or even intersected by the fault, sustained nil to only light damage. South of San Francisco, the concrete block gravity-arch dam of the Crystal Springs Reservoir was sited only 100-200 yards from the fault, with the reservoir astride the fault. The dam was virtually undamaged by the event, and the nearby San Andreas earthen dam, whose abutment was intersected by the fault rupture, was also virtually undamaged, although surrounding structures sustained significant damage or were destroyed (Lawson et al, 1908).

The earthquake and resulting fires caused an estimated 3,000 deaths and $524 ($1906) million in property loss. Fires that ignited in San Francisco soon after the onset of the earthquake burned for three days because of the lack of water to control them. Damage in San Francisco was devastating, with 28,000 buildings destroyed, although eighty percent (80%) of the damage was due to the fire, rather than the shaking. Fires also intensified the loss at Fort Bragg and Santa Rosa. Although Santa Rosa lies about 30 kilometers from the San Andreas Fault, damage to property was severe, and 50 people were killed.

2.2.2 Ignitions and Fire

Despite the strong shaking, the vast majority of the damage in the entire earthquake, and especially in San Francisco, was due to fire. Scawthorn and O'Rourke (1989) compiled data on the 52 known ignitions in the City of San Francisco.

The conflagration following the 1906 earthquake was a complex fire, actually consisting of several separate major fires that grew together until there was one large burnt area, comprising the northeast quadrant of the city and destroying over 28,000 buildings. The progress of these fires has generally been divided into four "periods" (NBFU, Bowlen Outline), although actual times for these periods differ between sources. Generally, the periods comprise the following times:

- From the earthquake until mid or late in Day 1, when most of the south of Market section had been destroyed, but the higher value north of Market section still remained largely intact.

- The night of Day 1 and the early hours of Day 2, when the fire invaded the north of Market section, progressing from the west.

- Continued progress of the fire to the north, and a bit to the south, during the remainder of Day 2.

- During Day 3, the last day, when the fire progressed almost entirely to the north, around Telegraph Hill, and burnt down to the Bay.

Figure 2-2 shows some of the ignitions in the Central Business District, around the foot of Market and Mission Sts., and their spread during the first day, while Figure 2-3 shows the final extent of the burnt area. Figure 2-4 to Figure 2-6 show scenes during and after the fire.

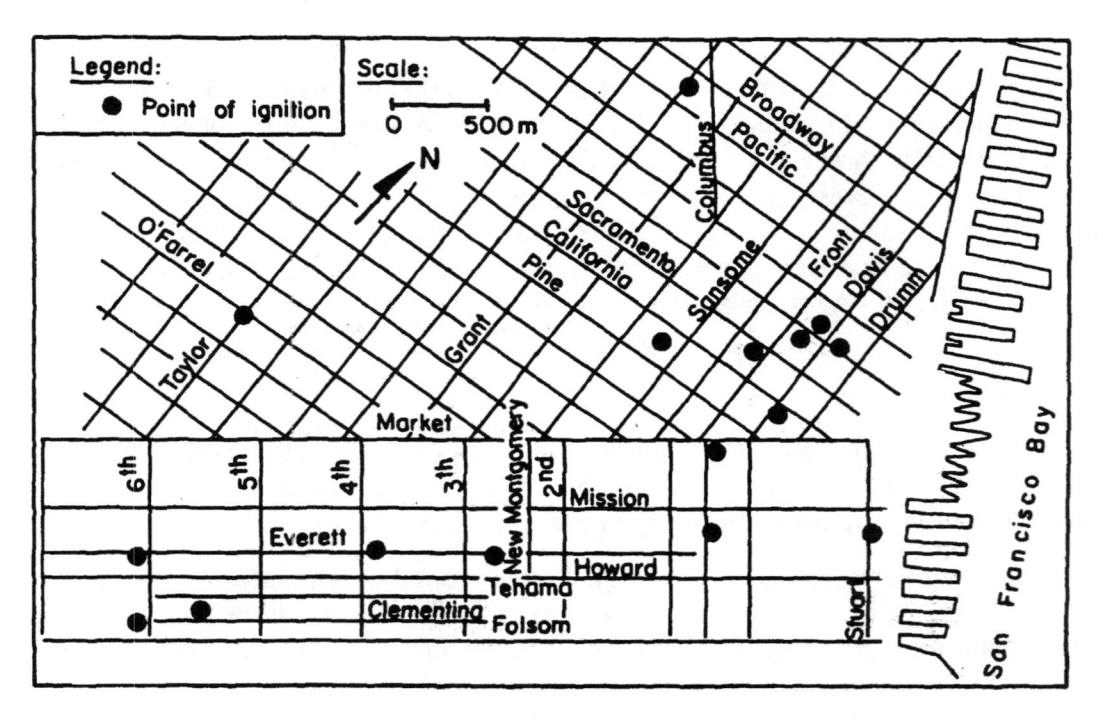

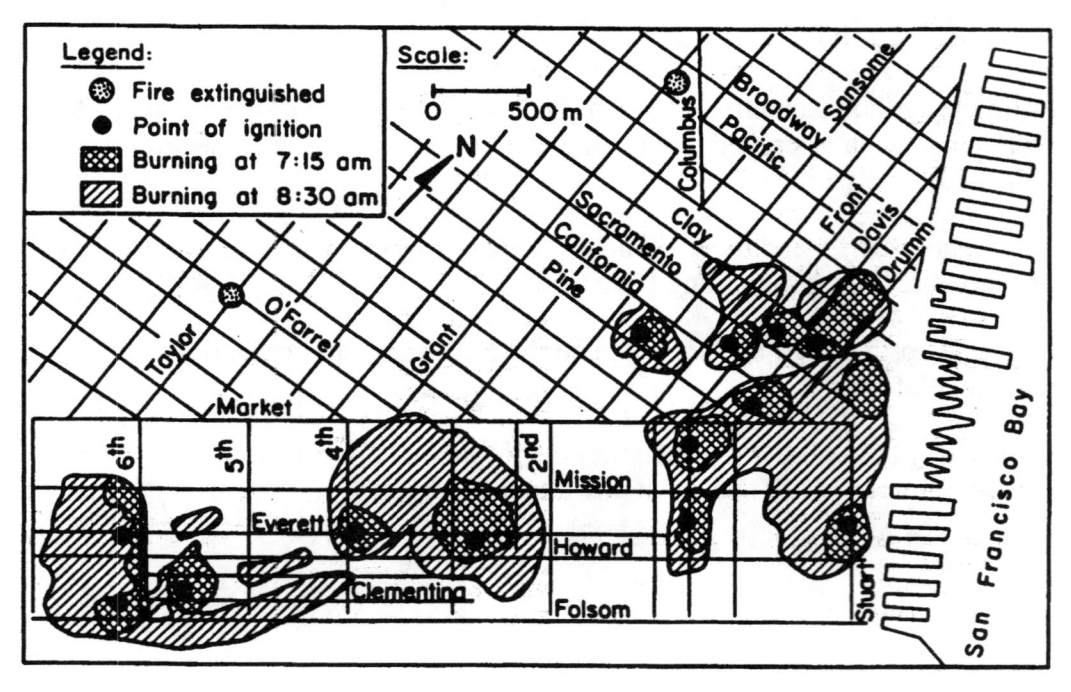

Figure 2-2 San Francisco 1906 Fire – Ignitions at 6:00 am (top) and Spread at 7:15 am and 8:30 am (bottom), Central Business District

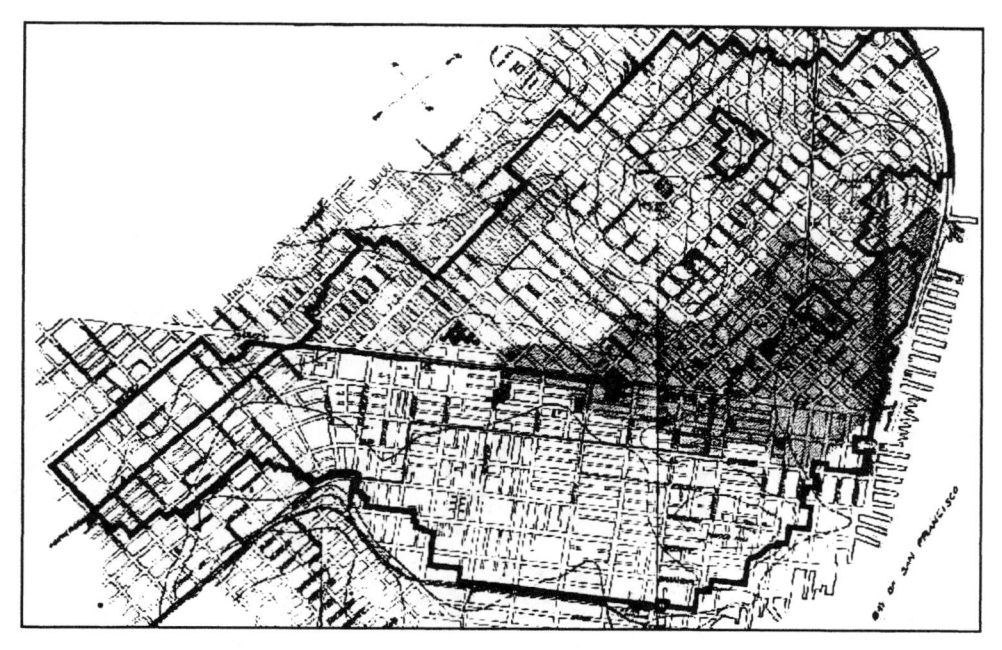

Figure 2-3 San Francisco 1906 Fire – final extents in solid line; many collapsed masonry buildings in shaded area.

Figure 2-4 San Francisco 1906 collapsed building, fire in background

Figure 2-5 View down Sacramento St. during San Francisco 1906 fire

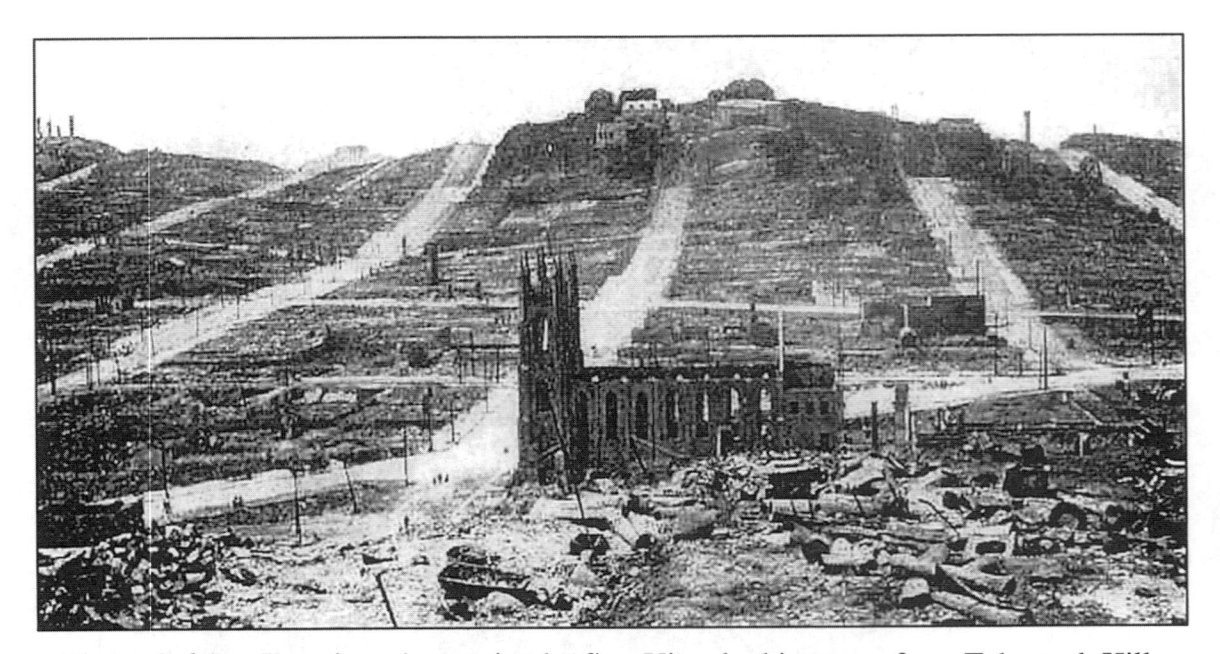

Figure 2-6 San Francisco destruction by fire: View looking west from Telegraph Hill, showing unburned houses on summit of Russian Hill. St. Francis Roman Catholic Church, with excellent brick walls in foreground.

Fire Department Response

The San Francisco Fire Department in 1905 protected approximately 400,000 persons occupying an urbanized area of approximately 21 square miles. The department consisted of a total of 585 full paid fire force personnel (resident within the city and on duty at all times), commanded by Chief Dennis T. Sullivan and deployed in 57 companies (38 engine, 1 hose, 10 ladder, 1 hose tower, and 7 chemical) (NBFU, 1905). The distribution of these companies was well conceived, being centered about the congested high value district (i.e., the Central Business District or CBD, known in San Francisco as the Financial District), with 24 engine, 8 ladder, 1 water tower and 7 chemical companies within 2 miles of the center of the CBD. All but two of the 38 steam engine companies dated from 1890 or later, and were rated at an average of 680 gallons per minute (gpm), although the eight engines tested in 1905 averaged only about 70% of their rated capacity, and the "ability of the men handling the engines was in general below a proper standard". The rated pumping capacity of the 38 first line and 15 relief and reserve engines totaled 35,100 gpm. In summary, the department was rated by the National Board of Fire Underwriters (NBFU, 1905) as efficient, well organized and, in general, adequate. The NBFU however concluded in 1905 that

> "...In fact, San Francisco has violated all underwriting traditions and precedent by not burning up. That it has not done so is largely due to tile vigilance of the fire department, which cannot be relied upon indefinitely to stave off the inevitable."

Within moments after the earthquake, Chief Dennis T. Sullivan was seriously injured due to the damage to the fire station where he was sleeping, and afterwards died. Ten fire stations sustained major damage (Tobriner, personal communication) although the earthquake seriously disabled no engines and all went into service (NBFU, 1906). Street passage was in general not a problem, and a number of fires were quickly suppressed, although many more could not be responded to. That is (NBFU, 1906),

> "...fires in all parts of the city, some caused directly by earthquake, some indirectly, prevented an early mobilization of fire engines and apparatus in the valuable business district, where other original fires had started and were gaining headway".

The NBFU Conflagration Report (NBFU, 1906) concluded

> "the lack of regular means of communication and the absence of water in the burning district made anything like systematic action impossible: but it is quite likely that during the early hours of the fire the result would not have been otherwise, even had not of these abnormal conditions existed" [sic].

That is, the NBFU concluded that even under normal conditions the multiple simultaneous fires would have probably overwhelmed a much larger department, such as New York's, which had three times the apparatus (NBFU, 1905). Nevertheless, Bowlen (n.d.) concluded that by 1 PM (i.e., about 8 hours after the earthquake)

> "the fire department, except that it was without its leader, was in fairly good shape, that is the men and horses were in good trim for firefighting, the apparatus

was in shape and could be worked where there was water. There is not one report of an engine or man going out of commission during the early hours of the fire, and the department was hard at work all the time, even though there was little to show for its effort"

2.2.3 Water System Performance

Several factors contributed to the initial ignitions rapidly growing out of control. While the weather was relatively hot and dry, undoubtedly the primary factor leading to the conflagration was the failure of the water system (Scawthorn and O'Rourke, 1989). In summary, in 1906 water to San Francisco was supplied from three reservoirs located 5 to 10 miles south of San Francisco as well as a watershed located about 35 miles southeast of San Francisco. The three reservoirs south of San Francisco were the San Andreas, Crystal Springs, and Pilarcitos Reservoirs. Three transmission pipelines conveyed water from these three reservoirs to a second series of smaller reservoirs within the city limits. All three of these pipelines were damaged to varying degrees. The Pilarcitos pipeline was so heavily damaged that it was abandoned; The San Andreas pipeline was broken at one location, and took 62 hours to restore to service. The Crystal Springs pipeline was broken at many locations, taking 28 days to restore to service. Compounding the damage to these transmission pipelines was pervasive areas of distorted ground within the City resulting in about 300 breaks in local city distribution water mains. As the local distribution mains were connected to some local city reservoirs, it was the damage to the distribution mains that was the primary reason for lack of water supply for firefighting.

Water was distributed throughout the city by means of trunk and distribution pipelines.

Figure 2-7 presents a map of the damage to the 1906 San Francisco water supply. At the time of the earthquake, there was a combined volume of 88.7 billion liters in the San Andreas, Crystal Springs, and Pilarcitos Reservoirs. These reservoirs, coupled with the Alameda watershed on the east side of the San Francisco Bay, supplied all water for the city of San Francisco in 1906. Transmission pipelines conveying water from the southern reservoirs were built mainly of riveted wrought iron. Within the city limits, there were approximately 711 km of distribution piping at the time of the earthquake, of which roughly 18.5 and 66.5 km were wrought and cast iron trunk lines, respectively. These lines were larger than or equal to 400 mm in diameter. The bulk of the system had been constructed during the years of 1870 to 1906.

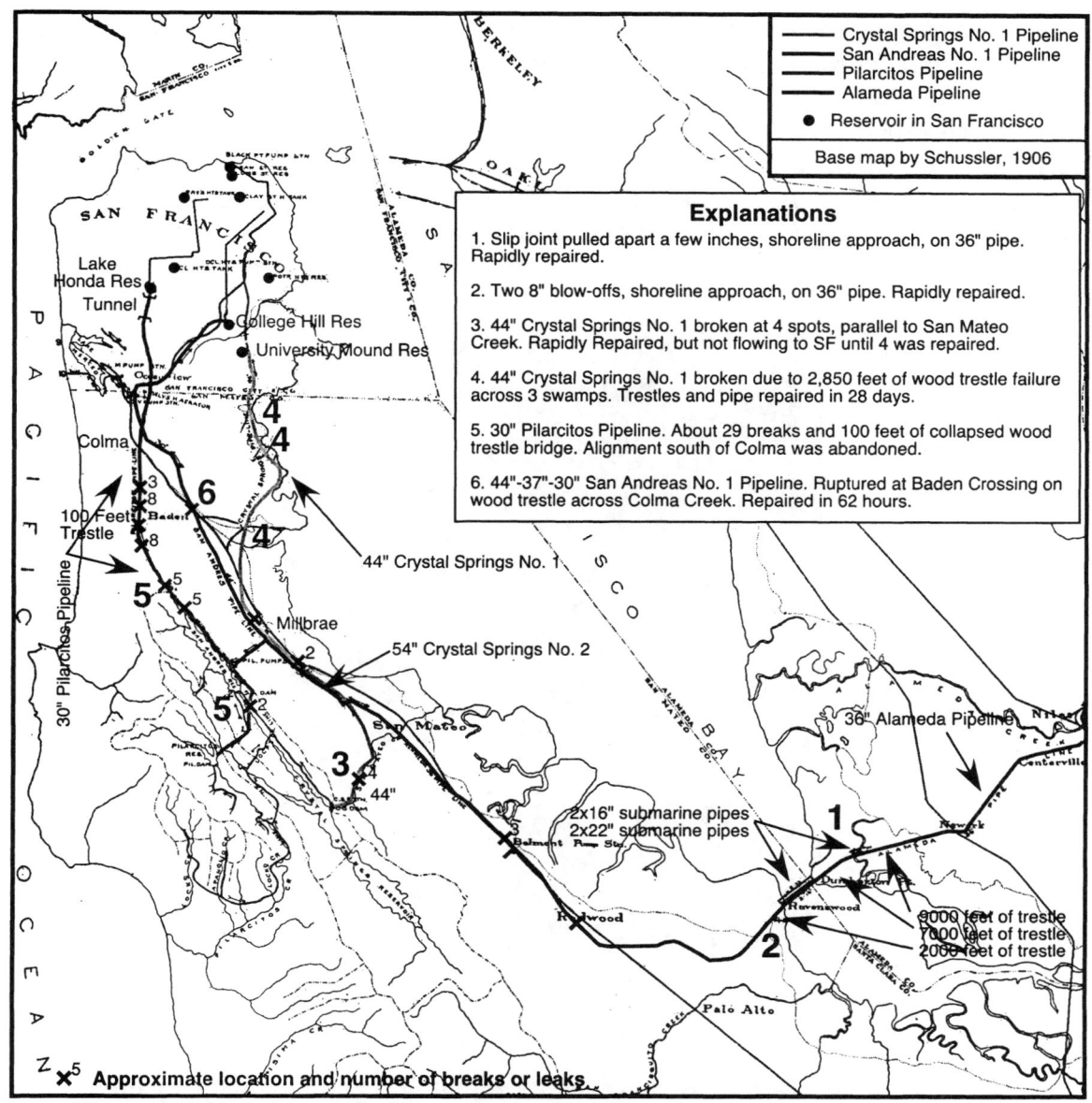

Figure 2-7 Damage to the Transmission Pipelines Serving San Francisco, 1906

Superimposed on Figure 2-7 are the approximate locations of transmission pipeline damage caused by the earthquake. Flow from all transmission pipelines stopped shortly after the earthquake. Because telephone service was out, emergency control information had to be obtained by dispatching personnel into the field where maintenance crews reported on the damage. Right lateral strike-slip movement along the San Andreas Fault ruptured a 750 mm diameter wrought iron pipeline conveying water from the Pilarcitos to Lake Honda Reservoir, Figure 2-8. Over 29 breaks were reported north of the San Andreas Reservoir, where the pipeline was constructed parallel to the San Andreas Fault. Fault movement near the San Andreas Reservoir was measured as 3.6 to 5.6 m. Pipeline ruptures were caused by tensile and compressive deformation of the line. Over three months were required to reconstruct the pipeline. Within 16 hours after the earthquake, repairs were made to that part of the Pilarcitos Conduit that was located within the city limits. Water then was pumped from Lake Merced through the Pilarcitos line into Lake

Honda at a rate of approximately 25 million liters per day. Dynamic distortion of bridges was responsible for rupturing a 925 mm diameter wrought iron pipeline conveying water from the San Andreas to College Hill Reservoir, and for rupturing at three swamp crossings an 1100-mm diameter wrought iron pipeline conveying water from the Crystal Springs to University Mound Reservoir. The wooden trestle bridges all were damaged by strong ground shaking, with no damage or misalignment observed in their timber pile foundations. Approximately three days were required to repair the 925-mm diameter pipe, and over a month was required to restore the 1100-mm diameter Crystal Springs Pipeline.

Figure 2-8 Failure of Pilarcitos 30 inch pipeline. Pipe telescoped and was thrown sideways on the wooden bridge crossing Small Frawley Canyon. San Andreas fault crosses pipe diagonally.

Figure 2-9(a) is a map of the 1904 water supply within the San Francisco City limits. There were nine reservoirs and storage tanks, for a total capacity of 354 million liters. Approximately 92% of this total, or 325 million liters, were contained in the Lake Honda, College Hill, and University Mound Reservoirs. These reservoirs and the pipelines linking them with various parts of the city were the backbone of fire protection. All trunk lines, 400 mm or larger in diameter, are plotted in Figure 2-9(a). Trunk lines are shown connected to the Lake Honda, College Hill, University Mound, Francisco Street, and Clay Street Reservoirs; all other reservoirs were connected to piping 300 mm or less in diameter. Superimposed on Figure 2-9(a) are the zones of lateral spreading caused by soil liquefaction, as delineated by Youd and Hoose (1978). Breaks in the pipeline trunk system crossing these zones are plotted from records provided by Schussler (1906) and Manson (1908). It can be seen that multiple ruptures of the pipeline trunk systems from the College Hill and University Mound Reservoirs occurred in the zones of large ground deformation, thereby cutting off supply of over 56% of the total stored water to the Mission and downtown districts of San Francisco. Two pipelines, 400 and 500 mm in diameter, were broken by liquefaction induced lateral spreading and settlement across Valencia Street north of the College Hill Reservoir. These broken pipes emptied the reservoir of 53 million liters, thereby depriving fire fighters of water for the burning Mission District of San Francisco. Figure 2-9(b) shows a map of the San Francisco water supply and area burned during the tire. All trunk lines of the College Hill and University Mound Reservoirs downstream of the pipeline ruptures are removed from this figure to show the impact and lack of hydraulic conductivity caused by severing these conduits. With the College Hill and University Mound Reservoirs cut off, only the Clay Street Tank and the Lombard and Francisco Street Reservoirs were within the zone of most intense fire, and therefore capable of providing water directly to fight the blaze. The combined capacity of these reservoirs was only 21 million liters, or 6% of the system capacity. The usefulness of such limited supply was further diminished by breaks in service connections, caused by widespread subsidence, burning and collapsing buildings. Schussler identifies service line breaks as a major source of lost pressure and water. There were roughly 23,200 breaks in service lines, between 15 and 100 mm in diameter. Fallen rubble and collapsed structures often prevented firemen from closing valves on distribution mains to diminish water and pressure losses in areas of broken mains and services. As is evident in Figure 2-9(b), the Lake Honda Reservoir was able to provide a continuous supply of water to the western portion of the city. The fire eventually was stopped along a line roughly parallel to Van Ness Avenue, where water still was available from the Lake Honda Reservoir. Moreover, the southern and southeastern extent of the fire is bounded by areas south and southeast of the trunk system ruptures. It is likely that these unburnt areas had water from the University Mound Reservoir. Recognition of the critical role played by damage to the water system led to the construction of San Francisco's Auxiliary Water Supply System, which is described later in this chapter.

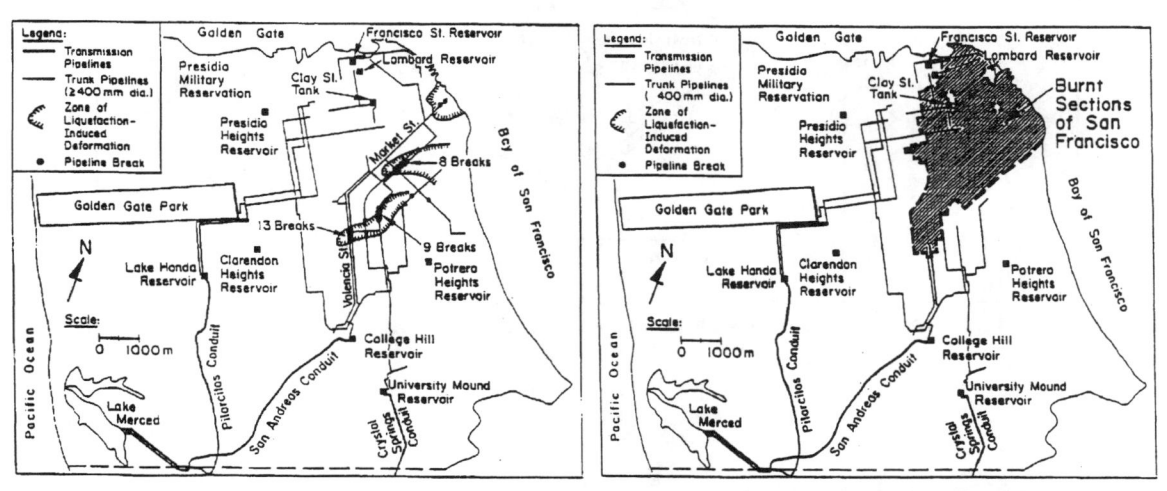

Figure 2-9 (a) map of the 1904 San Francisco water system (b) showing final burnt area

2.3 The 1923 Tokyo Earthquake and Fires

This section summarizes fire following aspects of the 1923 Kanto earthquake for Tokyo and nearby cities. The 1923 earthquake was followed by the largest urban conflagration in history, in which approximately 140,000 persons perished. The earthquake and conflagration are first briefly summarized, followed by a discussion of the effects on water supply components, and a qualitative analysis of the contribution of the water supply damage to the occurrence of the conflagration.

2.3.1 Seismological and Overall Damage Aspects:

The M 7.9 Kanto earthquake (Usami, 1987) occurred at 11:58 AM local time September 1, 1923, with an epicenter located just offshore in Sagami Bay, at 139.5 E, 35.1 N. Damage was extensive, with major crustal movements (maximum 2 m. uplift), significant numbers of engineered buildings sustaining structural damage (Freeman. 1932), destroying approximately 128,000 houses and damaging another 126,000 (Kanai, 1983), with extensive liquefaction (Hamada et al, 1992) (Figure 2-10) and land sliding (ASCE, 1929). Shaking intensity has been estimated at JMA 6 (equivalent to MM1 IX) (Kanai, 1983) (Figure 2-11). A major tsunami (4-6 m. in height) affected the Miura and Boso peninsulas, destroying 868 houses (Kanai, 1983; Hamada et al 1992).

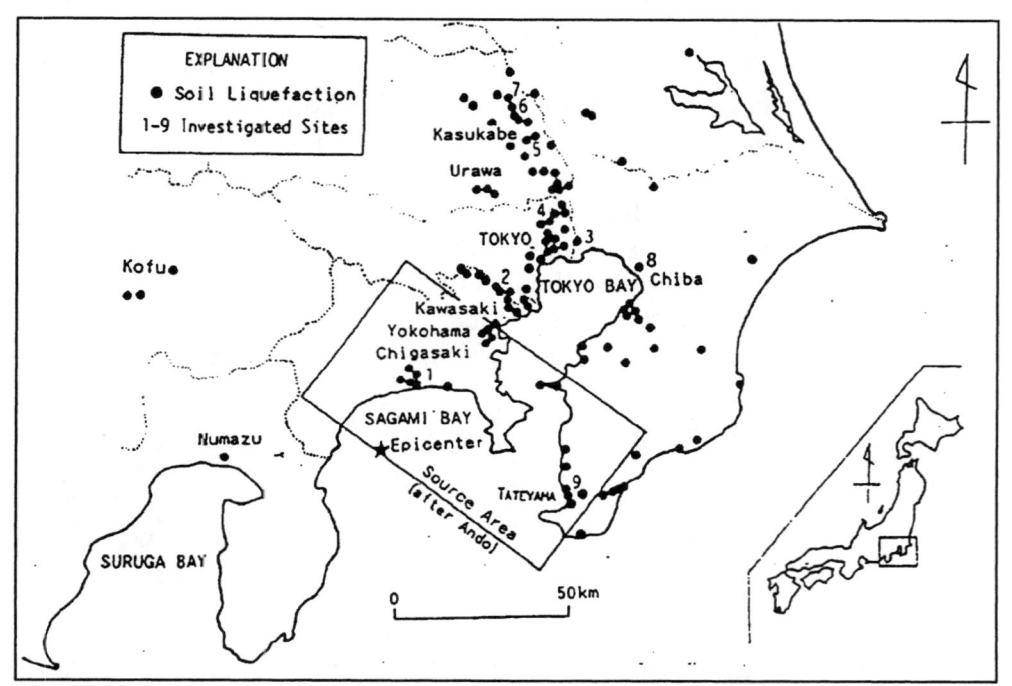

Figure 2-10 Kanto and Chubu Districts showing areas principally affected by liquefaction

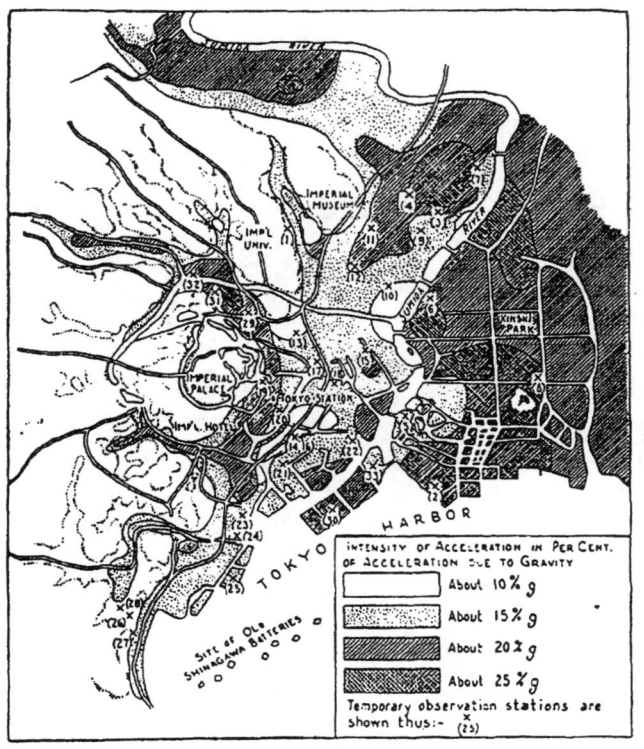

Figure 2-11 Distribution on earthquake intensity in Tokyo

2.3.2 Ignitions and Fire:

Tokyo had long been recognized as a major conflagration hazard, due to a dense urban aggregation of wood buildings. Due to a recent dry period and nearby typhoon, meteorological conditions were particularly adverse at the time of the earthquake, with hot (approximately 26 C, 80 F) dry winds of approximately 12.5 meters per second (28 mph) at the time of the earthquake. Winds grew continuously all day, reaching a maximum of 21 meters per second (48 mph) at 11 pm that evening. The fire occurred just prior to lunchtime, and numerous small charcoal braziers were lit for the noontime meal, resulting in approximately 277 outbreaks of fire, about 133 of which spread (Okamoto, 1984), Figure 2-12. The result was a major conflagration with rapid fire spread, Figure 2-13, which burned for several days causing approximately 140,000 deaths and destroying approximately 447,000 houses by fire (Kanai, 1983; Hamada et al 1992).

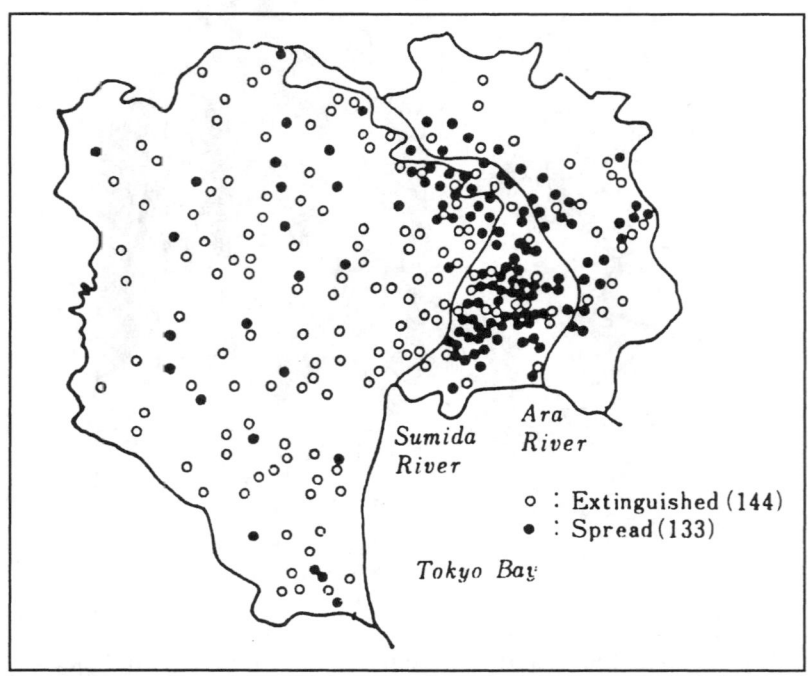

Figure 2-12 Outbreak of fires in the city of Tokyo (Kanto earthquake of 1923)

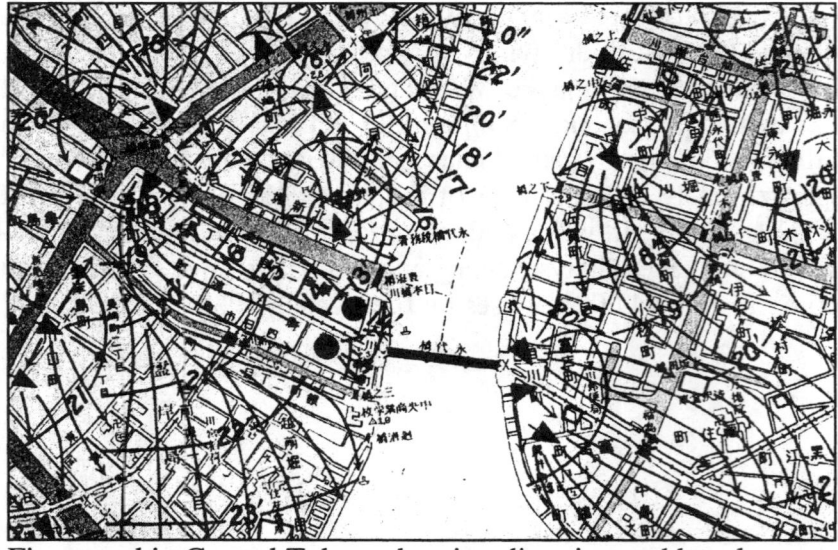

Figure 2-13 Firespread in Central Tokyo, showing direction and hourly progress of flame front

2.3.3 Damage to Water Supply Components:

Water supply systems for the cities of Tokyo, Yokohama, Kawasaki and Yokosuka were all generally similar in configuration, drawing their supply from rivers emanating from the mountains surrounding the Kanto plain, conveying the water by gravity to terminal stilling basins via concrete or cast iron aqueducts, and thence via distribution systems composed mainly of cast iron pipe. The performance of each component is detailed in an unpublished report by ASCE (1929). Particularly relevant was the performance of the distribution system within the urbanized areas.

Tokyo's distribution system totaled 723 miles in length, and included 255,000 lineal feet of 16 to 60 inch diameter cast iron (CI) trunk line pipe, and secondary mains from 4 inch to 14 inch, totaling about 700,000 lineal feet. No specific information is available regarding damage to the Tokyo distribution system, except that cast iron pipes throughout Tokyo suffered in places, especially in filled ground and along riverbanks.

Yokohama's distribution system totaled 170 miles of CI pipe, ranging in size from 4 inch to 36 inch. All pipe trenches were re-excavated in order to re-caulk joints - in general, smaller pipe was more damaged than larger, and fittings, tees and elbows were the worst damaged. Wherever the pipe crossed bridges, pipes were broken. There were 83,600 services to houses - 80% of these houses were burned, and their services were destroyed, while services to the 20% unburned were undamaged.

Yokosuka's distribution system comprised 117,000 feet of pipe, about 50% of which was on reclaimed land. About 50% of joints required re-caulking. Pipe damage was greatest to smaller diameter pipe and to fittings etc, repeating the experience in Yokohama. Burned houses in Yokosuka resulted in destroyed service pipes. Distribution systems were constructed of cast iron. The pipes sustained substantial damage, mostly in smaller diameters, and at tees, elbows and other fittings, and mostly in softer ground.

A major impact was the additional damage to the distribution system as the fire grew in size, due to damage to service pipelines, as the buildings burned.

2.3.4 Impacts of Water Supply Damage on Firefighting

Analyses are not available as to whether Tokyo and Yokohama could have defended themselves against conflagration if meteorological conditions had been more favorable. In the actual event, conditions were extremely unfavorable - perhaps several hundreds ignitions occurred almost immediately, due to the lunch hour timing of the earthquake, at which time thousands of small grills were being employed. The ignitions were fanned by high winds, and grew rapidly in the densely built up neighborhoods of almost exclusively wooden buildings, which had been made more flammable by a recent dry period.

Tokyo and Yokohama, particularly the areas most heavily burnt, are low-lying. Damage to the distribution systems in these areas was heaviest, so that hydrants were probably often dry. As the fire grew, it impacted the water system in two mutually exacerbating ways:

- **Demand**: In general, fire growth is exponential, for a plentiful fuel supply. While the perimeter, which must be defended, will grow as the square root of the area of the fire, in general the net result is that the firefighting water demands increase exponentially over time.

- **Supply**: As the fire grows in area, relatively more of the distribution system is available for supply, so that it would be expected that water supply increases. However, buildings within the fire are collapsing, breaking their service connections. In general, the net result is that the supply capacity of the distribution system is actually decreasing as the fire grows, due to the increased drain on the system due to hundreds or thousands of broken services.

This situation is typical of urban conflagrations, and undoubtedly existed in Tokyo, irrespective of the initial seismic damage to the distribution system.

The portions of Tokyo and Yokohama most heavily burnt are low-lying, with numerous canals and access to waterways. Therefore, secondary emergency water supply should have been available to the fire fighters. However, it is likely the equipment of the time was limited in capacity, and could not furnish the volume of water required to contain the large fires that quickly developed. Therefore, the primary factor leading to the conflagrations in Tokyo and Yokohama can be conjectured as not only due to the seismic damage to the water supply system, but also the rapid growth of numerous simultaneous ignitions under a situation of adverse meteorological and dense wood building conditions. These conditions would have overwhelmed the fire service, and an undamaged distribution system, even had there been no earthquake. As in San Francisco in 1906, and in the Oakland Hills in 1991, the spread of the fire caused damage to the water system (broken services), negatively impacting the capacity of the water system to supply flows at hydrants.

2.3.5 Summary of the 1923 Kanto earthquake

The 1923 Kanto earthquake resulted in strong shaking and widespread permanent ground deformations. In general, however, this did not severely damage the main water supplies to Tokyo and neighboring cities. The distribution systems, on the other hand, sustained numerous breaks, primarily in smaller pipes in low-lying ground. Notwithstanding the failure of the distribution system, the conflagrations which developed, and which were the main agent of damage in this catastrophe, were primarily due to non-seismic factors. These included a conflagration-prone built environment and an extremely adverse ignition scenario and meteorological conditions. Under these circumstances, it is likely the fire service would have proved inadequate, even had there been no earthquake-induced water system damage.

2.4 The 1989 Loma Prieta, California Earthquake and Fires

This section summarizes aspects of a significant fire following the 1989 Loma Prieta earthquake. The earthquake and damage are first briefly summarized, followed by a discussion of the effects on water supply components, and an analysis of the interaction of water supply damage to the occurrence of fires.

2.4.1 Seismological and Overall Damage Aspects

On October 17, 1989 at 5:04 pm local time, an M, 7.1 earthquake occurred due to approximately 40 km rupture along the San Andreas (or adjacent) fault. The epicenter of the 20 second earthquake was located near Loma Prieta in the Santa Cruz mountains about 16 km northeast of Santa Cruz, 30 km south of San Jose and about 100 km south of San Francisco. Major damage included the collapse of the elevated Cypress Street section of Interstate 880 in Oakland, the collapse of a section of the San Francisco-Oakland Bay Bridge, multiple building collapses in San Francisco's Marina district, and the collapse of several structures in Santa Cruz and in small rural towns in the epicentral region. Direct damage and business interruption losses were estimated as high as $6 billion. Human losses were 62 people dead, 3,700 people reported injured, and over 12,000 displaced. At least 18,000 homes were damaged, 960 were destroyed and over 2,500 other buildings were damaged and 145 destroyed.

2.4.2 Water Supply and Fire Protection in the City of San Francisco

San Francisco possesses three water supply systems, the first of which is typical of water systems in any other city. This system is called the Municipal Water Supply System (MWSS), owned and operated by the San Francisco Water Department (SFWD) and serving both fire fighting and municipal (potable water) uses. The two other systems are specifically dedicated to firefighting use and are owned and operated by the San Francisco Fire Department (SFFD). They were built following, and in direct response to, the 1906 earthquake and fire. These are the Auxiliary Water Supply System (AWSS), first developed following the 1906 earthquake and fire and extended periodically thereafter, Figure 2-14, and the Portable Water Supply System (PWSS), developed in the 1980's and primarily a truck-borne large diameter hose system, Figure 2-15.

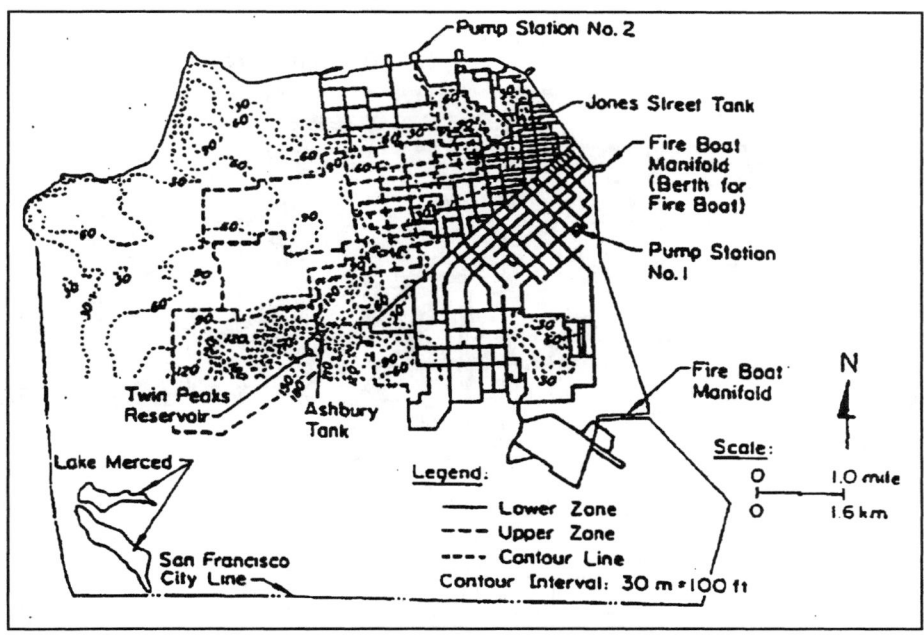

Figure 2-14 San Francisco AWSS Plan

The greatest damage to the MWSS in the 1989 Loma Prieta consisted of approximately 150 main breaks and service line leaks. Of the 102 main breaks, over 90 percent were in the Marina, Islais Creek and South of Market infirm areas. The significant loss of service occurred in the Marina area, where 67 main breaks and numerous service line leaks caused loss of pressure, Figure 2-16.

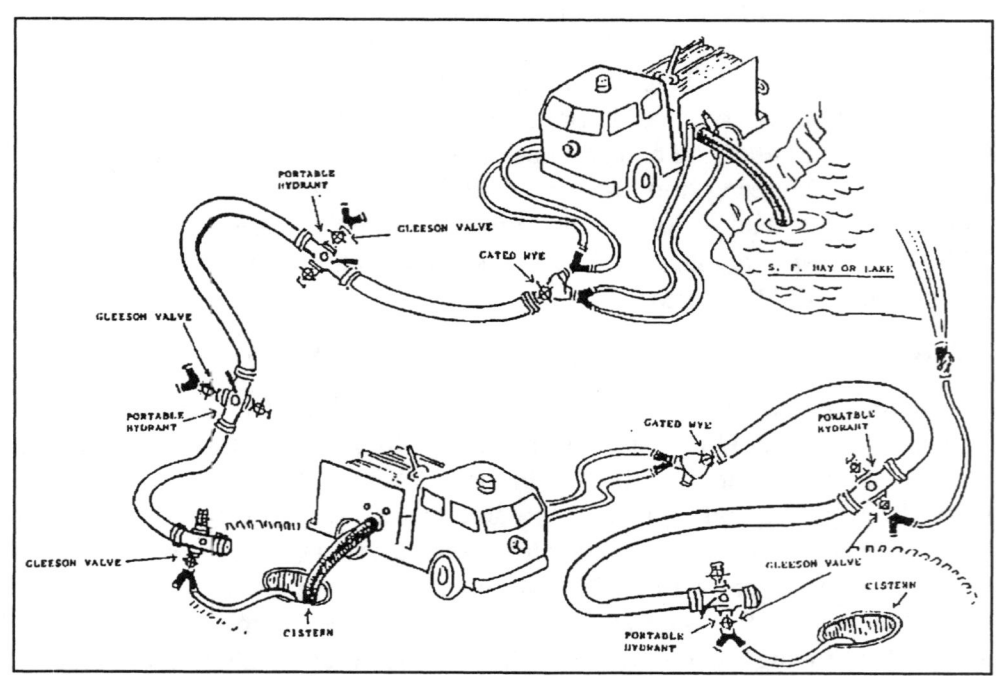

Figure 2-15 San Francisco PWSS

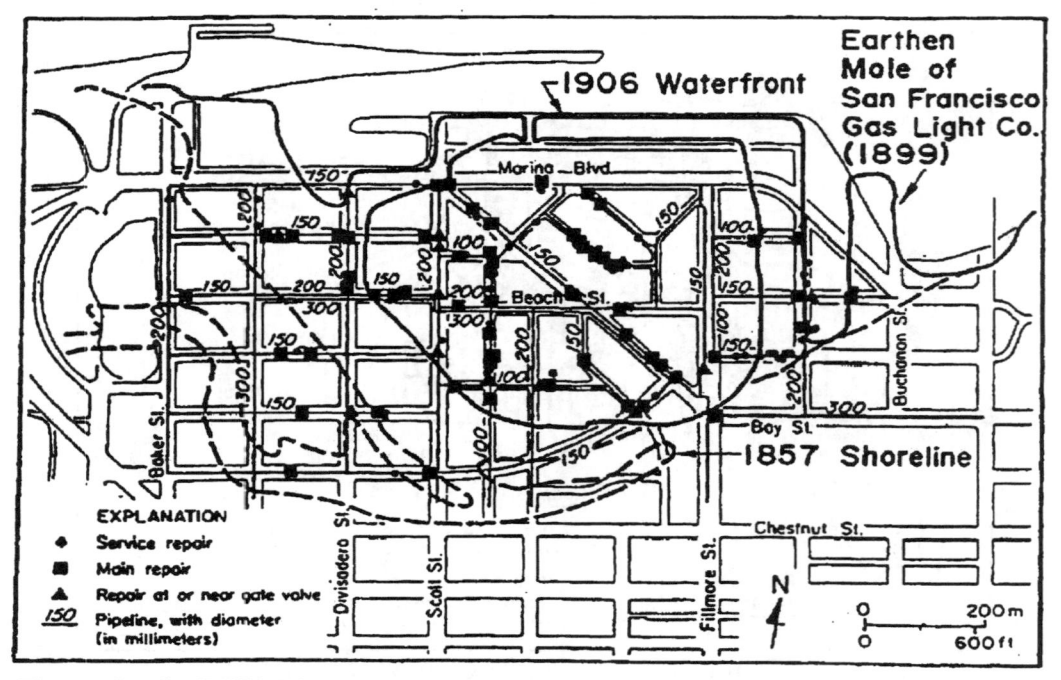

Figure 2-16 MWSS Pipe Breaks, Marina District, 1989 Loma Prieta Earthquake

The AWSS consists of several major components:

- Static Supplies: The main source of water under ordinary conditions is a 10 million gallon reservoir centrally located on Twin Peaks, the highest point within San Francisco (approximately 750 ft. elevation).

- Pump Stations: Because the Twin peaks supply may not be adequate under emergency conditions, two pump stations exist to supply salt water from San Francisco Bay - each has 10,000 gpm at 300 psi capacity.

Both pumps were originally steam powered but were converted to diesel power in the 1970's.

- Pipe Network: The AWSS supplies water to dedicated street hydrants by a special pipe network with a total length of approximately 120 miles. The pipe is bell and spigot, originally extra heavy cast iron (e.g., 1" wall thickness for 12" diameter), and more recent extensions are heavy ductile iron (e.g., .625" wall thickness for 12" diameter). Restraining rods connect pipe lengths across joints at all turns, tee joints, hills and other points of likely stress.

- Fireboat *Phoenix*[2]: The pipe network has manifold connections located at several points along the City's waterfront in order to permit the City fireboat *Phoenix* to act as an additional "pump station", drafting from San Francisco Bay and supplying the AWSS. The *Phoenix*'s pump capacity is 9,600 gpm at 150 psi, about the same as Pump Station No. 2.

- Cisterns: Lastly, in addition to the above components, San Francisco has 151 underground cisterns, again largely in the northeast quadrant of the City. These cisterns are typically built using concrete, 75,000 gallons capacity (about one hour supply for a typical fire department pumper).

The AWSS is a system remarkably well designed to furnish large amounts of water for firefighting purposes under normal conditions and contains many special features to increase reliability in the event of an earthquake.

The Loma Prieta earthquake resulted in only moderate damage for most of San Francisco, typically of MMI VI in rock- or firm-ground locations, although selected areas sustained much greater damage, perhaps as much as MMI IX in the Marina district. In the Marina district, 69 breaks in the domestic water supply and more than 50 service connections to water mains quickly dissipated all domestic water supply in the 40 blocks of the Marina district. The AWSS main serving the Marina district remained intact. However in

---

[2] Within days following the 1989 Loma Prieta earthquake, San Francisco Fire Department purchased a fireboat which Vancouver, B.C. had just discarded. Renamed the *Guardian*, the fireboat is arguably the largest in North America, with 20,000 Igpm pumping capacity. The *Phoenix* and the *Guardian* are both active as of this writing, with each alternately in service for one to several months, and the other in reserve. Both are stationed near the foot of Folsom St., close to the San Francisco-Oakland Bay Bridge.

locations other than the Marina, the AWSS sustained significant damage (see also Section 6.2.2.2.1):

- The most significant damage occurred on Seventh Street between Howard and Mission Streets, where a 12-inch main broke. This location is on the boundary of a filled-in swampy area and moreover, the AWSS pipe at this location crosses over a sewer line. Soil settlements in this area are thought to have occurred prior to and also as a result of the earthquake, causing the AWSS pipe over the sewer line to break.

- Other breaks included: (a) a break in an 8 inch hydrant branch, on Sixth Street between Folsom and Howard Streets (where the hydrant branch crossed up and over a sewer line) and (b) five 8 inch elbow breaks, four within filled-in swampy areas and including one on Bluxome Street where a portion of a building collapsed onto an AWSS hydrant.

Major leakage resulted from these breaks such that Jones Street tank (controlling pressure for the Lower Zone of the AWSS) had completely drained in approximately 15 minutes. Leakage continued so that first arriving engines at the Marina fire found only residual water when they connected to AWSS hydrants, as described above. Due to uncertainty as to the number and location of AWSS breaks, valves connecting the Upper Zone to the Lower Zone were not opened, and Pump Stations 1 and 2, although available, were not placed in-service immediately but only at 8 PM, following identification and isolation of broken mains. As a result all pressure in the AWSS Lower Zone was lost for several hours following the earthquake. The pump stations were operated at half capacity so as to fill the AWSS mains slowly out of concern for entrapped air, which was exhausted out of the Lower Zone through Jones Street tank (air could be heard exhausting from the tank). This operation continued until 10 PM when full pressure was restored and Jones Street tank had been filled with salt water.

Other damage was confined to:

- One 75,000-gallon cistern at Fifth and Harrison streets developed a leak at the cold joint between the roof and sidewall due to earthquake damage and lost 20 percent of its water, leaving 60,000 gallons for fire suppression purposes.

- Falling structures destroyed one High Pressure hydrant and damaged another.

2.4.3 Ignitions

Twenty-six fires occurred in San Francisco as a result of the earthquake, 11 on the 17th (SFFD, 1990). One of these fires occurred in the Marina District, and threatened to become a major conflagration. Firefighting efforts were severely hampered due to lack of MWSS and AWSS service to hydrants, due to the severe liquefaction and resulting pipe breakage in the Marina and elsewhere. Firefighters were forced to resort to drafting from nearby lagoons that however was inadequate, and the fire continued to grow. Deployment of San Francisco's PWSS in conjunction with the fireboat *Phoenix* provided the only adequate source of firefighting water, which was the only way the Marina fire was extinguished.

2.4.4 Marina Fire and use of the PWSS

A fire began in the four-story wood-frame building at 3701 Divisadero Street at the northwest corner of Beach Street. The building is a typical corner building in the Marina district. It was built in the 1920s and contained 21 apartments, with the ground floor being primarily a parking garage. The building's lower two floors had collapsed in the earthquake, and the third and fourth floors were leaning southward several feet. The fire was in the rear of the building and initially was small. This, combined with the confusion following the earthquake, resulted in a delayed report such that the first San Francisco Fire Department (SFFD) unit did not arrive until approximately 5:45 P.M. (all times estimated). The source of ignition has not been definitely determined as of this writing. Wind speed was virtually zero. Engine 41 closely followed arrival of SFFD Trucks 10 and 16. Based on the appearance of black smoke, the fire appeared to the officer in charge of E41 (Lt. P. Cornyn) to be a wood- structure-fueled fire. E41 connected to the AWSS hydrant directly in front of 3701 Divisadero (the building actually was leaning over the hydrant) and charged the pump, but found no water pressure, Due to radiant heat, E41 then withdrew across the street. At about 6:25 P.M., the building at the northeast corner (2080 Beach) ignited.

At about 6:30 pm 3701 Divisadero exploded (the fuel source was most likely leaking gas), flames shot 100 ft. or more into the air, buildings across both Divisadero and Beach streets were either burning or smoldering, and the fire had spread to the north and west neighboring buildings. At about this time, E41 and E10 were protecting the buildings across Divisadero and Beach streets. E41 was being supplied by E16 relaying from a hydrant at Scott and Beach streets, while E10 was being supplied by E21 drafting from a lagoon of the Palace of Fine Arts (located two blocks to the west) and relaying. By 6:50 pm, the fire was burning northward; E3, E36, E22, E31, E14, E25, HT25, and an additional fire reserve engine also were on the scene. E22 attempted to draft from the Marina lagoon, but was unable, due to low tide.

Because the fire was located only two blocks from the Marina, the fireboat *Phoenix* was called for, arriving at about 6:30 pm. At approximately the same time, PWSS hose tenders arrived at the scene and were able to connect to the *Phoenix*. In all, three PWSS hose tenders responded to the Marina fire (a fourth was in the shop being outfitted for service). Four major runs of hose (or portable water mains) were laid at the Marina fire, with some 6,000 ft. of 5-in. hose being deployed, using nine portable hydrants. The *Phoenix* pumped 6000 gpm at 180 psi for over 18 hours. Fire spread was stopped at about 7:45 pm. by master streams from the monitors on the hose tenders, as well as ladder pipes and hand lines.

Due to uncertainty as to the number and location of AWSS breaks, valves connecting the Upper Zone to the Lower Zone were not opened. Pump Stations 1 and 2, although available, were not placed in service immediately, but rather at 8 pm following identification and isolation of broken mains. As a result, all pressure in the AWSS Lower Zone was lost for several hours following the earthquake. The pump stations were operated at half capacity so as to fill the AWSS mains slowly, out of concern for entrapped air that was exhausted from the Lower Zone through the Jones Street Tank (air could be heard exhausting through the tank). This operation continued until 10 pm, when full pressure was restored and the Jones Street Tank had been filled with salt water. Figure 2-17 indicates the deployment of various SFFD apparatus, including the fireboat *Phoenix*, at about 6:24 pm.

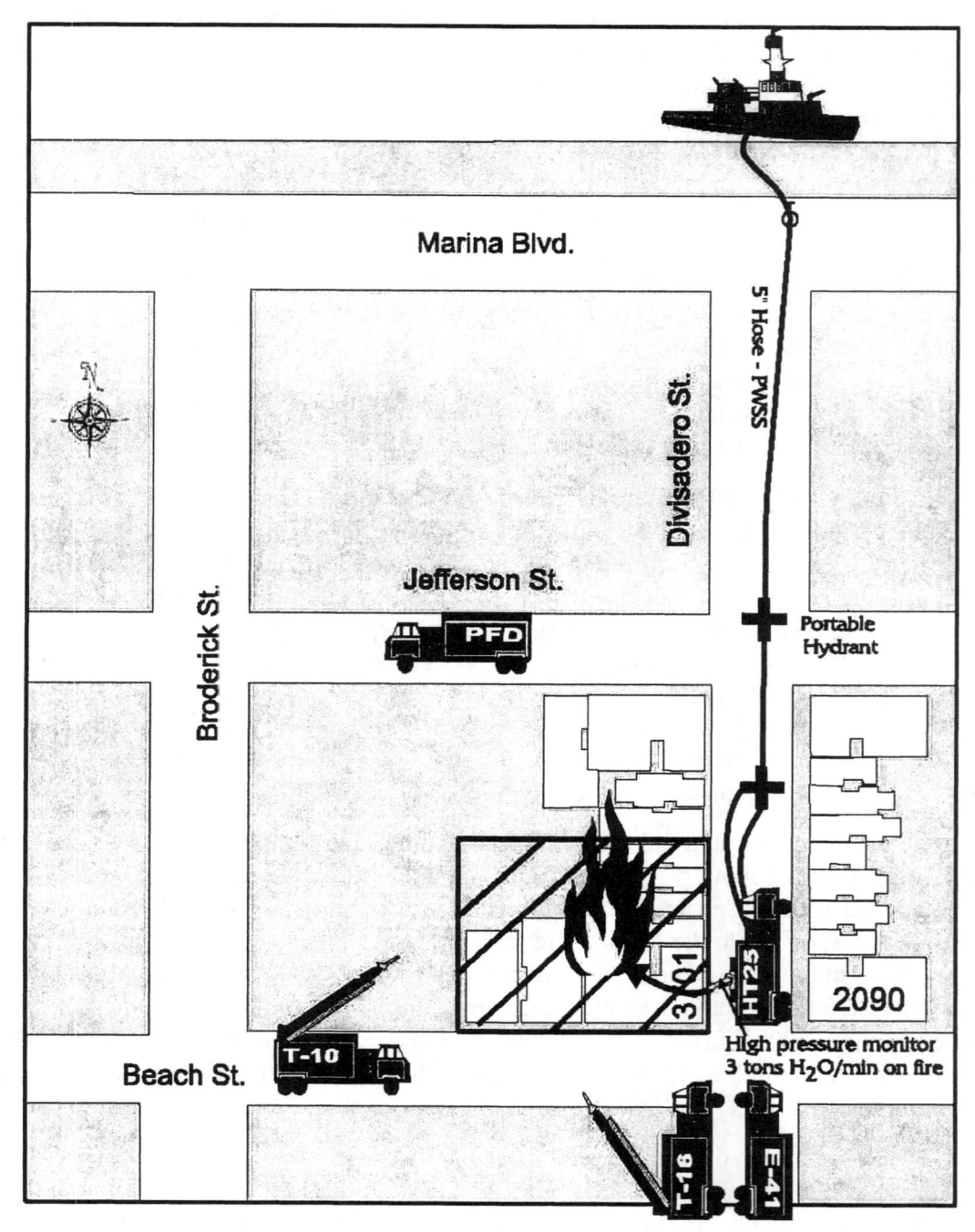

Figure 2-17 1989 Loma Prieta Earthquake, Marina Fire, SFFD deployment

2.4.5 Summary

The 1989 Loma Prieta earthquake provided a number of valuable observations and lessons, including:

- The Marina fire was potentially very severe - it was a very large fire in a dense neighborhood of wood frame construction - an unusually calm wind was a very fortuitous circumstance

- The fire was within 500 ft. of San Francisco Bay and the Pacific Ocean - the largest body of water on earth. However, these could not be drafted from by arriving fire engines, and the water was inaccessible.

- The MWSS system had over 400 million gallons of storage within San Francisco, but the numerous breaks in the Marina prevented adequate pressure or volume at Marina hydrants - elsewhere in the City, MWSS performance was generally satisfactory.

- The AWSS is designed for earthquake ground motions, and did not sustain damage in the Marina despite widespread liquefaction - nevertheless, it lost pressure in the Lower Zone due to breaks several miles away.

- The "backup to the backup" - that is, the PWSS backing up the AWSS which backs up the MWSS, provided firefighting water for extinguishment at the Marina fire. The PWSS' flexibility and portability proved adequate to the task.

2.5 The 1994 Northridge, California Earthquake and Fires

The January 17, 1994 M_w 6.8 Northridge earthquake was the largest earthquake to occur within a US city in more than 20 years, the previous event being the February 9, 1971 San Fernando M_w 6.7 earthquake, practically a duplicate of the Northridge event. The two events of similar magnitude had nearby epicenters, affected the same area, and occurred in the early morning in the winter. Both events caused over 100 ignitions, but no conflagrations. This section discusses the 1994 event.

2.5.1 Seismological and Overall Damage Aspects

The 4:31 AM January 17, 1994 M_w 6.8 earthquake was centered under the Northridge section of the San Fernando Valley area of the Los Angeles region. The event resulted in Modified Mercalli Intensity (MMI) shaking intensities greater than MMI VIII over approximately 700 square miles of the northern Los Angeles area. The population most heavily affected was in the San Fernando Valley, which is primarily protected by the Los Angeles City Fire Department.

2.5.2 Fire Service Resources

Table 2-2 lists fire departments significantly affected by the earthquake, and their summary statistics (see Scawthorn et al 1997, for additional detail).

Table 2-2 Fire Departments Affected by the 1994 Northridge Earthquake

Fire Department	Est. Popul. (thousands)	Area (Sq Miles)	No. Fire Stns.	No. FD Personnel	No. Engines
Los Angeles City	3,400	469	104	2,865	104
Los Angeles County	2,896	2,234	127	1,842	144
Ventura County	700	126	30	327	40 +/-
Santa Monica	97	8	4	100	5
Burbank	94	17	6	120	6
Pasadena	132	23	8	150	8
Glendale	166	30	9	167	9
South Pasadena	25	3	1	27	2
Beverly Hills	34	6	3	81	7
Culver City	41	5	3	66	5
Fillmore	12	2	1	9	1

As Sepponen (1997) reports:

> *At the time of the earthquake there were 788 LAFD personnel on duty. During the quake firefighters were injured, and several fire stations suffered major damage. Most of the city was without electrical power. Structural damage was noted in 35 station buildings. Although most were repaired quickly, Station 90 was closed for one month for roof repair, Station 70 was out of service for six months, and Station 78 was condemned and demolished. (Ward, 1995) Once equipment and personnel were safely outside of buildings, reconnaissance patrols were sent out as designated in the Earthquake Emergency Operational Plan. The timing of the earthquake at 4:31 a.m., was advantageous for the LAFD because their 24-hour shift change occurs at 6:30 a.m. each day. So soon after the earthquake occurred the new shift came to work and the other "old" shift continued to work due to the emergency, doubling the number of personnel available. This enabled reserve equipment to be easily put into service.*

2.5.3 Water System Performance

The Northridge earthquake affected the water supply for portions of the San Fernando Valley (Heubach, 1997). Breaks occurred in at least six trunk lines and a large number of leaks occurred at other locations. The Department of Water and Power estimated that the earthquake caused approximately 3,000 leaks, including two lines of the Los Angeles Aqueduct. The damage to the system resulted in dropping the water pressure to zero in some areas. On January 22nd, five days after the earthquake, between 40,000 and 60,000 customers were still without public water service, and another 40,000 were experiencing intermittent service.

2.5.4 Ignitions and Fires

Approximately 110 fires were reported as earthquake-related on January 17, as shown in Figure 2-1 and Table 2-3 (details of ignition location etc are provided in Scawthorn et al, 1997). The Northridge earthquake reportedly caused or was a contributing factor in 77 fires in the LAFD service area. The 77 fires were among a total of 161 fires that occurred on the day of the earthquake. The time line in Figure 2-19 shows all calls for assistance with fires on the day of the earthquake. Structure fires predominate (86%) the earthquake-related fires. More than 70% (66) of the earthquake-related fires occurred in single- or multiple-family residences, as might be expected from the building stock that is typical in the San Fernando Valley. The major cause of ignition was electric arcing as the result of a short circuit, although gas flame from an appliance is also a recurring source of ignition. The breakdown of the day's calls into dispatches categories is shown in Figure 2-20.

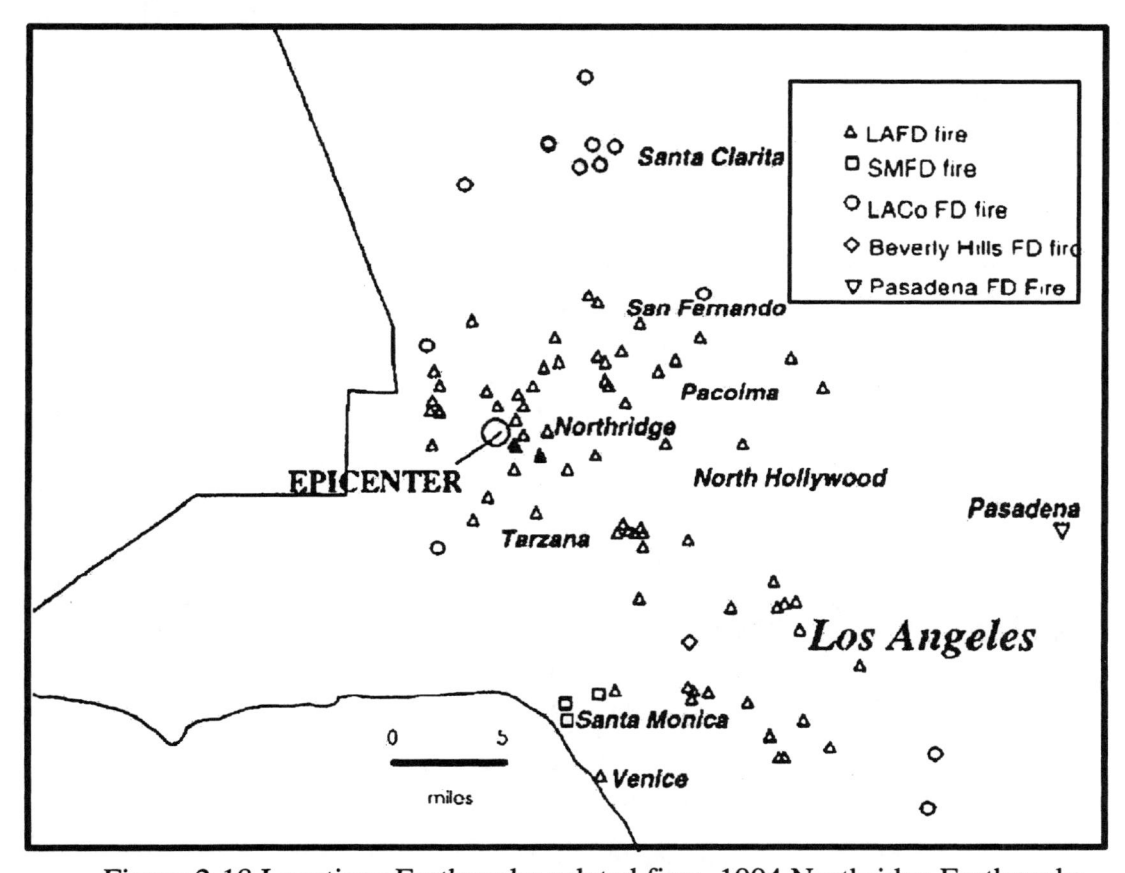

Figure 2-18 Locations Earthquake-related fires, 1994 Northridge Earthquake

Table 2-3 Fire Following the January 17, 1994 Northridge Earthquake

Jurisdiction	No. Earthquake-Related Fires
Los Angeles City	77
Los Angeles County	~15
Ventura County	~10
Santa Monica	4
Burbank	0
Pasadena	1
Glendale	0
South Pasadena	0
Beverly Hills	1
Culver City	0
Fillmore	2
TOTAL	**~110**

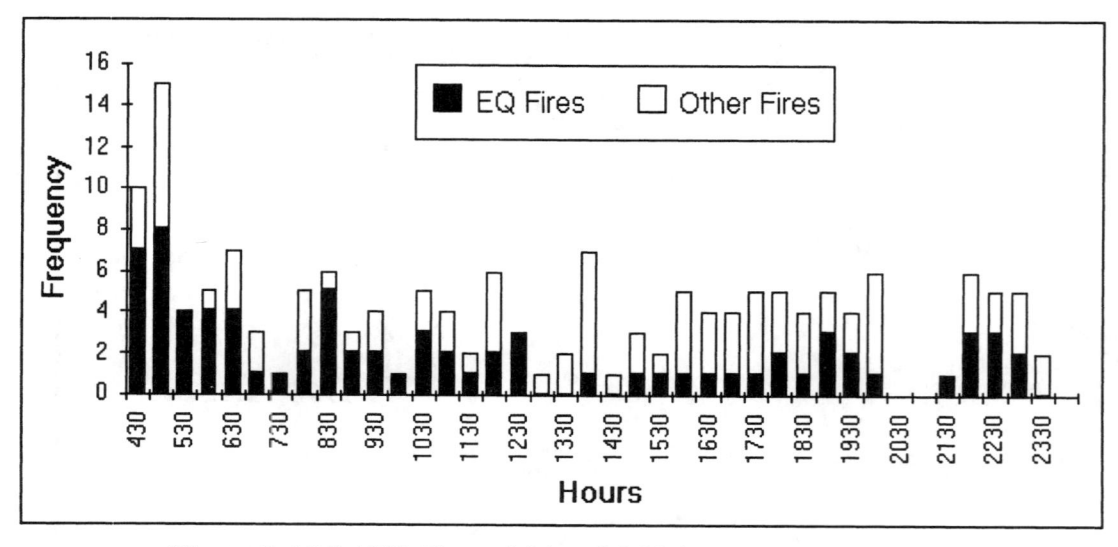

Figure 2-19 LAFD Fires, 4:31 to 24:00 hrs, January 17, 1994

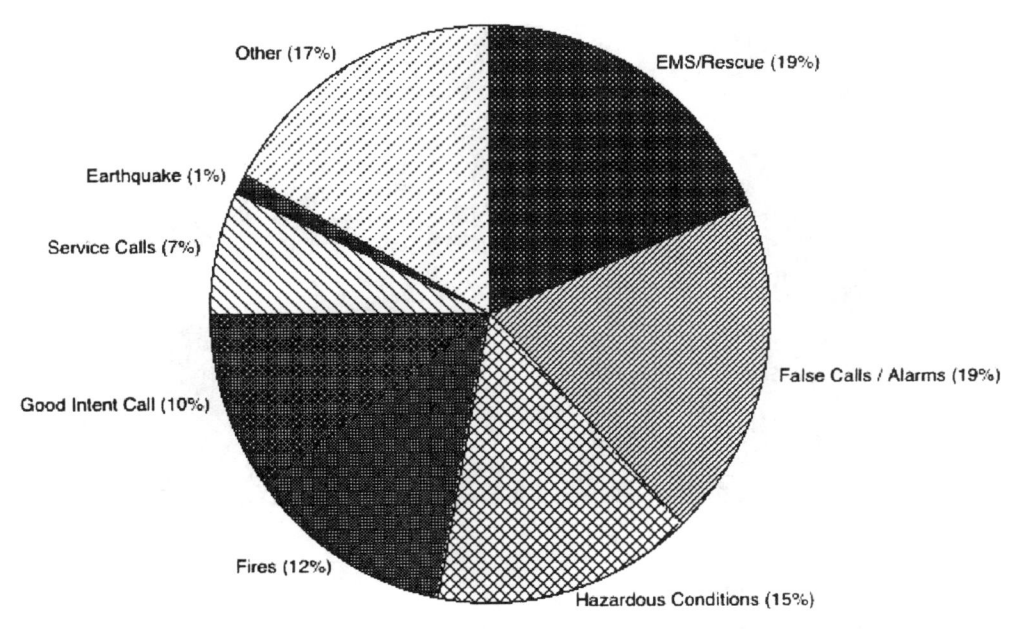

Figure 2-20 LAFD Incident Response Types, 1308 Incidents
4:31 to 24:00 hrs, January 17, 1994

Scawthorn et al (1997) have documented a number of specific fires and fire department operations, as well as all ignitions, in this event. The North Balboa Boulevard fire is summarized in the next section, as being of particular interest.

2.5.5 North Balboa Blvd. Fire

2.5.5.1 *Site Description*

This fire scene is located in the Granada Hills area of the San Fernando Valley. It is a residential area with one- and two-story wood-frame single-family dwellings, many with swimming pools, Figure 2-21 (top). A 48-inch water main under the street was broken, flooding the street and front yards of the homes, Figure 2-21 (bottom).

2.5.5.2 *Location/Ignition/Cause.*

A broken 20-inch gas main under Balboa Boulevard was ignited by the driver of a nearby stalled pick-up truck who was attempting to start the vehicle. Electric arcing in the ignition system ignited a large gas cloud, creating a fireball and igniting two dwellings on the east side of Balboa and three on the west side. Radiant heat from the gas fire was a major factor in the spread of fire. Wind was 15 to 20 mph from the northeast. Ignition occurred about 20 minutes after the earthquake struck. A total of five homes were destroyed, with minor damage to four others.

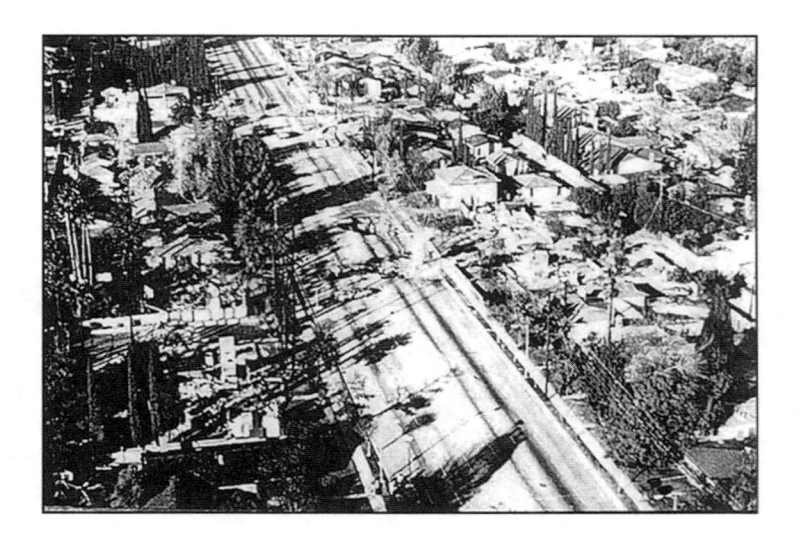

Figure 2-21 North Balboa Blvd. Incident, 1994 Northridge Earthquake
(top) Aerial View, (bottom) Surface View

2.5.5.3 *Fire Department Operations:*

Fire fighters from Engine 8 and Engine 18 were out on district survey, saw the fire, and responded. Engine 74 responded to a radio request for assistance from Engine 8. Fire fighters from Engine 8 arrived first and found Balboa Boulevard impassable due to the water flowing from the broken water main. Captain Rust took Engine 8 around the streets parallel to Balboa Boulevard (Paso Robles and McLennan Avenues) and cross streets (Lorilland and Halsey) to check the fire hydrants for water. They were dry. Engine 8 fire fighters entered the alley west of Balboa to protect the structures on that side of the fire. They located a swimming pool behind a home on Paso Robles and used it as a water source. Water from this swimming pool was also supplied to Engine 18 at the south end of the alley. Engine 18 fire fighters entered the alley west of Balboa and set up to protect the homes at the south end. The heat from the fires was intense and forced firefighters to operate from protected areas.

A Los Angeles County brush fire hand crew arrived on the scene as a mutual aid resource and was directed to cut and remove combustible shrubs, trees, fences, etc. around homes exposed to the fire.

Fire fighters from Engine 74 arrived on the scene, checked the hydrants on the north side of the fire, found them dry, and entered the alley east of the fire. The alley was impassable due to debris from collapsed block walls. Resident volunteers removed the debris, and Engine 74 fire fighters proceeded south to use a swimming pool for a water source. Engine 74 fire fighters extinguished a fire in the attic of an exposed one-story dwelling and continued to direct water streams on the exterior of this building.

A group of local citizen volunteers formed a "bucket brigade" on the northeast side of the fire using a swimming pool for a water source. They protected the house exposed to the fire at that location.

Engine companies 8, 18, and 74 pumped water between 1 1/2 and 2 hours during the firefighting operation. It took about 2 hours for the natural gas leak fire to be reduced in size such that it presented a minimal threat from radiated heat.

The Incident Commander at the scene, Captain Rust, directed operations on the west side between his company and Engine 18, and coordinated efforts with Engine 74 on the east side. Heavy radio traffic use made radio communications very difficult.

An aerial photograph of the scene after the structure fires were extinguished is shown in Figure 2-21(b). Note that the ruptured gas main is still burning.

2.5.5.4 Water-related Aspects.

Breaks in the water mains rendered all surrounding fire hydrants inoperative. Fortunately, several homes in the area had swimming pools that were used as water supply sources. Engine 8 and Engine 74 fire fighters used their 1 1/2-inch siphon ejectors to draw water into their tanks. The 1 1/2-inch siphon ejector can supply water at 92 to 115 gpm. The swimming pools provided approximately 70 minutes of water flow. Hose layout as of 5:35 am, and water usage for the incident, are shown in Figure 2-22 and Table 2-4, respectively.

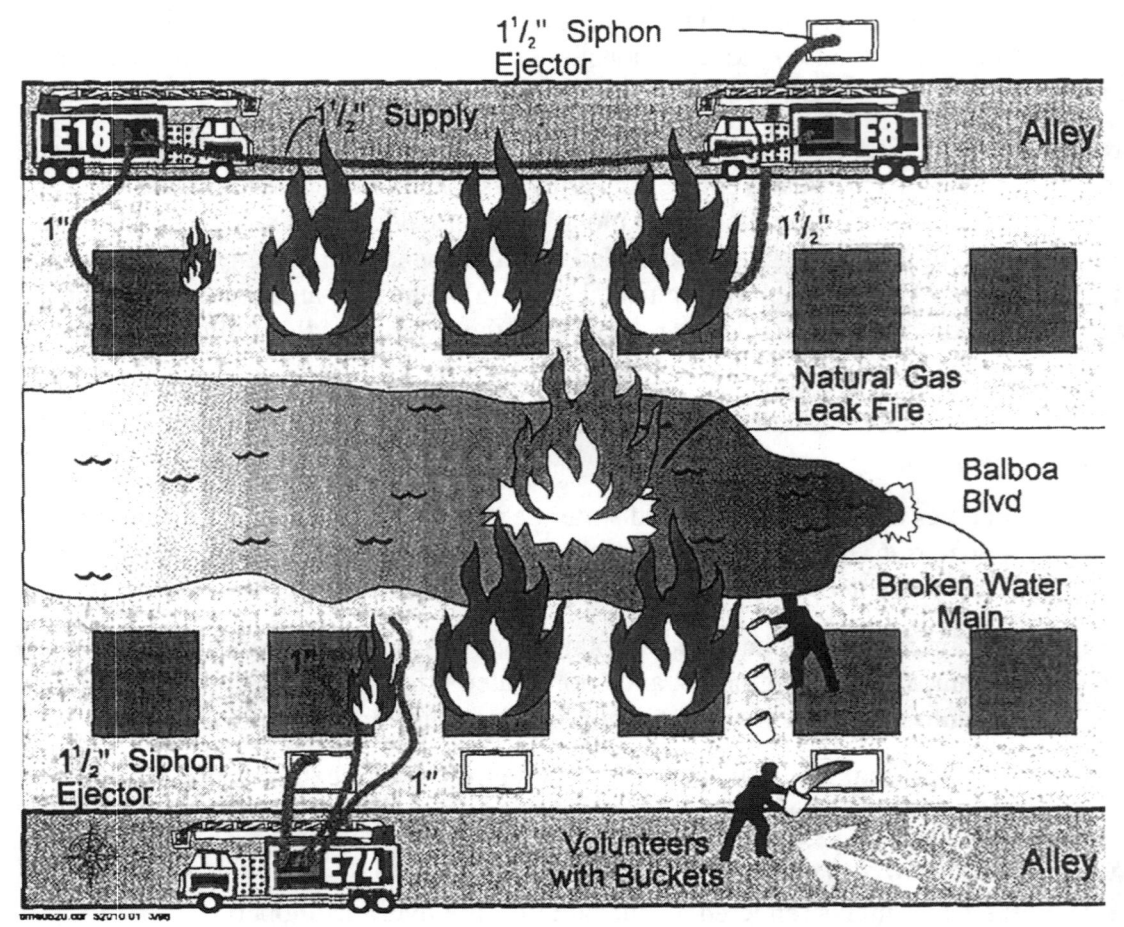

Figure 2-22 LAFD Deployment 5:35 am, No. Balboa Blvd., 1994 Northridge Earthquake

Table 2-4 Water Usage, Balboa Blvd. Fire

Engine 8	One 1 1/2-inch siphon ejector in pool supplying approx. 100 gpm
	One 1 1/2-inch supply line laid to Engine 18 for their water source
	One 1 1/2-inch tip line with spray tip - 125 gpm TOTAL: 8,750 gallons
Engine 18	One 1 1/2-inch supply line in to fill tank
	One 1-inch line with spray tip - 25 gpm TOTAL; 1,750 gallons
Engine 74	One 1 1/2-inch siphon ejector in pool supplying approx. 100 gpm Two 1-inch lines/spray tips 50 gpm TOTAL: 3,500 gallons
Total Estimated Water Employed to Control/Extinguish Fires: 14,000 Gallons	

2.6 The 1995 Hanshin (Kobe), Japan Earthquake and Fires

2.6.1 Seismological and Overall Damage Aspects

The 5:46 AM January 17, 1995 Mw 6.9 (JMA M7.2) Hanshin (official name: Hyogo-ken Nambu) earthquake was centered under the northern tip of Awaji island near Kobe, in the Kansai region of Japan. The event resulted in Modified Mercalli Intensity (MMI) shaking intensities greater than MMI VIII did over approximately 400 square km of the Kobe-Ashiya-Nishinomiya area. Population of the affected area (MMI VIII or greater) is approximately 2 million.

2.6.2 Fire Service Resources

The Kobe Fire Department (KFD) is a modern, well-trained fire response agency, organized into Prevention, Suppression, and General Affairs sections, and a Fire Academy. The city is served by 1,298 uniformed personnel. Equipment includes two helicopters, two fireboats, and 196 vehicles.

2.6.3 Ignitions and Fires

Approximately 100 fires broke out within minutes, primarily in densely built-up, low-rise areas of the central city, which comprise mixed residential-commercial occupancies, predominantly of wood construction. Within 1 to 2 hours, several large conflagrations had developed. There were a total of 108 fires reported in Kobe on January 17 (Kobe FD, 1995), the majority being in the wards of Higashi Nada, Nada, Hyogo, Nagata, and Suma. Fire response was hampered by extreme traffic congestion, and collapsed houses, buildings, and rubble in the streets. Because of the numerous collapses, many areas were inaccessible to vehicles. Table 2-5 lists time of ignitions for Kobe and other cities, and Figure 2-23 is an aerial view of one of the burnt areas.

Table 2-5 Post-Earthquake Fire Ignitions, Jan. 17, 1995 Hanshin Earthquake

Kobe City Ward	~6:00	~7:00	~8:00	~9:00	~24:00	1/17 Total	1/18 Total	1/19 Total	1/17~19 Total
Higashi-nada	10	1	2	1	3	17	2	4	23
Nada	13	0	1	1	2	17	2	0	19
Chuo	8	4	2	1	5	20	3	3	26
Hyogo	11	0	2	2	2	17	4	3	24
Nagata	13	1	0	0	3	17	1	4	22
Suma	4	4	0	4	1	13	2	1	16
Tarumi	0	0	0	0	6	6	0	0	6
Kita	0	0	0	0	1	1	0	0	1
Nishi	1	0	0	0	0	1	0	0	1
Kobe City Total	**60**	**10**	**7**	**9**	**23**	**109**	**14**	**15**	**138**

Cities other than Kobe	~6:00	~7:00	~8:00	~9:00	~24:00	1/17 Total	1/18 Total	1/19 Total	1/17~19 Total
Ashiya	4	4	1	0	0	9	2	2	13
Nishinomiya	11	11	1	1	10	34	4	3	41
Takarazuka	2	0	0	0	2	4	-	-	4
Itami	2	2	2	1	0	7	-	-	7
Kawanishi	1	1	0	0	0	2	-	-	2
Amagasaki	3	2	1	0	2	8	-	-	8
Awajicho	1	1	0	0	0	2	-	-	2
Osaka	7	4	1	1	2	15	-	-	15
Toyonaka	3	1	0	1	0	5	-	-	5
Suita	1	1	0	0	0	2	-	-	2
Other Cities Total	**35**	**27**	**6**	**4**	**16**	**88**	**6**	**5**	**99**

Note: "-" means no report.

Figure 2-23 Aerial View Burnt Area, 1995 Kobe Earthquake

2.6.4 Water System Performance

Firewater is primarily from the city water system, served by gravity from about 90 tanks and reservoirs. Of these, 22 sites include dual tanks, with one tank having a seismic shutoff valve so that, in the event of an earthquake, one tank's contents is conserved for preserving and emergency supply of drinking water (or fire fighting) purposes. In this event, all but one valve functioned properly, conserving 30,000 cubic meters of water, which, however, could not be delivered because of approximately 1,750 breaks in the underground piping. Kobe has approximately 23,500 fire hydrants, typically flush-mounted (i.e., under a steel plate in the sidewalk or street) with one 150-millimeter-diameter hose connection. The city has provided underground storage of water for disaster fire fighting in 968 cisterns, mostly ranging from 10,000-liter to 40,000-liter capacity, sufficient for up to about a 10-minute supply of a pumper. All engines carry hard suction, so that additional water can be drafted from Osaka Bay or the several streams running through Kobe, although this strategy was not particularly effective in this earthquake.

Kobe sustained approximately 1,750 leaks and breaks in its underground distribution system. Water for fire-fighting purposes was available for 2 to 3 hours, including the use of underground cisterns. Subsequently, water was available only from tanker trucks. KFD attempted to supply water with a fireboat and relay system, but this was unsuccessful due to the relatively small hose used by KFD. In an overflight of the area at about 5:00 p.m. on January 17, Scawthorn observed all of the larger fires (about eight in all) from an altitude of less than 300 meters. No fire streams were observed, and all fires were burning freely—several with flames 6 meters or more in height. No fire apparatus were observed in the vicinity of the large fires, although fire apparatus could be seen at other locations (their activities were unclear from the air). Some residents formed bucket brigades (with sewer water) to try to control the flames.

2.6.5 Comparative Analysis: Hanshin and Northridge events

Selected aspects of the 1994 Northridge and 1995 Hanshin earthquakes are compared in Table 2-6. Several key observations include:

Earthquakes in urban areas continue to cause multiple simultaneous ignitions, and degrade emergency response due to impaired communications, transportation and water supply

These events are replicable, as shown by comparison of the 1971 San Fernando and 1994 Northridge events (Scawthorn, et al, 1995), and by comparison of the ignition rates and other factors in the Northridge and Hanshin events, providing some validation for simulation modeling and projections for larger events

Under adverse conditions, large conflagrations are possible in modern cities, as shown by events in California (i.e., the 1991 East Bay Hills Fire, and the 1993 Southern California wildfires), and by the Hanshin earthquake in Japan.

Table 2-6 Hanshin and Northridge Earthquakes: Comparative Analysis

Aspect	Factor	Northridge	Hanshin
Event	Magnitude (M_w)	6.8	6.9
	Date (winter)	Jan 17	Jan 17
	Time	0431	0546
Region	Population (MMI 8)	1~1.5 million	2 million
	Density (pop/sq km)	1,000~1,500	4,000
Ignitions	Number (total)	110	108
	Structural Fires	86%	97%
	Rate (MMI 7) pop/Ignition:	14,719	13,676
Response	FD Communications	manual dispatch	
	Resources (pop/firefighter):	1,338	1,138
	Stations	104	26 (Kobe)
	Traffic Congestion	Minor	Major
	Mutual Aid	Available - not needed	after 10 hrs
Water	Water System Damage	Some	Total?
	Cisterns	Swimming Pools	946, mostly 10 to 40 tons (10 mins)
Wind		Calm	Minor
Gas	Automatic Shut-offs	? few %	70% - ineffective due to structural collapse
Spread		Minor	Major: 5,000 bldgs.

2.7 The 1999 Marmara, Turkey, Earthquake and Fires

The 1999 Marmara, Turkey earthquake was a catastrophic event claiming over 17,000 lives. It did not, however, result in any conflagrations, due primarily to the fact that the overwhelming building material in the region is non-combustible reinforced concrete, or masonry. It did result in a major fire at a petroleum refinery, which burned for several days and which was only suppressed by special fire brigades brought in from other countries.

2.7.1 Seismological and Overall Damage Aspects

The M_w 7.4 Kocaeli (Izmit) earthquake occurred at 3:10 am local time 17 August 1999 on the east-west trending north strand of the North Anatolian Fault Zone (NAFZ), about 100 km. SE of Istanbul (Scawthorn, 2000). The 125 km long fault and high damage area follows or is close to the south shore of Izmit Bay, and has predominantly 2.2 m right lateral displacement, from Adapazari in the east to Yalova in the west. Significant vertical fault scarps of as much as 2 m occur at several locations. Peak ground accelerations of approximately 0.4g were recorded near the fault, and liquefaction and subsidence were observed on the shores of Izmit Bay and Lake Sapanca. Several million persons live in the Izmit region, which has experienced rapid growth and heavy industrialization in the last two decades. The predominant building type is mid-rise non-ductile RC frames with hollow clay tile infill, thousands of which collapsed in a *pancake* mode. Estimated population requiring short to long-term shelter was 200,000. Lifelines generally performed well, with the exception of underground piping in the heavily affected areas, where major damage was reported (Tang, 2000). Electric power, highways, rail and telephone were generally functional within several days following the earthquake, and the pipelines of the Thames Water Project (the regional water supply and transmission system) suffered only leaks and was able to be kept in service for several days after the earthquake, before being shutdown for repairs (Eidinger at al, 2002). Fires occurred in a number of collapsed buildings but were generally confined to building of origin. Two fires broke out at the Tupras oil refinery, which burned for several days, and which are of especial interest.

2.7.2 Tupras petroleum refinery fire

The most widely publicized and spectacular damage to any industrial facility occurred at the massive USD 3.5 billion petroleum refinery near Korfez, owned by the state-owned oil company, Tupras. The damage was at the tank farm, where fires burned out of control for several days (Johnson et al, 2000). The first fire was initiated in a floating roof naphtha tank. Naphtha is a highly volatile material with a low flashpoint, and is easily ignited. The speculation is that the sloshing of naphtha in the tank caused the floating roof to breach its seal, allowing naphtha to spill. Sparks from the friction between the steel roof and tank wall likely ignited the naphtha.

The refinery receives its entire water supply through a dedicated pipeline from Lake Sapanca, some 45 km to the east. Due to multiple breaks in the pipeline, the refinery quickly lost all water and all fire-fighting capabilities. As the fire spread to additional tanks, aircraft attempted to douse the fires by dropping foam. After a few days, the refinery used diesel pumps to draw water directly from Izmit Bay to fight the fire, along with the aerial foam attack. The fires were finally declared under control on Sunday, some 5 days after the earthquake.

While the fire was burning out of control, an area within 2 to 3 miles of the refinery was evacuated, including some areas where search and rescue operations were taking place in collapsed buildings. Train service was disrupted in the area because of the fire.

The fire and heat eventually consumed numerous tanks in the tank farm. It was reported that at least 17 tanks were considered to be total losses. These tanks were generally buckled by the intense heat, with one tank expanding as if ready to explode.

The other area of severe and spectacular damage in the refinery occurred in one of their three crude units, when a 90-meter high reinforced concrete heater stack collapsed. The break appeared to occur at about the height of the large diameter heater duct. The cause of the failure was not immediately obvious from sifting through the rubble of the stack.

The top of the stack fell into the unit, destroying the heater, while the bottom portion fell into a pipeway running around the perimeter of the unit. The destroyed pipeway was heavily congested with piping from all over the refinery. It is likely to take several months to identify, isolate, and repair damaged piping in this area.

One of the pipes broken by the stack collapse was a naphtha line from the original burning naphtha tank in the tank farm. A fire started when the collapse occurred, and although it was extinguished relatively quickly, it flared up several times because of the new fuel from the broken pipe. The supply could not be stopped because the two block valves were at the tank, inaccessible because of the fire, and downstream from the crude unit.

Figure 2-24 Tupras Refinery damage due to fire following earthquake (a) Burnt and collapsed tank; (b) Top of stack collapsed into unit, severely damaging heater unit. Top of stack is in foreground of photo.

2.8 The 2000 Napa Earthquake

The City of Napa was shaken by a moderate-sized earthquake (M_L 5.2, M_W 5.0) on September 3, 2000 at 1:36 am local time. Recorded peak ground accelerations in the Napa area were as high as 0.49g, with recorded peak ground velocity of 15 inches per second (Eidinger, 2001).

The epicenter of the earthquake was at 38.3770 degrees latitude, 122.3137 degrees longitude. The hypocentral depth was 9.4 km. The epicenter was located 3 miles west south west of Yountville, and 9 miles northwest of central Napa. The earthquake occurred on an unnamed fault. The earthquake had a strike-slip mechanism.

Most of the developed area of the City of Napa is in an alluvial valley. The Napa River runs northwest to southeast through the city and the river is flood prone. No widespread evidence of liquefaction or landslides is known to have occurred in developed areas. There was no fault rupture in the developed area. In one section of town with known high ground water table (Westport), liquefaction was not known to have occurred, and there was no readily observed increased evidence of failure of buried utilities in that area.

The City of Napa operates a water distribution system. Raw water is from two dams on the north and east of town and the North Bay Aqueduct. There are three water treatment plants. Typical summer time usage is in the range of 20 million gallons per day. There are about 350 miles of buried pipeline in the water system.

The earthquake caused 22 pipe breaks and leaks in the water system. These 22 do not include numerous leaks on the customer side of the water meter. 18 of these repairs were reported on September 3; 4 more repairs were reported by September 8. The 22 pipe repairs include: 18 mains (4 inch to 8 inch diameter) and 4 0.75-inch service connections (up to the meter). Damage to mains included: holes in pipe (9); cracks or splits in barrel (9). Repair methods for mains used full circle clamps (7), bell clamps (3), cut-in sections of pipe (7), cut-in new fitting (1). Repair methods for the four damaged laterals included new laterals (3) and a clamp (1). At some locations where pipes were damaged, the City hooked up some customers using flex hose attached to hose bibs. This was reportedly done to more rapidly restore water supply to a few houses. All customers had some sort of water supply restored by 11:30 pm September 3. Of the 22 pipe repairs, no known damage could be attributed to obvious ground settlements or lateral spreads.

There was one fire ignition caused by the earthquake. It happened in a hotel, reportedly due to falling items that somehow created a fire ignition requiring fire department response. The City of Napa fire department responded and had this fire controlled before it spread (reportedly a room and contents fire).

The City of Napa and the County of Napa fire departments and law enforcement agencies responded to about 400 non-fire related incidents following the earthquake. These included about 30 medical incidents.

There were 30 natural gas leaks that the City of Napa Fire Department and Pacific Gas & Electricity (PG&E) responded to. These were primarily (possibly all) related to water heaters within buildings that toppled or slid, breaking their gas flex hose connection. There were no fire ignitions from these leaks.

The Fire Department also reported several instances of minor hazardous material spills; the Napa County Hazardous Materials Team responded to these incidents.

The City of Napa obtained mutual aid for law enforcement and fire department strike teams. The law enforcement aid was released by the morning of September 3. Five fire department strike teams (25 engines) were obtained through mutual aid to supplement the City of Napa fire department (10 engines); the fire department mutual aid teams were released the night of September 3.

2.9 Summary Observations

The accumulation of experience based on observations of the above events, and others which space does not permit discussing here, leads to the conclusion that the potential exists for large conflagrations following a major earthquake in an urban area, particularly in a region with a large wood building stock. Under adverse meteorological and other conditions, these conflagrations may burn for several days, replicating the events of 1906 in San Francisco, and 1923 in Tokyo. Extensive, well-drilled mutual aid systems are required, in order to mobilize large resources in response, but the deployment of these resources will be hampered by transportation difficulties and, perhaps most tellingly, failure of firefighting water supplies. Improvements in planning and infrastructure are absolutely essential to forestall this potential.

2.10 Credits

Figures 2-1, 2-2, O'Rourke, Beaujon and Scawthorn, 1992. Figure 2-3, USGS, 1907. Figure 2-4, NOAA. Figure 2-5, Arnold Genthe. Figure 2-6, USGS, 1907. Figure 2-7, Eidinger, 2003. Figure 2-8, Schussler 1906. Figure 2-9, Scawthorn and O'Rourke, 1989. Figure 2-10, Hamada et al, 1992. Figure 2-11, Scawthorn, 1997a. Figure 2-12, Okamoto, 1984. Figure 2-13, Japanese govt. report. Figure 2-14, 2-16, O'Rourke et al, 1992. Figure 2-15, Blackburn. Figure 2-17, Scawthorn. Figure 2-18, 2-19, 2-20, 2-22, Table 2-2, 2-3, 2-4, Scawthorn et al 1997. Figure 2-24, Gayle Johnson, EQE International. Table 2-5, Sekizawa 1997.

2.11 References

ASCE, 1929. Special Committee on Effects of Earthquakes on Engineering Structures with Special Reference to the Japanese Earthquake of September I, 1923, American Society of Civil Engineers San Francisco.

Bowlen, Fred J., Batt. Chf., S.F.F.D., Outline of the History of the San Francisco Fire, Original Manuscript

Freeman, J.R., 1932. Earthquake Damage and Earthquake Insurance, McGraw-Hill, New York.

Eidinger, J., Ed., 2001, Gujarat (Kutch) India M 7.7 Earthquake of January 26, 2001 and Napa M 5.2 Earthquake of September 3, 2000, American Society of Civil Engineers, Technical Council on Lifeline Earthquake Engineering, Monograph No. 19, June, 2001.

Eidinger, J., 2003, Economics of Seismic Retrofit of Water Transmission and Distribution Systems, 6th US Conference on Lifeline Earthquake Engineering, American Society of Civil Engineers, August.

Hamada, M., Wakamatsu, K. and Yasuda, S., 1992. Liquefaction-Induced Ground Deformations During the 1923 Kanto Earthquake: in Case Studies of Liquefaction and Lifeline Performance During Past Earthquakes, v. 1, Japanese Case Studies, M. Hamada and T.D. O'Rourke, editors, NCEER Tech. Report. NCEER 92-0001, National Center for Earthquake Engineering Research, SUNY-Buffalo, Buffalo.

Heubach, W. 1997. The 1994 Northridge, California Earthquake Water Supply Effects, in Ballantyne and Crouse (1997)

Johnson, G.S., Ascheim, M. and Sezen, H. 2000. Performance of Industrial Facilities, Chapter in Scawthorn, 2000.

Kanai, K., Engineering Seismology, U. Tokyo Press, Tokyo,1983.

Lawson, A. C. et al, 1908. The California Earthquake of April 18, 1906, Report of the State Earthquake Investigation Commission, in two volumes and atlas, Vol I, Carnegie Institution of Washington, Washington (reprinted 1969).

Manson, M., 1908. Report on an Auxiliary Water Supply System for Fire Protection for San Francisco, California, Board of Public Works, San Francisco, CA.

NBFU, 1933. Report on the Southern California Earthquake of March 10, 1933, National Board of Fire Underwriters, N.Y.

NBFU, 1905. Committee of Twenty, Report of National Board of Fire Underwriters by its Committee of Twenty on the City of San Francisco, Cal., Henry Evans, Chmn., July 1905.

NBFU, 1906. Reed, S. Albert, The San Francisco Conflagration of April, 1906, Special Report to National Board of Fire Underwriters Committee of Twenty, May.

NIST. 1997. Fire Hazards and Mitigation Measures Associated with Seismic Damage of Water Heaters, Report NIST GCR 97-732, M.P and T. T. Soong., Building and Fire Research Laboratory, Mroz, National Institute of Standards and Technology Gaithersburg, Maryland 20899

O'Rourke, T.D., Pease, J.W., and Stewart, H.F., 1992. Lifeline Performance And Ground Deformation During The Earthquake, USGS Prof. Paper 1551-F, The Loma Prieta, California Earthquake of October 17, 1989 - Marina District, T.D. O'Rourke, ed., Strong Ground Motion and Ground Failure, T.L. Holzer, Coord.

O'Rourke, T., Beaujon, P.A. and Scawthorn, C., 1992. Large Ground Deformations and their effects on Lifeline Facilities: 1906 San Francisco earthquake, *in* Case Studies of Liquefaction and Lifeline Performance During Past Earthquakes, Vol. 2, United States Case Studies, Technical Report NCEER-92-0002, February 17.

Okamoto, S. 1984. Introduction to Earthquake Engineering, U. Tokyo Press, Tokyo.

Scawthorn, C. 2002. Fire Following Earthquake, chapter in Earthquake Engineering Handbook, Chen, W.F. and Scawthorn, C. editors, CRC Press, Boca Raton.

Scawthom, C., K.A. Porter, F. T. Blackbum, 1992. Performance of Emergency Response Services After the Earthquake, USGS Prof. Paper 1551-F, The Loma Prieta, California Earthquake of October 17, 1989 - Marina District, T.D. O'Rourke, ed., Strong Ground Motion and Ground Failure, T.L. Holzer, Coord.

Scawthorn, C. (editor) 2000. The Marmara, Turkey Earthquake of August 17, 1999: Reconnaissance Report, MCEER Tech. Rpt. MCEER-00-0001, Multidisciplinary Center for Earthquake Engineering Research, State University of New York, Buffalo.

Scawthorn, C. 1997a. The 1923 Kanto, Japan, Earthquake, in Ballantyne and Crouse.

Scawthorn, C. 1997b. The 1989 Loma Prieta, California Earthquake Water Supply Effects, in Ballantyne and Crouse.

Scawthorn, C. and T.D. O'Rourke. 1997. The 1906 San Francisco, California Earthquake, in Ballantyne and Crouse.

Scawthorn, C, A.D. Cowell and F. Borden, 1997. Fire-Related Aspects of the Northridge Earthquake. Report prepared for the Building and Fire Research Laboratory, National Institute of Standards and Technology, NIST-GCR-98-743, Gaithersburg MD 20899.

Scawthorn, C., Cowell, A. et al, 1995. Fire-related Aspects of the January 17, 1994 Northridge Earthquake, Ch. 8, *Spectra*, Earthquake Engg. Research Inst., Oakland CA.

Scawthorn, C. and T.D. O'Rourke. 1989. Effects of Ground Failure on Water Supply and Fire Following Earthquake: The 1906 San Francisco Earthquake, Proceedings, 2nd U.S. - Japan Workshop on Large Ground Deformation, July, Buffalo.

Scawthorn, C. et al., 1985 Fire-related aspects of the 24 April 1984 Morgan Hill Earthquake, Earthquake Spectra, v. 1, n. 3, Earthquake Engg. Research Inst., Oakland CA.

Scawthorn, C. and Donelan, J., 1983. Fire-related aspects of the Coalinga Earthquake, Chapter in the Earthquake Engineering Research Institute report on the Coalinga Earthquake of May 2, 1983, Berkeley, CA.

Schiff, A. (editor) 1997. Northridge Earthquake: Lifeline Performance and Post-Earthquake Response, Report NIST GCR 97-712, National Institute of Standards and Technology Building and Fire Research Laboratory, Gaithersburg, MD 20899

Schussler, H.,1906. The Water Supply of San Francisco, California, Spring Valley Water Company.

Sekizawa, A. 1997. Post-Earthquake Fires and Firefighting Activities in The Early Stage in The 1995 Great Hanshin Earthquake, paper presented at Thirteenth Meeting of the UJNR Panel on Fire Research and Safety, March 13-20, 1996, NIST Tech. Report NISTIR 6030, Building and Fire Research Laboratory, National Institute of Standards and Technology, Gaithersburg, MD 20899.

Sepponen, C. 1997. Fire Department Emergency Response, Ch. 16 in Schiff (1997)

SFFD. 1990. Report on the Operations of the San Francisco Fire Department Following the Earthquake and Fire of October 17, 1989, San Francisco Fire Department.

Tobriner, Prof. Stephen, U. Calif. at Berkeley, Personal Communication, based on various data including Bowlen Outline.

Usami, T. 1996. Nihon Higai Jishin Soran (List of Damaging Japanese Earthquakes), U. Tokyo Press, Tokyo.

USGS. 1907. The San Francisco Earthquake and Fire of April 18, 1906 and their effects on structures and structural materials, Bull. No. 324, Washington.

Youd, T. L. and Hoose, S. N. 1978. Historic Ground Failures in Northern California Associated with Earthquakes, Professional Paper 993, USGS, Washington.

3 Non Earthquake Fires and Conflagrations

3.1 Introduction

This Chapter discusses selected U.S. and foreign non-earthquakes large fires and conflagrations, with relevance to the fires following problem. The purpose of this Chapter is several-fold, including:

- To document the record of large fires and conflagrations, to demonstrate that it is a recurring problem. Not all persons, even in the fire service, understand that, typically under adverse meteorological conditions, a large fire can develop in a city that places demands on the fire service resources greatly exceeding its capacity. Conflagrations are seen as 'ancient history' and, indeed, conflagrations date back to the ancient world. Interestingly, however, the US has a long and rich record of conflagrations, including quite recent events such as the 1991 East Bay Hills fire in which over 3,000 structures were destroyed in a few hours. Thus, it is imperative that concerned parties, particularly the fire service, not lose sight of the fact that constant vigilance is still required to prevent conflagrations.

- To provide data and narratives for specific events, in order that engineers, fire service personnel and emergency planners can understand the nature of what they can be confronted with.

- To compare and contrast non-earthquake and earthquake conflagrations so as to emphasize that, in selected cases such as in large cities with large wood building inventories such as in the US, Canada, Japan, New Zealand and elsewhere, that the multiple simultaneous ignitions that occur following a large earthquake have the potential to grow into a mass conflagration, of catastrophic proportions.

Among the great fires of modern history, the Great Fire of London in 1666, and the 1871 Chicago come to mind as the most destructive, with the loss of 13,000 and 17,000 structures, respectively. Lesser known perhaps (and not discussed in detail here) is the destruction by arson of the city of Moscow by order of the Russian governor, Rostopchin, in order to deny it to Napoleon. Almost 12,000 structures burned to the ground.

Table 3-1 lists selected North American large urban fires and conflagrations, while Table 3-2 lists selected North American wildland fires. In the next sections we discuss selected events, generally providing an overview of the event, and data on the ignitions, performance of the water system, and overall result.

Table 3-1 Selected North American Large Fires and Conflagrations[3]

Year	Location	incident	Loss Year$	Loss 2002$
1835	New York, NY	500 bldgs. covering 13 acres	$ 15	$ 2,467
1838	Charleston, SC	1,158 bldgs.	na	na
1839	New York, NY	conflagration	$ 10	$ 1,434
1845	Pittsburgh, PA	1000 bldgs.	$ 4	$ 468
1845	Quebec	1,500 bldgs.	na	na
1845	Quebec	1,300 bldgs.	na	na
1845	New York, NY	302 bldgs.	$ 6	$ 702
1848	Albany, NY	300 bldgs.	$ 3	$ 317
1849	St. Louis, MO	425 bldgs.; 27 steamships	$ 4	$ 408
1849	San Francisco, CA	conflagration	$ 12	$ 1,225
1851	San Francisco, CA	2500 bldgs.	$ 4	$ 382
1861	Charleston, SC	conflagration	$ 10	$ 681
1866	Portland, ME	1500 bldgs.	$ 1	$ 58
1871	Chicago ,IL	17,430 bldgs.; 250 killed	$ 196	$ 9,532
1871	Peshtigo, WI	bldgs.,1.3 million acres; 1500 killed	$ 75	$ 3,647
1872	Boston, MA	776 bldgs.; 13 killed	$ 75	$ 3,527
1874	Chicago, IL	conflagration	$ 5	$ 220
1889	Seattle, WA	conflagration	$ 5	$ 133
1889	Spokane, WA	conflagration	$ 6	$ 160
1889	Boston, MA	52 bldgs.; 4 killed	$ 4	$ 107
1889	Lynn, MA	conflagration	$ 5	$ 133
1892	Milwaukee, WI	conflagration	$ 6	$ 145
1900	Hoboken, NJ	steamships, 326 lives	$ 5	$ 85
1900	Hull, PQ	800 acres burned. spread to Ottawa	$ 5	$ 83

[3] Loss is given in $ millions for the year of the incident, and in dollars adjusted to 2002, using the Consumer Price Index (extrapolated prior to 1913).

FIRE FOLLOWING EARTHQUAKE

Year	Location	incident	Loss Year$	Loss 2002$
1901	Jacksonville, FL	1700 bldgs.	$ 11	$ 197
1902	Paterson, NJ	525 bldgs.	$ 6	$ 95
1904	Baltimore, MD	80 city blocks	$ 50	$ 812
1904	Toronto, Ontario	104 bldgs.	$ 12	$ 195
1906	San Francisco, CA	28,000 bldgs.	$ 350	$ 5,324
1908	Chelsea, MA	3,500 bldgs.	$ 12	$ 171
1911	Bangor, ME	55 acres of bldgs. burned	$ 3	$ 41
1912	Houston, TX	140 bldgs.	$ 12	$ 150
1913	Hot Springs, AR	518 bldgs.	$ 2	$ 23
1914	Salem, MA	1,600 bldgs.	$ 14	$ 140
1916	Paris, TX	1,440 bldgs.	$ 11	$ 106
1916	Nashville, TN	648 bldgs.	$ 2	$ 14
1916	Augusta, GA	682 bldgs.	$ 4	$ 41
1916	Black Tom, NJ	ship explosion	$ 4	$ 38
1917	Atlanta, GA	1,938 bldgs.	$ 6	$ 47
1917	Halifax, NS	2,000 killed, munitions ship exploded.	$ 5	$ 43
1918	Minnesota	forest fire 599 lives lost	$ 25	$ 179
1920	Grandview, TX	136 bldgs.	$ 2	$ 10
1922	Norfolk VA.	100 bldgs.	$ 1	$ 6
1922	New York (Arverne)	141 bldgs.	$ 2	$ 12
1922	New Bern, NC	40 blocks destroyed.	$ 2	$ 9
1922	Astoria, OR	30 blocks destroyed	$ 10	$ 59
1922	Northern Ontario	forest fires, 44 lives lost.	$ 6	$ 36
1923	Berkeley, CA	640 bldgs.	$ 6	$ 36
1925	Shreveport, LA	196 bldgs.	$ 1	$ 6
1926	Newport, AR	280 bldgs., 1 life lost	$ 2	$ 8
1926	Lake Denmark, NJ	munitions plant	$ 93	$ 520
1928	Fall River, MA	107 bldgs., 1 life lost	$ 3	$ 15

Year	Location	incident	Loss Year$	Loss 2002$
1929	Mill Valley, CA	130 bldgs.	$ 2	$ 9
1930	Nashua, NH	350 bldgs.	$ 2	$ 12
1931	Norfolk, VA	60 bldgs.	$ 1	$ 8
1931	Spencer, IA	39 bldgs.	$ 1	$ 6
1932	Coney Island, NY	4 blocks destroyed	$ 2	$ 14
1933	Ellsworth, ME	127 bldgs.	$ 1	$ 10
1933	Auburn, ME	250 bldgs.	$ 2	$ 12
1934	Chicago, IL	stockyard conflagration	$ 5	$ 35
1936	Bandon, OR	400 bldgs., 13 lives lost	$ 1	$ 9
1937	Cincinnati, OH	conflagration during flood	$ 1	$ 9
1938	New London, CT	conflagration during hurricane	$ 1	$ 7
1940	Camden, NJ	factory, several city blocks	$ 2	$ 14
1941	Marshfield, MA	450 summer resort bldgs.	$ 1	$ 7
1942	New York, NY	U.S.S. Lafayette (Normandie)	$ 60	$ 382
1942	Kewanee, IL	2 business blocks	$ 2	$ 11
1947	Texas City, TX	ship explosion; 468 killed	$ 67	$ 312
1950	Rimouski, Quebec	346 bldgs.	$ 16	$ 68
1950	Cabano, Quebec	151 bldgs.	$ 6	$ 26
1961	Los Angeles, CA	484 bldgs., several firefighters killed	na	na
1970	San Diego, CA	382 bldgs., 175,000 A., 5 killed	na	na
1977	Santa Barbara, CA	234 bldgs.	na	na
1980	San Bernardino, CA	325 bldgs., 23,000 A., 4 killed	na	na
1990	Santa Barbara, CA	641 bldgs., 1 killed	na	na
1991	Oakland, CA	3,354 bldgs., 25 killed	$ 1,500	$ 2,020
1993	Southern, CA	1,000 bldgs., 3 killed	$ 500+	$ 713+
2003	Southern, CA	2,722 bldgs., 734,000 A., 21 killed	$ 2,000±	$ 2,000±

Table 3-2 Selected North American Wildland Fires

Date	Name of Fire	Location	Acres	Deaths	Structs
1825	Miramichi, Maine	NB/ME	3,000,000		
1871	Peshtigo	WI/MI	3,780,000	1,500	
1881	Michigan	MI	1,000,000	169	
1894	Hinckley	MI	Undetermined	416	
1894	Wisconsin	WI	Several Million	some	
1902	Yacoult	WA	1,000,000 +	38	
1903	Adirondack	NM	637,000		
1910	Great Idaho	ID	3,000,000	85	
1933	Tillamook	OR	311,000	1	
1947	Maine	ME	205,678	16	
1949	Mann Gulch	MN	4,339	13	
1967	Sundance	ID	56,000		
1970	Laguna	CA	175,425		382
1977	Sycamore	CA	805		234
1980	Panorama	CA	23,600		325
1987	Siege of 87'	CA	640,000		
1988	Yellowstone	MN	1,585,000		
1988	Canyon Creek	MN	250,000		
1990	Painted Cave	CA	4,900		641
1990	Dude Fire	AZ	24,174	6	63
1991	Oakland Hills	CA	1,500	25	3,500
1992	Foothills Fire	ID	257,000	1	
1994	South Canyon Fire	CO	1,856	14	
1994	Idaho City	ID	154,000	1	
1996	Cox Wells	ID	219,000		
1996	Millers Reach	AK	37,336		344
1997	Inowak	AK	610,000		
1998	Volusia Complex	CO	111,130		
1998	Flagler/St. John	CO	94,656		
1996	Dunn Glen	MN	288,220		
1996	Big Bear Complex	CA	140,947		
1996	Kirk Complex	CA	86,700		
2000	Cerro Grande	NM	47,650		235

3.2 Historic Conflagrations

3.2.1　64 Rome

Perhaps history's first major, and most famous[4], urban conflagration, is the destruction of Rome by fire in July 64 AD. It is well summarized in the following:

During the night of July 18, 64 AD, fire broke out in the merchant area of the city of Rome. Fanned by summer winds, the flames quickly spread through the dry, wooden structures of the Imperial City. Soon the fire took on a life of its own consuming all in its path for six days and seven nights. When the conflagration finally ran its course it left seventy percent of the city in smoldering ruins…from the ashes of the fire rose a more spectacular Rome. A city made of marble and stone with wide streets, pedestrian arcades and ample supplies of water to quell any future blaze. The debris from the fire was used to fill the malaria-ridden marshes that had plagued the city for generations. By the sixth day enormous demolitions had confronted the raging flames with bare ground and open sky, and the fire was finally stamped out…but…flames broke out again in the more open regions of the city. Here there were fewer casualties…of Rome's fourteen districts only four remained intact. Three were leveled to the ground. The other seven were reduced to a few scorched and mangled ruins.(Source: http://www.ibiscom.com/rome.htm)

Even in this ancient event, almost all elements of modern conflagrations are seen to exist:

- Dry, windy conditions
- Narrow, winding streets
- Dense wooden construction
- Inadequate water supply
- Firefighting via building demolition (fuel removal)
- Rekindles
- Reconstruction with fire-resistant materials, wide fire lanes and an improved water supply

These essential elements of the urban conflagration will be seen to repeat themselves time and again, down to the modern day.

3.2.2　1666 London

This event is included because it is regarded by many as the impetus for the formation of the fire service and modern property insurance industry. In this sense, it is worth putting this event in perspective. London in 1666 was one of the leading capitals of Western civilization, with a population of about three to four hundred thousand (Hanson, 2001).

---

[4] Emperor Nero is of course accused of 'fiddling while Rome burned" - these rumors have never been confirmed. In fact, Nero rushed to Rome from his palace in Antium (Anzio) and ran about the city all that first night without his guards directing efforts to quell the blaze. Rome's population in this era is variously estimated to have been from 0.5 to 1.5 million (http://www.mmdtkw.org/VOstia.html)

However, it was also a medieval city, with a long history of conflagrations (798, 982, 989, 1212 – the latter fire killed perhaps as many as 10,000). The city in 1661 was mostly built of wood, with a narrow irregular street pattern, much of which survives to this day, albeit hardly as congested as in 1666. The narrow streets barely separated blocks of wood buildings, due to cantilevered upper stories, which almost met over the narrow lanes and paths.

On Sept. 2, 1666, the Great Fire of London broke out in a baker's shop on Pudding Lane in the City. The fire, driven by high winds, was only brought under control five days later. It had destroyed about 436 acres, or 80% of the city within its walls, Figure 3-1. This is less than a square mile – given that the total number of buildings destroyed was 13,000, this gives some idea of the density of wood buildings feeding the fire. Losses included St. Paul's Cathedral, more than 80 other churches, the Royal Exchange, and the halls of 44 craft and trade guilds. While very few deaths were attributed to the fire, the total may have been much higher (Hanson, 2001).

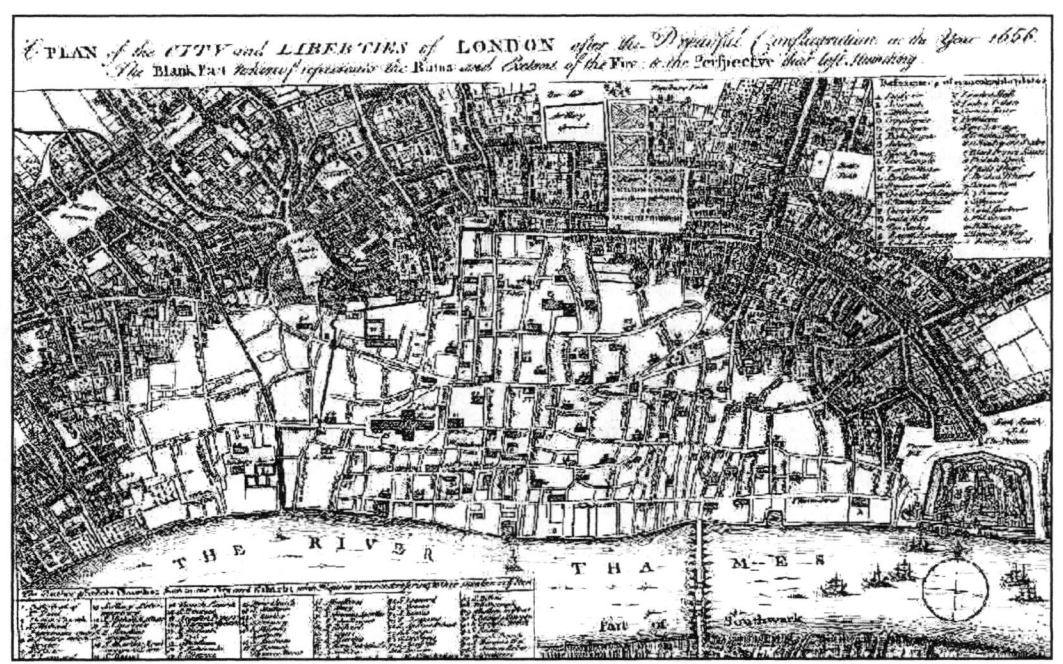

Figure 3-1 Great Fire of London 1661

3.2.3　1871 Chicago

The Chicago fire of 1871 was the largest urban fire in history to that time. It began at 9.45 o'clock Sunday night, October 8, when flames were discovered in a small stable in the rear of a house on the corner of De Koven and Jefferson streets[5]. A very strong wind was blowing from the south-southwest, and it had been rainless and very dry for weeks. As the fire grew, the wind increased and fire whirls were observed. Hot cinders, firebrands and black, stifling smoke, were driven fiercely at fleeing people who had their clothes burned off their backs. The fire was fueled not only by the wood buildings but also by numerous furniture factories, lumberyards, wood railroad rolling stock, and large wooden grain elevators. Tar and gas works then added to the fuel, and then the explosion of the South Side gas works.

Attempts were made during the night to create firebreaks using explosives – the blowing up of the Merchants' Insurance Company building, using kegs of powder, was momentarily successful, but wind-driven flames crossed the firebreak and extended into the financial district. Cook County Court House was situated in the middle of a large square, so that it was thought it was safe, being surrounded by firebreak on all sides. However, large wind-driven firebrands ignited its wooden dome, and it was soon fully involved. The entire fire lasted only eighteen hours, but the entire business portion of the city was obliterated, Figure 3-2. The radiant heat from the fire was such that thousands had to wade into the lake up to their necks (Goodsell and Goodsell, 1871).

Over 17,000 buildings were destroyed and 250 people killed. The progress of the fire was directly related to the strong wind, Figure 3-4. It should be noted that simultaneous with the Chicago fire was the Peshtigo Wisconsin wildland fire, which burnt over 2,000 sq. miles and killed 1,500 people. Both fires were driven by extremely dry hot winds coming off the prairies.

---

[5] *Mrs. O'Leary's cow* is reputed to have kicked over a lantern, although this has been debated ever since.

FIRE FOLLOWING EARTHQUAKE

Figure 3-2 Ruins of Chicago Fire of 1871

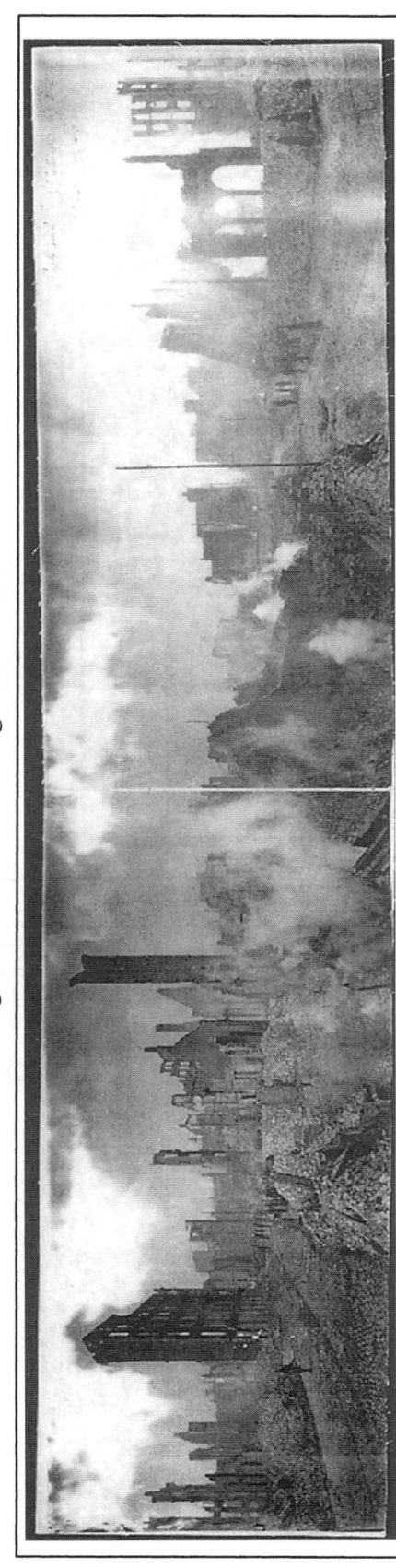

Figure 3-3 Baltimore Fire, 1904

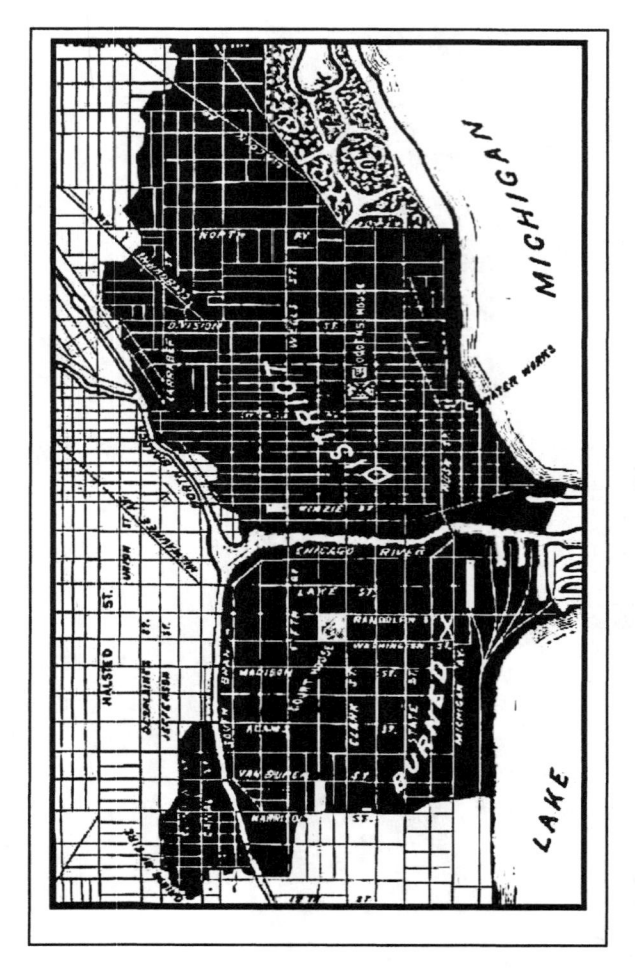

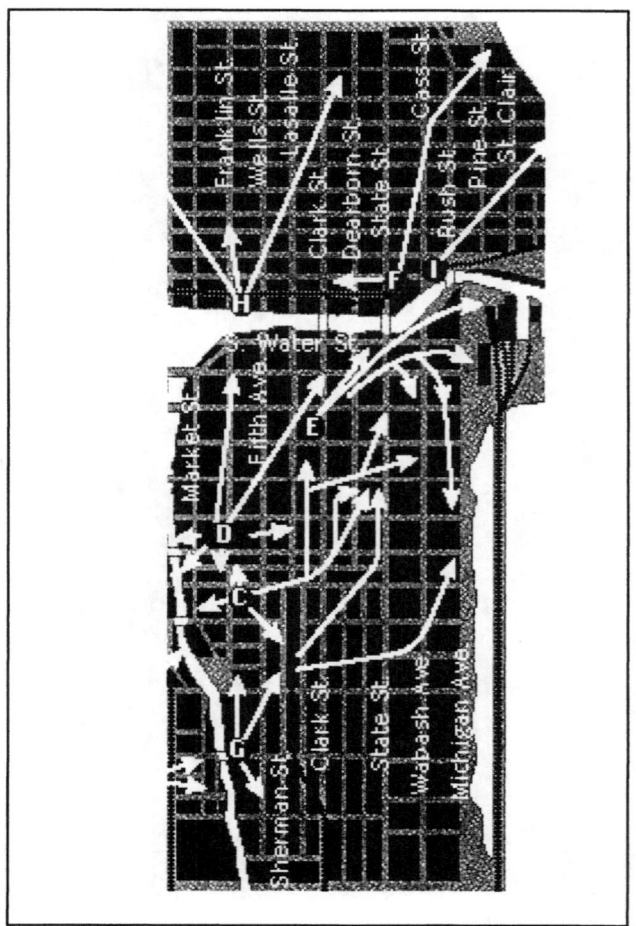

Fire	Time (Oct. 8/9)
A. O'Leary barn	8:30 pm
B. Bateham's Mills	10:00 pm
C. Parmelee's Stables	11:30 pm
D. Gas Works,	12:00 midnight
Conley's Patch	12:20 am
E. Court House	1:30 am
F. Wright's Stables	2:30 am
G. Polk Street	2:30 am
H. Northwestern Elevator	7:00 am
I. Galena Elevator	7:00 am

Figure 3-4 Chicago Fire 1871: (l) Extent of Burnt area; (r) path and time of flamefront arrival

3.3 Early 20th C. Conflagrations

3.3.1 1904 Baltimore

On Feb. 7th 1904 at 10:48am a fire was reported in the basement of a commercial building in downtown Baltimore MD. A subsequent explosion ignited exposures and at 11:40am a general alarm responded the entire Baltimore fire department to the scene. Whipped by the wind, the fire spread northeast through the financial district and then southeast through the docks. By early afternoon additional apparatus was brought in by train from Washington DC but due to differing hose sizes, the apparatus could not connect to hydrants. Because the fire had spread to 12 city blocks, dynamiting of building to create firebreaks was attempted, but this only caused more ignitions. The fire was finally brought under control after 31 hours, with the combined efforts of 1,231 firefighters, 57 Engines, 9 Trucks, 2 Hose Companies, 1 Fire Boat, one Police Boat, several Tug Boats, brought in from as far away as Washington, Wilmington, Philadelphia, New York and other areas. The City of Baltimore furnished 460 firefighters, 24 Engines, 8 Trucks, and several Boats. The fire destroyed the entire downtown portion of Baltimore, Figure 3-3, claiming 1,526 buildings in more than seventy city blocks, or over 140 acres. One life was lost – a firefighter who died of pneumonia.

3.3.2 1923 Berkeley

Berkeley California is located in the East Bay part of the San Francisco Bay Area, on the flat lands along the Bay shore and extending into the East Bay hills, which rise to about 1,000 ft. The region is subject to a phenomenon known as, *Diablo winds*, which are the local name for a wind technically known as a *foehn* – a hot dry wind coming down off a mountain range, rising in temperature as it descends (adiabatic compression)6. On September 17, 1922, a Diablo wind condition existed and a fire began about 2 pm in wildlands in the hills above the City. The winds drove it into the hill portion of the City, which was residential with many houses of a highly combustible brown shingle style – wood shingle siding and roofing. The fire department immediately made a request to neighboring cities and even to San Francisco across the Bay. The large city of Oakland to the immediate south had 13 fires of its own at the time, and could not respond.

San Francisco responded by ferrying four engines, two hose wagons, two chemicals, as well as dispatching a fireboat[7]. The engines were 750-gallon pumpers, hose wagons each carried 2,000 feet of hose, chemicals had double 100-gallon tanks. Engine companies consist of 6 men each; chemical companies, 2 men each; fire boat crew, 7 men. The fireboat left the San Francisco water front at 4:15 p.m., proceeded to the Berkeley wharf and worked until 10pm that night, when the winds had abated and the fire was brought under control.

---

[6] Foehn winds are known in various regions as Chinooks (Pacific Northwest), Santa Anas (southern California), etc.

[7] http://www.sfmuseum.org/oakfire/berkeley.html

Figure 3-5 Berkeley 1923

Figure 3-6 Berkeley 1923 – North of U.C. Berkeley Campus

FIRE FOLLOWING EARTHQUAKE

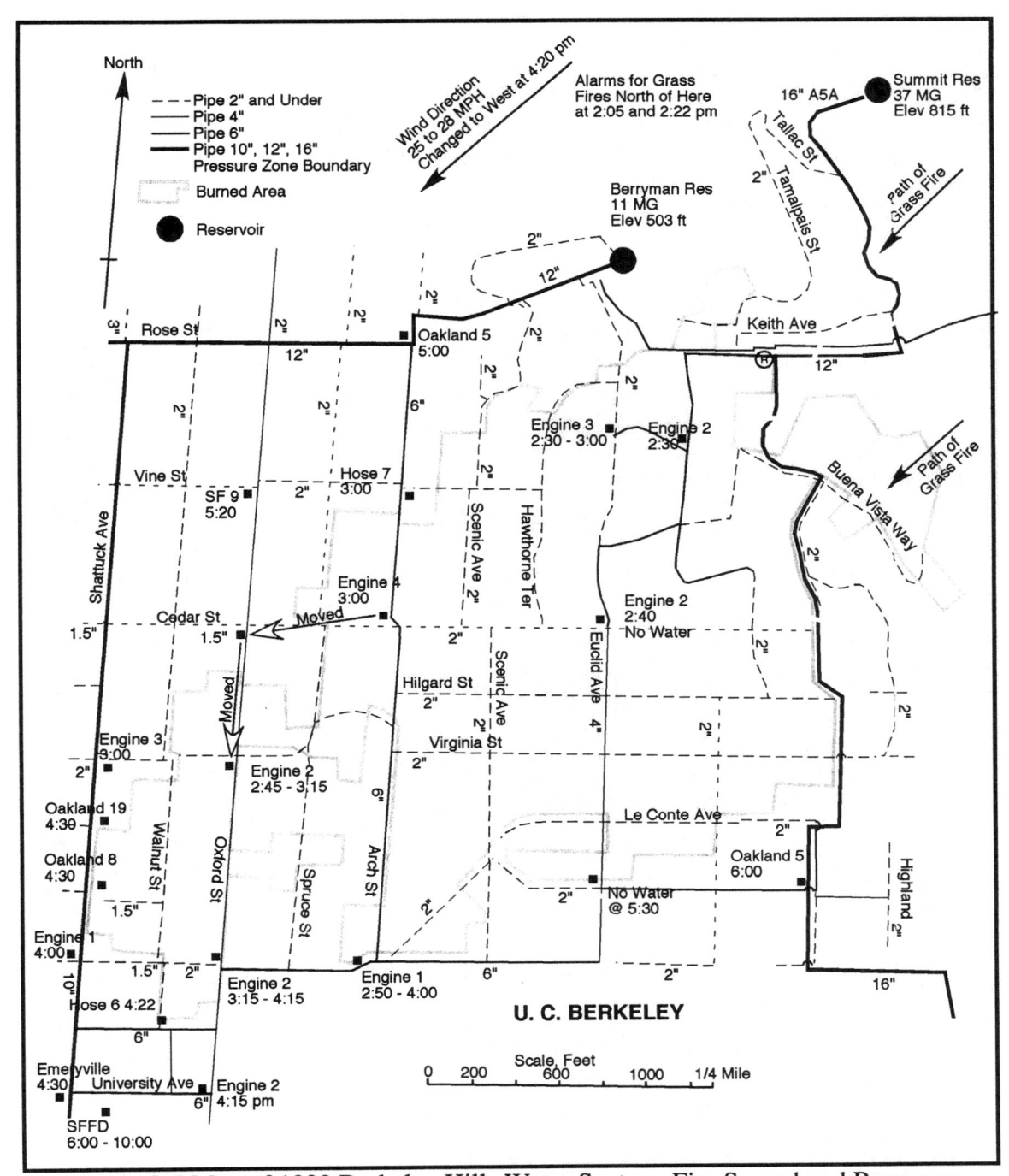

Figure 3-7 Map of 1923 Berkeley Hills Water System, Fire Spread and Response

A total of 584 buildings were wholly destroyed and about 30 others were seriously damaged, Figure 3-5 and Figure 3-6. Only 7 buildings were masonry, the remaining of wood frame construction. Only about 8% of the buildings had fire retardant roof coverings. Many buildings had close set backs between each other (under 5 feet), although streets were moderately wide (60 feet). The total loss was estimated at $10,000,000 ($1923). The fire rendered 4,000 persons homeless. The fire is significant in light of the 1991 East Bay Hills fire, discussed in Section 3.3.4.

At the time of the fire, the local water utility had substantial infrastructure in place. Two large local reservoirs had been built in 1884 and 1891; these reservoirs were (at that time) filled from local springs and run-off, although these same reservoirs are now (2003) filled with treated surface water from sources over a hundred miles away. At an elevation higher than the ultimate burned area, there was a 37,000,000-gallon water reservoir that provided water into the "A5A" pressure zone, Figure 3-7. At about mid-elevation in the burn area, there was a 24,000,000-gallon reservoir (since downsized to 15 MG and operated at 11 MG) that provided water into the "A2A" pressure zone. The middle elevation A4B pressure zone got all its water via a pressure regulator from the higher elevation A5A zone. There were water pipelines along every city street that ultimately burned. In an after-incident analysis of why this fire could not be better controlled, several observations were made by the water utility:

- First, the fire spread from uphill (elevation 650 feet) to downhill (elevation 200 feet) locations.

- Second, there was plenty of water in local storage tanks at the time of the fire; these water sources were not depleted in the course of fighting the fire.

- Third, the street layout was fairly regular, with water pipe service on every street.

- Fourth, water distribution pipes in the burn area were commonly 2" diameter cast iron or redwood pipes

- Fifth, immediately adjacent to the burn area the pipes included 10", 12" and 16" diameter pipes.

- Sixth, the fire continued to burn downhill towards the central business district of Berkeley until about 5:00 pm that day, when a combination of factors worked together to halt the fire.

The factors that halted the fire were as follows:

- The winds were dying down by 5:00 pm, so the natural spread of the fire was slowed;

- As the burned area expanded down hill, the ability for fire fighters to apply water onto the fire greatly increased, as the water distribution system for the central business district of Berkeley included many large (10" and larger) diameter pipelines, thereby providing a much better water supply to control the fire at those locations. The use of a fireboat at the Berkeley shoreline some 4 miles to the westernmost bounds of the fire would have been decidedly ineffective.

In the years following this fire, the local water utility (East Bay Water, subsequently merged into the East Bay Municipal Utility District) decided that use of 2" diameter pipeline for city service was no longer acceptable, and undertook a large effort to replace every 2" diameter pipeline with new pipelines of at least 4" diameter or larger. This was an expensive effort, with the cost borne in general rates; apparently the decision to not replace 4" pipeline with larger pipes was in part a consideration to the overall cost; this decision would prove costly again in 1991 (see Section 3.5). In response o this fire, for the area seen in Figure 3-7, many improvements were made including the following: the 12" pipe to feed Berryman reservoir was upgraded to 24" in 1927; the bulk of the 2" pipeline seen in Figure 3-7 was replaced by 6" or 8" pipes by 1925; a few 2" pipes were replaced with 4" pipes by 1937. The 6" and larger pipelines (originally installed about 1901) seen in the figure are largely still in service in 2003. The single regulator station to provide water into the A4B pressure zone (middle level zone between the yellow lines in the figure) has been supplemented by two other regulators at geographically dispersed areas. It is interesting to note that the active Hayward fault cuts through the area in the figure on a north-north-west direction, more or less parallel to the 16" Summit pipeline; it remains to be seen how well this area does in future earthquakes.

In 1951, the NFBU rated the risk of fire in the area burned in 1923 as follows: "water supply varies from poor to fair"; about 1 in 50 new roof coverings are combustible; however there are still some districts with a large proportion of wooden shingle roofs; … with consideration of high winds and periods of low humidity, the conflagration hazard in the closely built residential districts is fairly severe; … steps taken to prevent grass and brush fires in the hills [will] help to keep the conflagration hazard down". The 1932 version of the NFBU report for the residential parts of Berkeley rated the risk of conflagration as "severe".

The rapid spread of this 1923 fire can be summarized as being caused by four principal factors:

- An ignition located outside the urban area where there were no water supplies to rapidly control it

- High winds (20 to 28 mph) spread the fire into the City

- The building inventory was comprised of mostly wooden buildings, most with shingle wall or roof coverings

- The inability of fire departments to get adequate flows out of hydrants along streets with 2" diameter pipelines

In the years following the 1923 fire, there were various panels established to examine and make recommendations to reduce the risk of fires in the Berkeley – Oakland Hills. Fire risk in this area remains a continuing high hazard, with 14 separate fires of at least 3 acres in size from 1923 to 1990, including the 1971 Claremont Canyon fire that burned 40 structures. Recommendations to remove shingle roof coverings, or at least not allow any news ones to be built, were sometimes recommended but not implemented. Some improvements in the 1980s were made to the hillside water systems including those in neighboring Oakland, albeit that these improvements were somewhat compromised by anti-development protests. Minor firebreaks were maintained. Section 3.5 examines how some of these mitigation measures fared in the 1991 Oakland Hills firestorm.

3.4 WW2 experience

Air attacks played a major role in World War 2, the best known of which are London, Coventry, Pearl Harbor, Hamburg, Dresden, Tokyo, Hiroshima and Nagasaki. While many fires were caused by air attack in London and Coventry, these were not incendiary attacks. Hamburg, Dresden and Japanese air attacks were intentional incendiary attacks however, and provide useful information so that this section summarizes some key elements of this experience, relevant to the problem of fire following earthquake.

3.4.1 Hamburg

The air campaign in the European theatre during WW2 was two-pronged – due to unsustainable daylight bombing losses early in the war, the RAF resorted to nighttime bombing especially of urban areas while the US, when it undertook operations, confined itself largely to daylight bombing of industrial targets. On the night of May 30, 1942, the RAF mounted its first "thousand plane" raid against Cologne and two nights later struck Essen with almost equal force. On three nights in late July and early August 1943 the RAF struck Hamburg, the second largest city in Germany and Germany's largest port and shipbuilding center. The attack was a combination of high explosive and incendiary bombs – the high explosives to destroy water mains (and kill firefighters). Three thousand blockbusters (4,000 lb. bombs), 1,200 land mines, 25,000 other high explosives, 3 million incendiaries, and 80,000 phosphorus bombs were employed. After the area caught fire, the air above the bombed area became extremely hot and rose rapidly. Cold air then rushed in at ground level at velocities exceeding 100 mph, such that trees were uprooted and a standing person could not withstand the wind forces. This phenomenon, termed a *firestorm*, had not typically been seen in peacetime conflagrations, with the exception of Tokyo in 1923 and possibly Chicago in 1871. About one third of the houses of the city were destroyed and German estimates show 60,000 to 100,000 people killed (USSBS, 1945), out of a population of 2 million (one million subsequently fled the city).

3.4.2 Dresden

Dresden is a major cultural and inland port city in eastern Germany, which had been beyond the reach of the strategic bombing campaign and therefore unbombed for most of the war. In February 1945 it was attacked in a nighttime incendiary campaign, in support of Russian forces advancing from the east. On the 13th February 1945, 773 RAF Lancasters bombed Dresden. During the next two days the USAAF sent over 527 heavy bombers to follow up the RAF attack. Dresden was nearly totally destroyed in the ensuing firestorm, in which at least 135,000 were killed out of a population of about 650,000.

3.4.3 Japan

While 2.7 million tons of bombs were dropped in the European theatre of operations (1.4 million within Germany), about 160,000 tons were dropped on Japan, or 11% of that dropped on Germany. Up until March 1945 only 7,000 tons had been dropped on Japan, typically from 30,000 feet where the accuracy was such that target area hits averaged less than 10 percent.

> *On 9 March 1945, a basic revision in the method of B-29 attack was instituted. It was decided to bomb the four principal Japanese cities at night from altitudes averaging 7,000 feet. Japanese weakness in night fighters and antiaircraft made this program feasible. Incendiaries were used instead of high-explosive bombs and the lower altitude permitted a substantial increase in bomb load per plane. One thousand six hundred and sixty-seven tons of bombs were dropped on Tokyo in the first attack. The chosen areas were saturated. Fifteen square miles of Tokyo's most densely populated area were burned to the ground. The weight and intensity of this attack caught the Japanese by surprise. No subsequent urban area attack was equally destructive. Two days later, an attack of similar magnitude on Nagoya destroyed 2 square miles. In a period of 10 days starting 9 March, a total of 1,595 sorties delivered 9,373 tons of bombs against Tokyo, Nagoya, Osake, and Kobe destroying 31 square miles of those cities ... In the aggregate, 104,000 tons of bombs were directed at 66 urban areas... some 40 percent of the built-up area of the 66 cities was destroyed...30 percent of the entire urban population of Japan lost their homes... 185,000 casualties were suffered in the initial attack on Tokyo of 9 March 1945[8]. Casualties in many extremely destructive attacks were comparatively low. Yokohama, a city of 900,000 population, was 47 percent destroyed in a single attack lasting less than an hour. The fatalities suffered were less than 5,000... The Japanese had constructed extensive firebreaks by tearing down all houses along selected streets or natural barriers. The total number of buildings torn down in this program, as reported by the Japanese, amounted to 615,000 as against 2,510,000 destroyed by the air attacks themselves. These firebreaks did not effectively stop the spread of fire, as incendiaries were dropped on both sides of the breaks. They did, however, constitute avenues of escape for the civilian population. (USSBS, 1946).*

On 6 August and 9 August 1945, the first two atomic bombs to be used for military purposes were dropped on Hiroshima and Nagasaki respectively. One hundred thousand people were killed, 6 square miles or over 50 percent of the built-up areas of the two cities were destroyed.

> *About 20 minutes after the detonation of the nuclear bomb at Hiroshima, a mass fire developed showing many characteristics usually associated with firestorms. A wind blew toward the burning area of the city from all directions, reaching a maximum velocity of 30 to 40 miles per hour about 2 to 3 hours after the explosion, decreasing to light or moderate and variable in direction about 6*

---

[8] The official Japanese count [for the Tokyo attack of March 9~10] found 83,793 dead and 40,918 injured. A total of 267,171 buildings were destroyed leaving one million people homeless. 15.8 square miles of the city had been burned to the ground, including 18% of the industrial area, and 63% of the commercial center. (http://www.danshistory.com/ww2/ww2faq.shtml#5)

hours after...No definite firestorm occurred at Nagasaki, although the velocity of the southwest wind blowing between the hills increased to 35 miles an hour when the conflagration had become well established, perhaps about 2 hours after the explosion. (Glasstone and Dolan, 1977)

3.4.4 Analysis

Incendiary attacks were a significant aspect in both the European and Pacific campaigns. The nature of the attacks, with generalized high explosive weapons causing building and water supply damage over a wide area, impairing transportation and communications, and then followed by multiple simultaneous ignitions, is closely replicated by a major earthquake. In Germany, two major firestorms resulted, while in Japan the only firestorm was in Tokyo (which had previously had a firestorm following the 1923 earthquake). In all incidents, tens of thousands of ignitions probably occurred, perhaps two orders of magnitude more ignitions than would be expected in a large earthquake.

Based on World War II experience with mass fires resulting from air raids on Germany and Japan, the minimum requirements for a fire storm to develop are considered by some authorities to be the following: (1) at least 8 pounds of combustibles per square foot of fire area, (2) at least half of the structures in the area on fire simultaneously, (3) a wind of less than 8 miles per hour at the time, and (4) a minimum burning area of about half a square mile. Highrise buildings do not lend themselves to formation of firestorms because of the vertical dispersion of the combustible material and the baffle effects of the structures. (Glasstone and Dolan, 1977)

While there is precedent for a firestorm following an earthquake (1923 Tokyo), a firestorm following a major US earthquake is unlikely, while a firestorm following a major Japanese earthquake is still a possibility. However, mass fires, or conflagrations, are definite possible in US cities, as shown by the 1991 East Bay Hills fire discussed next.

3.5 1991 East Bay Hills

3.5.1 Overview

The 1991 East Bay Hills fire in the eastern portion of the San Francisco Bay area resulted in the largest dollar fire loss[9] in United States history. The fire began as a grass fire of unknown origin at the urban-wildland interface on Oct. 19, and was suppressed by that afternoon. Sunday morning, Oakland fire department companies were overhauling the fire, to extinguish hot spots and prevent rekindles. Despite their efforts, hot dry extreme winds rekindled the fire at about 11am, which quickly became a conflagration, with the fire advancing at one point at 132 ft. per min (40 m/s), Figure 3-9. Twenty-five lives were lost and 3,354 structures were destroyed with a value of $1.5 billion (Routley, 1991).

This section discusses this important event at length, first describing the event including the devastatingly rapid fire spread. Performance of the water system is next discussed,

---

[9] In absolute terms – however, if the 1906 San Francisco fire is inflated to equivalent dollars, it dwarfs the 1991 event.

followed by comments on fire flow guidelines. The section concludes with a discussion of alternative fire protection strategies for high-risk hillside areas.

3.5.2 Fire Spread

Figure 3-8 charts the progression of the fire in terms of structures burned as a function of time after ignition. For the first 30 minutes of the spread, a structure was being burned down at a rate of almost one every three seconds. The apparent slow down of the spread of the fire at about 12 noon to 1 pm reflects that at this time the fire had to traverse the large fire breaks afforded by Highways 24 and 13. Upon examination of the fire department response, there was ample human and equipment resources to control the fire before it spread much beyond the first 2,000 dwelling structures. The lack of good water supply plus lack of coordinated response may have contributed to the loss of the last 1,000 structures.

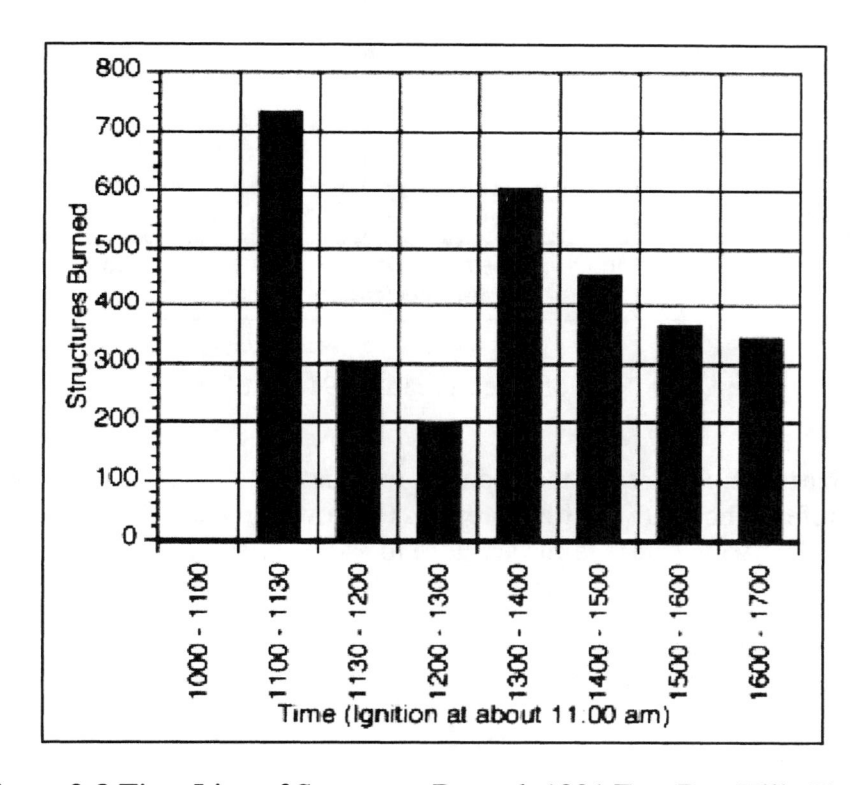

Figure 3-8 Time Line of Structures Burned, 1991 East Bay Hills Fire

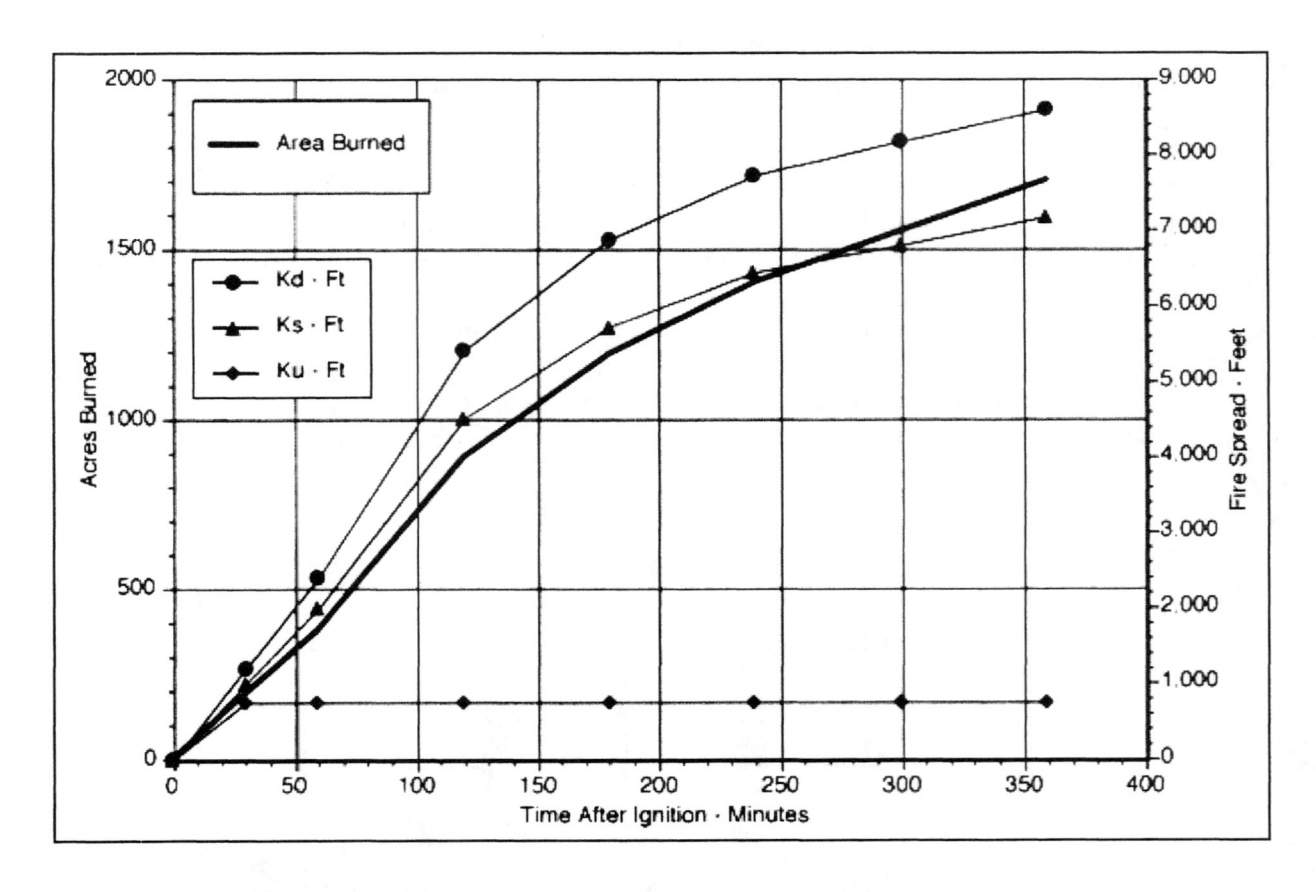

Figure 3-9 Rate of Fire Spread, 1991 East Bay Hills Fire

Figure 3-9 charts the progress of the fire with regards to the area burned. The terms "Kd, Ks and Ku" refer to how fire the fire spread in the downhill, side-hill and up-hill directions, respectively. These terms are used to establish empirical rules that can be used for developing analytical fire spread models such as those presented in Chapter 4.

The 1991 Oakland Hills fire was controlled when a similar combination of factors that controlled the 1923 Berkeley fire occurred:

- The *Diablo* wind conditions abated
- The fire spread to pressure zones with large quantities of water available to control the fire from further spread
- A large (but not necessarily well organized) fire service response had been mobilized

The spread of the fire was the result of a number of factors, including:

- A high fire risk created by low humidity and Diablo winds. There was a highly unusual winter freeze in the area (several days with temperatures ranging from 10°F to 30°F) in December of 1990, which killed quantities of the lighter brush and eucalyptus. It was not common at the time to aggressively remove dead brush, resulting in large amounts of highly combustible natural fuels

- Limited separation between natural fuels and structures in many cases

- Unregulated use of wood shingles as roof and siding material (some of the lessons learned from the 1923 Berkeley fire were largely ignored)

- Steep terrain (making it hard for initial fire service responders to control the fire before it spread

- Homes overhanging hillsides in some areas

- Narrow roads and limited access in several locations. On some of the winding roads and hairpin turns, portions were so narrow that two automobiles had difficulty in passing, Figure 3-11. As Routley (1991) observes: *Two major developments occurred after 1970:*

  Hiller Highland, a neighborhood of 340 densely built two-story condominiums and townhouses to the hills. New, wider access roads were built to serve this development, but the heavy overgrowth of trees was maintained. A single connection was provided to Charing Cross Road, one of the older hillside streets.

  The second major development was the Parkwood Apartments, a complex of seven buildings, containing 456 apartments, that was built at the bottom of Temescal Canyon, just north of the Caldecott Tunnel entrance. The buildings had three and four stories of wood-frame apartments, built on top of open parking areas, and were terraced into the lower levels of the Canyon. This exclusive development had only one access point, controlled by a security gate, connecting to the Highway 24 frontage road.

- The role of the water supply system is detailed in Section 3.5.4.1 below.

Many of these factors had been foreseen, and risk reduction measures had been studied and recommended for several years, but only a few were implemented (Routley, 1991).

Fire spread was very rapid, due to high winds as well as *branding* (burning debris carried aloft by the rising air from the fire and carried great distances downwind – when the burning firebrands land on flammable material, such as wood shake roofs, they trigger new ignitions). Branding was responsible for the spread of the fire across Highway 24, an eight-lane freeway and natural firebreak:

At 1202 hours, Oakland Engine 2, which was responding to the main fire, reported to the IC that it had discovered a vegetation fire in the Temescal Recreation Area, west of the interchange of Highways 13 and 24. This area is 2,000 feet beyond the face of the burning hills and 400 feet lower in elevation. Flying brands were dropping into this area and within minutes were beginning to ignite a line of trees that border the west side of Lake Temescal. This area was in the direct path of the fire's convective thermal column, which dried and preheated the ignitable fuels. (Routley, 1991)

> *At 11:45 am to 12:00 noon, flying brands were occasionally falling into the upper Rockridge district of Oakland, located southeast of the Highway 24 – Highway 13 interchange. At the time, residents responded by using garden hoses to easily extinguish these sources. By 2:00 pm that afternoon, this entire area of several hundred houses was evacuated as the fire continued its spread. This particular area of Oakland had been built largely from 1909 to 1935, with many 4" and 6" cast iron pipelines.*

The fire triggered a very large (some would say efficient, but not too effective) response by the California Office of Emergency Services (OES) mutual aid system. In the largest response ever recorded – 440 engine companies and 1,539 personnel from 250 agencies. Eighty-eight strike teams were mobilized, many within the first several hours and some from more than 350 miles away. However, this response was most of all too much, too late and with too little coordination. About 75% of the loss occurred within the first 2 hours of the spread of the initial ignition, and this fire spread was through water pressure zones of modern design (design for 1,500 gpm fire flows). The spread of the fire south and west of Highways 24 and 13 could probably have been much better controlled had there been better coordination of mutual aid response and a better water system in that area.

In the area near the Hiller Highlands development, the water supply system more than fully met modern fire codes (1,500 gpm fire flows), and yet the area still suffered a complete loss:

> *At approximately the same time that the fire was jumping across the freeway interchange, it was also sweeping through the 340-unit Hiller Highlands development. There were no fire suppression forces there to make a stand, because all of the companies that had been committed to the left flank of the fire had been overrun or were retreating from the overwhelming fire front. The fire arrived well ahead of the mutual aid strike teams that had been ordered to protect the area. The fire swept over the crest of the hill from Charing Cross Road into the tightly packed two-story townhomes, as residents rushed to escape the flames by driving down Hiller Road to Tunnel Road. Several residents of this development were unable to escape ahead of the flames and died in their homes or on the roadways. (Routley, 1991)*

As the massive mutual aid response gathered, it should have been available and prepared to prevent further losses. However, the large mutual aid response revealed additional problems, including:

- Shortage of command officers handicapped initial implementation of ICS. As Routley (1991) reports:

The Oakland Fire Department [had] been severely impacted by budget reductions during the past decade and [had] lost approximately 40 percent of its on-duty staffing. At least 10 companies were discontinued and the remaining companies operate with reduced staffing. The hardest hit area was the command level, which was reduced from an Assistant Chief and five Battalion Chiefs on each shift to an Assistant Chief and two Battalion Chiefs. The Assistant Chief now serves as both the Shift Commander and as Battalion 2. The Chief's Operator (Battalion Aide) positions were also cut, leaving the three on-duty command officers with no support staff.

- Multiple commands developed as additional agencies became involved. Unified Command was however implemented after several hours.

- Radio channels and Communications Center became overwhelmed by the situation.

To help understand the role of water supply on the actual spread of the fire, as well as to see what lessons could be learned, the East Bay Municipal Utility District undertook a comprehensive examination of the performance of its water system immediately before, during and after the firestorm of October 20, 1991. Section 3.3.4.1 provides a summary of the findings.

- Initially, responding mutual aid units were unable to hook up to Oakland hydrants. When California adopted a standard 2 1/2 inch threaded connection for all hydrants, the cities of Oakland, Berkeley, San Francisco and more than 20 other East Bay cities opted to maintain their 3-inch connections and to keep a supply of adapters on hand for mutual aid units. Fire departments in the area normally carry adapters on their apparatus, but the plan called for adapters to be obtained from the warehouse to meet incoming mutual aid strike teams at staging areas. Since this fire occurred on a Sunday, there was a delay in obtaining the adapters until off-duty personnel could open the warehouse and send them to the scene on supply trucks.

- In general, however, water service in the Hills area is not very robust. It is an older system with smaller pipe (some dating back to the 1890s [Eidinger, personal communication]), many cascading pressure zones fed by limited capacity pumps, and has limited tank capacities. Fire demand on this system greatly exceeded fire service design requirements. Two additional factors further degraded water supply:

- Early in the incident, electric power failed so than the normal water system booster pumps were ineffectual, which eliminated any capacity to re-supply emptied tanks. Even if electric power had been available, however, pumping capacity was much less than the extraordinary demands being placed on the system by the fire.

- As hundreds and then thousands of structures were destroyed by the fire, each structure's collapse broke the plumbing, resulting in an open service line. While each service line is typically 5/8 inch diameter, one hundred service breaks is approximately the equivalent of a 6 inch main, which might result in an equivalent flow of many hundreds of gallons per minute being lost[10].

In the final analysis however, lack of adequate water supply was not a crucial factor in this conflagration, since the wind-driven flames greatly exceeded the fire service's on-site capability to apply water, if they'd had it.

One interesting aspect of the water supply was the utilization of the San Francisco Fire Department's Portable Water Supply System (PWSS), discussed above in the Loma Prieta earthquake context. Despite communications deficiencies during the East Bay Hills incident, mutual aid SFFD strike teams were able to maintain contact with their own Communications Center and requested additional support from their own Department. Two PWSS Hose Tenders were activated by reassigning ladder company crews and responded to the staging area at the Claremont Hotel. These units are loaded with 5-inch hose and are equipped to establish emergency aboveground water supply systems in the event of an earthquake. Three water tenders and a communications vehicle from the Department of Public Works were also dispatched to Oakland. In response to further requests for assistance, San Francisco called in 85 additional off-duty personnel and sent them to Oakland by bus to reinforce the companies at the Claremont Hotel and in the Rockridge district (Routley, 1991). On the hill above the Claremont Hotel, the fire was advancing through a mixture of modem homes and stately mansions, working its way down the hillside. With the hotel protected by master streams poised for action, with these streams coming from the robust pipeline network in the A5A pressure zone, the crews climbed the hill to engage the fire on the upper streets, but found that the hydrants on the upper streets were dry. They returned to the bottom and started up again, hand-stretching a 5-inch hoseline to support an offensive attack on the fire. The fire was also successfully held at Alvarado Road.

In the Rockridge district, PWSS 5-inch hose and portable hydrants were used to bring a strong water supply from robust A2A pressure zone into the area where SFFD units were operating (A4B pressure zone). Several companies were able to obtain water from the portable hydrants, and the fire did not progress past this line. At this location, the driver of a local area TV news truck insisted that he be allowed to cross the line into the fire area; he somehow managed to get officials to allow him to do so; he then drove his truck up the hill, drive over the 5" hose, and broke it.

The result was one of the worst conflagrations in US history, Figure 3-12, which clearly demonstrates the continuing conflagration potential in modern US cities, at least under adverse meteorological and/or other conditions.

---

[10] The pipe size equivalence is based on area only, and neglects hydraulic effects, which if considered, reduce the one hundred, break equivalent to a smaller pipe, and less flow. However, many lost hundreds of gallons per minute, or a significant drop in pressure, will still occur in any case given one hundred service breaks.

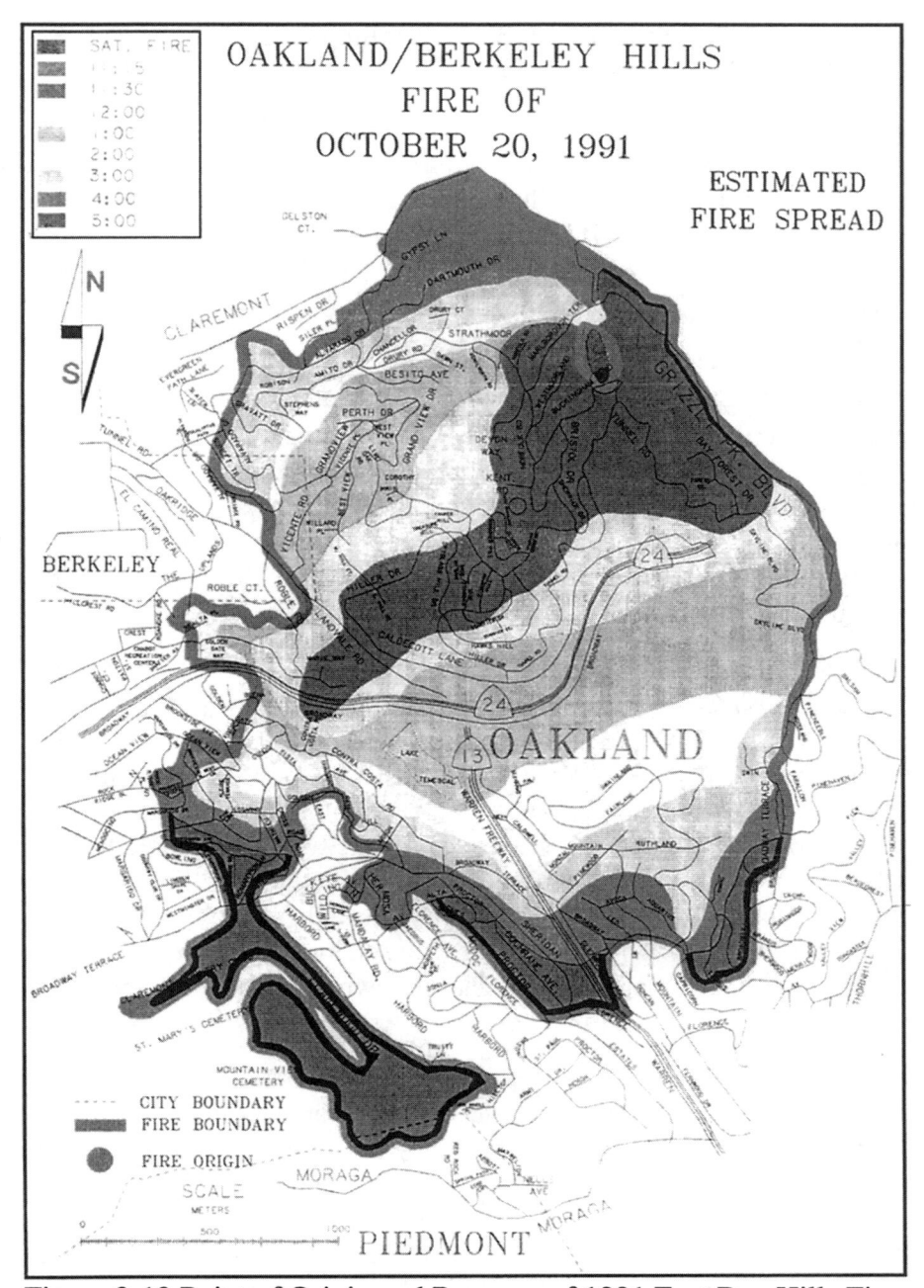

Figure 3-10 Point of Origin and Progress of 1991 East Bay Hills Fire: contours show origin at 11am, and progressive extent of fire at 11:15, 11:30, 12:00, 1:00pm, 2:00 pm, 3:00 pm, 4:00 pm and 5:00 pm

3.5.3 Water System Performance in the Oakland Hills Firestorm

The water system that was involved in the 1991 Oakland Hills Firestorm is owned by the East Bay Municipal Utility District (EBMUD). EBMUD has had a history of large fire conflagrations in its service area, including the 1923 Berkeley fire (614 structures burned), the 1970 Claremont Canyon fire (40 structures burned) and the 1991 Oakland

Hills firestorm (over 3,000 structures, or 3,400 dwelling units burned). This Section discusses how the water system performed in this event.

At the time of the 1991 fire, EBMUD maintained a large and robust system. Typical system-wide delivery was over 200 million gallons per day, peaking to about 300 million gallons on hot summer days. EBMUD had 6 treatment plants, 175 potable water storage tanks, 139 pumping plants, and 3,700 miles of distribution pipelines in service. Overall, EBMUD had 125 different pressure zones. About 75% of all customers received their water via gravity flow and 25% via pumped flow. Eleven of these pressure zones were involved in this fire.

The oldest parts of the EBMUD pipe distribution system within the fire area dated back to the early 1900s. The oldest reservoir is the Dingee reservoir, placed in service in 1894. Age of pipelines was important in considering the performance of the water distribution system during the firestorm in that the hydraulic carrying capacity of unlined cast iron had become degraded performance over time. The net result was that old cast iron pipe, which was still in use on older neighborhoods involved in the firestorm, often had a flow capacity less than half that of when it was originally installed.

In 1970, a fire in the Claremont Canyon destroyed some 40 houses. The location of this 1970 fire was a few hundred yards north of the ultimate extent of the 1991 Oakland Hills fire. Alarmed by the high fire risk in the hills, EBMUD immediately responded in 1971 by constructing three new water tanks at the highest elevations in these hills; all these tanks supplied some water in the 1991 fire. Reflecting upon the "urgent" need to get more water supply in the hills prior to the 1971 fire season, EBMUD selected tanks with the fastest construction schedule: small redwood tanks; with the recognition that these were to be temporary structures. At the time of the 1991 fire, these redwood tanks were still in service. After the 1991 fire, coupled with ongoing seismic threat tot he water system, EBMUD decided to replace these small isolated redwood tanks with newer designs, as well as improve their hydraulic connections to the rest of the water system.

Another fire in the Berkeley hills occurred in December 1980, devastating 5 homes that bordered Tilden Park, an open grassland and woodland area, being the same location as the ignition in the 1923 fire. On June 22, 1981, the so-called Atlas Peak fire occurred near Napa, California, resulting in the loss of 146 structures. In 1981, a "blue ribbon" panel had convened to study ways in which the fire risk could be reduced in the Berkeley Oakland Hills (EBRPD 1982). The panel studied the various aspects of fire risk, and recommended, amongst other things, to install a 21-mile long fuel break in the Berkeley and Oakland Hills, dividing the open lands in the east from the urbanized hill areas to the west. This fuel break proved to be of little or no value in the ensuing 1991 Oakland Hills firestorm, as the fire ignition occurred on the western urbanized side of the fuel break. Of concern to the panel was that the fuel break be constructed in an environmentally sensitive way and in an manner so as in no way to damage unique and sensitive natural plant communities; this served to increase the cost of the fuel break by four to six times higher than constructing the fuel break by mechanical means.

As a secondary outcome of the panel, it was recognized by the Oakland and Berkley fire chiefs that the water system in the urbanized hill areas had been built over decades to varying standards. The fire chiefs made recommendations to improve the water system. The fire chiefs of Oakland and Berkeley finally approved a new fire flow standard for the high-risk area of the Oakland – Berkeley Hills: namely a 1,500 gpm fire flow rate for 2 hours. In the mid-1980s, EBMUD designed retrofits to the 9 highest elevation pressure zones that were ultimately involved in the 1991 fire, to meet the following standards: in-zone water storage for each zone; storage in each zone equal to 1.5 maximum demand days plus an additional 1,500 gpm x 2 hours = 180,000 gallons; pipeline upgrades to provide 1,500 gpm to each hydrant in the hills. Recognizing the extent of the upgrades, an Environmental Impact Report was prepared. Public comments were vehemently against the proposed upgrades EBMUD wished to install. Of the more than 20 written letters from local citizens available for review, all but one were dramatically against the proposed water system upgrades. Citizen comments ranged from modest complaints due to the necessity of digging up streets to put in larger diameter pipes; to other complaints such as: "the new tanks will reduce the value of my property". One citizen's written report said that she remembered the 1970 Claremont Canyon fire, and she would appreciate anything EBMUD could do to improve the water system. In response to these comments, EBMUD ultimately installed smaller water storage tanks in the hills.

The 1981/82 Blue Ribbon panel did not signify the area south of Highway 24 and west of Highway 13 as a "high fire risk" area. Presumably, this was because the area was "protected" by the large firebreaks afforded by the highways. As will be described below, this proved to be unfortunate, as the water system in that area could not provide adequate fire flows to adequately protect structures in this conflagration.

Another contributing factor to the water supply situation in the hills was an unrelated landslide that had taken place several years before the fire. This landslide broke a 12" diameter pipeline that supplied water from a large tank to an area that was ultimately involved in the 1991 fire. The City of Oakland ultimately decided not to repair the street that went through this landslide. This forced EBMUD to put in place a "temporary fix" to provide water to those people cut off from their normal supply. This "temporary fix" was a pressure regulator to drop water into the isolated neighborhood from a distant and higher elevation tank; this fix could provide a few hundred gpm flow, but not the few thousand gpm flow originally available via the 12" pipeline. Needless to say, this "temporary fix" was still in service at the time of the 1991 fire.

The following observations can be made with regards to EBMUD's actions before the fire:

- In the 1923 Berkeley fire, 2" diameter pipeline was a contributing factor. EBMUD replaced all of its 2" pipeline with larger pipeline. For economic reasons, 4" diameter pipeline was still considered to be acceptable. Fore aesthetic reasons, initial efforts to limit the use of shingle roof coverings went largely unheeded.

- After the 1980 fire, a "blue ribbon" panel was convened. 14 miles of fire break were improved and 7 new miles were constructed The Berkeley and Oakland fire chiefs agreed to adopt a new "high fire risk" standard of 1,500 gpm for 2 hour fire flows, in higher elevation hill areas considered to have high fire risk. The water utility re-designed the existing system. Due to the construction and visual impacts posed on local citizens by this new infrastructure, the ultimately upgraded system was not quite as good as intended by design.

- The lower elevation portions of the water system that were involved in the 1991 fire (south of Highway 24) were not upgraded to the higher fire flow standards. Presumably, the fire risk was considered moderate, in that there were 4 lane and 8 lane highways to serve as firebreaks to protect this area. EBMUD, like most other water districts, sets its fire flow standards based upon what the local fire chief requests; as fire standards have evolved over time, EBMUD does not have a formal internal policy to upgrade its older parts of its water system to meet newer standards; this occasionally does get done, but often only when there is an unusually event to trigger the process, such as construction of a major new school in the area. Like any public utility, EBMUD considers the cost of new infrastructure in context of their typical annual capital budget; if the utility wants to build more, then it needs to charge more. Ultimately, it is the citizens of the area who decide as to how much they are willing to spend for infrastructure; and they often decide based on their perception of the extra benefits they get versus the cost. There is little doubt that had the lower elevation portions of the water system had been upgraded to provide 1,500 gpm fire flows, then arguably many hundreds of structures would have been saved. As of 2003, the water system infrastructure in the lower elevations of the firestorm area have been upgraded, at local user's extra cost, but this begs the question as to whether other places should also be similarly upgraded, and if so, at how much cost versus how much benefit.

3.5.3.1 Water System Response - North of Highway 24 (Area Upgraded to 1,500 gpm)

For the three-hour period from 8:00 am to 11:00 am, Sunday October 20, 1991, water usage in this area ranged from 1,000 to 2,000 gpm; this is normal for a Sunday morning, representing a combination of irrigation and indoor uses. Between 11:00 and 12:00 noon, the amount of water used in the immediate fire ignition area increased to about 1,000 gpm, reflecting initial use by the first responding fire fighters. By 12 noon, many more fire fighters arrived in the area, and the amount of water used in the area jumped up from about 2,500 gpm to over 12,000 gpm. Part of this water was being used by fire fighters to fight the fire; while part of this water was lost through broken service connections, after houses had burned down. For the two-hour period from 12 noon to 2:00 pm, water usage averaged about 13,000 gpm.

Beginning at 2:00 pm, three of the water tanks in the area were emptied. Due to the drawdown of these reservoirs to empty, the total water used in the area dropped to about 8,000 gpm from 2:00 pm to 5:00 pm.

By 5:00 pm, the spread of the firestorm reached its maximum extent. From 6:00 pm through the night and into October 21, water usage averaged about 6,000 gpm.

3.5.3.2 Power Outages North of Highway 24

Power failures in PG&E's system temporarily shutdown five pumping plants that replenish the reservoirs. The interaction of water and power lifelines in addressing fire conflagrations is described in (Eidinger 1993). Based on analysis of actual pumping plant capacities, if no power outages had occurred, or sufficient on-line backup power had been immediately available, then the total water use in the firestorm area above Hiller Drive, between 12 noon and 2:00 pm, could have been increased by 1,250 gpm, to somewhat over 14,000 gpm. It is doubtful if this moderate increase in available water, along with other factors (fire fighting teams, fuel load, weather), would have had appreciable effect on halting the spread of the fire at this time.

3.5.3.3 Failed Service Connections North of Highway 24

As the fire spread through the area, structures were burned to the ground. In many cases, when a structure burned down, the service connection piping from that house into EBMUD's distribution main became exposed.

Beginning at about 3:00 pm on October 20, 1991, EBMUD personnel began closing off the taps off the distribution mains to these failed service connections. This effort lasted until about 7:00 pm that day, when it became too dark to continue operations. The effort was restarted at about 8:00 am October 21, 1991. By about 3:00 pm October 21, 1991, EBMUD personnel had closed off over 2,000 service connections.

Under normal conditions, a single 5/8 inch service connection can provide 20 gpm to a house. Many houses in the firestorm area have even larger service connections (3/4 and 1 inch connections are not uncommon - larger connections, from 2 - 6 inches, supply a few customers, such as schools and apartments). During the firestorm, lesser amounts of water will flow through an open service connection, owing to competing demands for water from fire fighters and other open service connections. Also, not all service connections were likely to be completely open due to the destruction of the house.

As an estimate of water lost through failed service connections, abandoned flowing hydrants, and other causes, we estimate that about one-third of all water used during peak water usage times (12 noon to 3:00 pm, October 20, 1991) was lost through failed service connections. This one-third estimate is based upon allowance for loss of water used in developing insurance rating guidelines for required fire flows[11].

3.5.3.4 Water System Response - South of Highway 24 and West of Highway 13 (Vintage Water System)

Due to the limitations of some of the old pipe distribution system in this area, it is not surprising that fire fighters reported lack of water at some locations. For example, Captain F. Baleria reported that:

> *"... connected to hydrants at Country Club and Beechwood, and Country Club and Bowling, and had barely enough water to supply one 1 1/2" hose line."*

Similarly, in another after action report, Captain G. A. Flom reported that:

---

[11] Standard Schedule for Grading Cities and Towns of the United States with Reference to their Fire Defenses and Physical Conditions, National Board of Fire Underwriters, 1956.

> *"... laid a hose from the hydrant [at Rockridge Blvd. South]... ...crews extinguished an apartment house on the east side of Margarido ... when the water failed. ... at this time, crews were in full retreat. ... restored water pressure ... enabled us to set up portable hydrants on Margarido and attack and suppress the house fire, saving the structure near the original hydrant."*

These reports show that there was intermittent supply of water in this area. This occurred because the pipeline network was largely made of pre-1930 4" and 6" tuberculated cast iron pipe; and severe overdrafting on the system. Reservoir supply to this area was never compromised: the tanks that serve this area were never drained with fill levels never less than 5,000,000 gallons.

Supplementary water was delivered using 5" diameter hose (San Francisco's portable water system), but these pipes failed when vehicles drove over the hoses; reportedly, fire department staff were unable to prevent vehicles from interested parties from entering the fireground area. When available, this supplementary water contributed less than 3% of the peak fire flows actually delivered.

3.5.3.5 *Overall Water Use South of Highway 24*

For the three-hour period from 9:00 am to 12 noon, October 20, 1991, water usage in this area averaged around 3,500 gpm. This amount of water usage reflects the amount of water used by residents in this area, for normal uses, such as consumption, irrigation, etc.

Between 1:00 pm and 4:00 pm, the amount of water used increased to about 9,000 gpm to 12,000 gpm, reflecting initial fire fighting efforts in the area.

Beginning around 4:00 pm, water usage increased to over 16,000 gpm. Water usage ranged from 16,000 gpm to over 20,000 gpm for the next ten hours.

Water usage decreased to about 14,000 gpm, from 2:00 am October 21 to about 7:00 am. Water usage declined to about 4,000 gpm by 2:00 pm October 21.

3.5.3.6 *Power Outages South of Highway 24*

Power failures in PG&E's system shutdown certain pumping plants in that replenish the reservoirs south of Highway 24. The Dingee pumping plant lost power at about 12:13 pm on October 20. PG&E power was restored to the Dingee pumping plant at 5:46 pm on October 20. If the Dingee pumping plant were not stopped due to the power outage, an additional 9,700 gpm flow would have been available to the B5A Dingee Zone. As the reservoirs that serve the Dingee zone were never emptied, this additional water would not have had much impact on increasing available fire flows in the area. No after action fire reports were obtained that suggested that there was a lack of water in the first lift served by this pumping plant. Instead, pipe size limitations effectively precluded use of higher flow rates in the area. This suggests that PG&E power outages at the main pumping plant serving this area had *no appreciable affect* on the firestorm outcome.

3.5.3.7 *Failed Service Connections South of Highway 24*

As the fire spread through the area, structures were burned to the ground. In many cases, when a structure burned down, the service connection piping from that structure, into EBMUD's distribution main became exposed.

Beginning at about 8:00 am on October 21, 1991, EBMUD personnel began closing off the taps off the distribution mains to these failed service connections in this area. The effect of closing these service connections, beginning at about 8:00 am October 21, resulted in a decrease in water usage from about 11,000 gpm to about 5,000 gpm over the next four hours. This decrease is also likely due to the cessation of fire fighting efforts in some areas. From this trend, we conclude that about 4,000 - 6,000 gpm was being lost through failed service connections, at a more or less steady rate, from 5:00 pm October 20 through 8:00 am October 21. This rate of water loss is about one-quarter to one-third of the total water used in this area during this time interval.

3.5.4 Fire Flow Guidelines for Water Systems in High Fire Risk Areas

3.5.4.1 *Pre-EBMUD Designs for Lower Elevation Areas*

Part of the area covered by the October 20, 1991 firestorm was built prior to the formation of the present day water utility (pre-1929). The Dingee reservoir and local water distribution system was built by Mr. Dingee, in the 1890s. The standards that Mr. Dingee used to build this system are not known; perhaps he tried for 500 gpm, but perhaps he just tried to connect as many customers to his water system as he could, for a reasonable cost.

3.5.4.2 *EBMUD Designs for Higher Elevation Areas*

Currently, the Oakland Fire Department has set 1,500 gpm for 2 hours as the required fire flow for new residential construction in the Oakland Hills area east of Highway 13. All EBMUD construction post-1973 for these areas included a minimum fire flow of 1,500 gpm for 2 hours. Almost all areas east of Highway 13 and north of Highway 24 complied with these fire flows at the time of the firestorm.

3.5.4.3 *Observations as to the Performance of the Water System*

Based upon the water use analysis, most of the areas within the firestorm had water systems that actually delivered at least the modern day fire flow requirement. Even so, these areas suffered almost 100% loss.

Due to high demands in the lower elevation old parts of the water system, water flow was interrupted to fire fighters located at some distances away from the pressure regulators feeding the area. For example, one fire fighting team drafted water essentially continuously throughout the day from the hydrant at Buena Vista Place, saving the houses on that street. The hydrant at Buena Vista Place is the first hydrant in the zone. Because of the small diameter, 1910 vintage cast iron pipe in this area, the distribution system could not simultaneously and continuously deliver water to hydrants further away from the pressure regulator, even though ample water was available in the reservoirs.

Based upon the huge extent of the October 20, 1991 firestorm, the EBMUD system ran out of water in certain areas, even in areas where the system completely met modern fire flow standards for high fire risk residential areas. Although the system was prepared for the fire it was designed to meet, it was not prepared to supply water for the October 20, 1991 firestorm. Substantial reservoir size increases, pumping plant size increases, and upgrades to 4 inch distribution piping would be required to supply water to many simultaneous fires within a single pressure zone. Such hardware improvements, well beyond that required by current standards, could have helped some fire fighting efforts - however, there may be other more cost effective approaches to improving total fire fighting capabilities, and reducing the risk of re-occurrence of this type of catastrophe.

3.5.5   Alternative Strategies for High Fire Risk Hillside Areas

1. Replace Older Small Diameter Pipes with Larger Pipes. In certain areas of the system, severe overdrafting led to the inability to deliver water to hydrants where needed, even though there was plenty of water available in reservoirs. Replacement of smaller diameter pipelines will alleviate this constraint. However, this is a costly retrofit, and will provide only local benefits. The people that will benefit from this type of upgrade may need to be assessed the associated costs.

2. Modification of Today's Fire Flow Standards. During the course of the October 20, 1991 firestorm, several of EBMUD's reservoirs were emptied. The reason that these reservoirs were emptied was not that the reservoirs were incorrectly designed - they all met the modern standard of 1,500 gpm for 2 hours plus an additional 1.5-day supply for the area's maximum demand rate. Rather, the prime reason that the reservoirs emptied was that the firestorm conflagration that occurred was far and beyond any size fire that was considered in design.

One option is to consider increasing today's fire flow standards from 1,500 gpm for 2 hours, to some higher flow rate, and some longer duration. This choice is an expensive one to implement, especially in the existing built up areas. Further, increasing the size of reservoirs will adversely affect water quality.

3. Construction Upgrades. A significant contributor to the rapid spread of the fire was airborne shingles from roofs. If all houses in the area had clay tile or other non-combustible roof, it is likely that the extent of the firestorm would have been considerably reduced.

Rules could be put into place to strictly enforce that new structures built in high-risk fire areas be built with tile or similar roofs.

It would also be highly desirable to rid the high fire risk areas of structures that already have shake roofs. This process can be mandated by City rules either at time of sale, at time of substantial structural improvements, or at times when re-roofing efforts are performed.

It is of some interest to note that the City of Oakland has occasionally insisted that new construction in the 1991 firestorm area include fire sprinklers. Actual installations require that the fire sprinkler system have a separate tap off the water main in the street, and no isolation valve on the homeowner's side of the connection is allowed. In practice, this type of fire sprinkler installation would have had negative effect in the 1991 Oakland Hills fire, in that indoor sprinklers would not stop a spreading fire; once sprinklered houses burn down, they would open two taps on the water mains, leading to further water losses and drop in system pressure, further aggravating the ability of fire departments to draft out of hydrants.

4. Vegetation Control. The high vegetative fuel load in the 1991 firestorm area contributed to the fire's rapid spreading. Strict control of vegetative fuel loads in high hazard fire areas would reduce future fire risks. Additional precautions should be taken to rid high hazard areas of fuel load caused by severe winter freezes. For the twelve years following this fire, the Oakland Fire department has more-or-less rigorously enforced the annual removal of dead and dying brush on private property in the high fire risk areas of the City; although little such effort has been done on publicly owned lands. As time passes and budgets are squeezed, and the collective memory of this firestorm abates, it remains to be seen if this effort will be maintained. For the period 1991 through 1997, the City of Oakland funded the effort to control vegetation growth by using grants from FEMA. In 1997, the residents in the affected fire zone area were asked, by ballot measure, to continue paying for these efforts through a special district assessment; this was voted down, and so the City continued to fund the effort through its general funds. One reason the local residents voted down the formation of a special assessment district was the sense that the City did nothing to address the fuel load on public agency-owned lands, thereby leading to the perception of an expensive, wasteful and ineffective vegetation control program.

5. Emergency Backup Power and Pumping. Several pumping plants lost PG&E power during the firestorm. If PG&E power had not been lost, at any time at any pumping plant, or if EBMUD had immediate use of emergency backup power, the amount of additional water available to fight the fire would have been incremental: perhaps an additional 1,650 gpm split between five higher elevation areas.

At the time of the fire, EBMUD had sufficient numbers of portable back-up generators and portable pumps to serve the affected pumping stations - however, due to the size and intensity of the fire, the time necessary to mobilize and install this equipment made them unusable for many hours into the firestorm.

The lower elevation areas with a lot of old smaller diameter pipe lost their source of water re-supply when PG&E power failed at the Claremont Center. However, these same zones had substantial intra- and inter-zone ties, so that they were able to draw water from remote reservoirs and pumping plants outside the firestorm area. Shortage of water due to pumping plants out of service for these areas was not a major problem to fire fighters.

Recognizing the continued threat of fires, including fires following earthquake, and the fact that EBMUD has about 140 pumping plants throughout its water system, and the fact that at the time of the fire, only two of these pumping plants had permanent on-site backup power supply, EBMUD has undertaken the following improvements:

- All pumping plants are now outfitted (or have been retrofitted) with quick connect couplings or bypass pumping tees, to allow the rapid deployment and use of either portable generators or portable pumps;

- EBMUD maintains a sufficient number of its own portable generators and pumps (more than 20) to deploy them to all pumping plants that might realistically be involved with concurrent fires;

- EBMUD has installed permanent emergency generators at a few large pumping plants (including three that were involved in the 1991 Oakland Hills firestorm).

- Within the older parts of the City of Oakland that were located within the firestorm area, a citizen-approved charge was made to replace older small diameter cast iron pipelines with newer lined pipelines with larger diameter and better long term hydraulic characteristics; and pressure regulating stations were upsized.

Even with these actions, there remain areas within EBMUD's service territory, as well as for other San Francisco Bay Area water districts, that still have many attributes that led to the 1991 fire conflagration. For example, there are still hundreds of miles of 2" to 4" diameter water mains in service in the hillside areas of the San Francisco Bay Area; and many of these areas have high fuel loads, limited size up-hill storage reservoirs, and narrow and winding streets.

Figure 3-11 1991 East Bay Hills fire - narrow roads in Hills area, and burnt abandoned automobiles indicative of the congestion during the evacuation.

Figure 3-12 West limit of burnt area, 1991 East Bay Hills fire, showing some houses burnt, others surviving

3.6 1993 Southern California

The Los Angeles metropolitan region is a highly built up metropolitan area prone to wildland fires. At the wildland-urban interface, these fires regularly cause substantial losses. The region has a population of almost 10 million spread over about 2,000 square miles, which is surrounded on three sides by mountains, creating the Los Angeles Basin – an important aspect of this topography are the *Santa Ana* winds - hot, dry high winds which blow off the mountains from arid interior regions. Combined with highly flammable brush in the mountains and wildland-urban interface, the Santa Anas create absolutely explosive fire conditions, which have resulted in major conflagrations at regular intervals during Los Angeles' history.

Beginning on October 27, 1993, Santa Ana conditions resulted in a series of fires over the next ten days that burnt over 300 square miles and destroyed over 1,000 structures in Southern California. The fires occurred in two series - for the first outbreaks, occurring October 27-29, causing $500 million in losses. At the height of the emergency, there were major fires burning in five Southern California counties with approximately 15,000 firefighters engaged in firefighting activities. Of the 14 fires in the region, the worst blaze was the Laguna Beach fire in Orange County that claimed 366 homes and was finally extinguished Sunday, October 31 after a five-day battle that blackened nearly 17,000 acres. The Altadena/Sierra Madre fire destroyed 151 homes in a 5,700-acre blaze. Although there were no deaths in the fires or among those fighting them, 67 firefighters were injured and 4 were seriously or critically burned.

The second outbreak of wild fires began November 1 much like the first and, in less than twenty-four hours, destroyed more than 300 homes in the Calabasas-Topanga-Malibu areas. Three people died and five were seriously injured as the fire raged through home after home in a matter of seconds. Firefighters, homeowners and media reporters literally had to run for their lives to escape the blazing wall of flame that consumed everything in the fire's path. Fanned by fifty mile-an-hour winds and fed by some of the most expensive residential real estate in the country, the fire increased the property losses by millions of dollars every minute. The west flank of the blazing inferno finally burnt down to the Pacific Ocean. At the same time, a few miles away, Pepperdine University was held at siege for several hours as fire fighters kept the flames at bay on the edge of the campus property. Resident students who could not be evacuated in time were sheltered in the gymnasium as a last resort protective measure. At nightfall, all aerial helicopter and airplane water support ended leaving the ground units on their own. It was the sole courage and skill of the already weary fire fighters and the dependability of the water system that eventually saved the campus (Goltz et al, 1993)

3.7 Highrise fires

Another important aspect of fires following earthquakes are those fires that may occur in a major highrise building, since not only are major property values at risk but, more importantly, it is conceivable that thousands of people could be trapped above the fire.

Highrise buildings are generally defined as buildings greater than 7 stories in height, based on this being the tallest height most fire department tall ladders can reach. There are about 10,000 highrise structure fires in the US each year, constituting roughly 10% of all structural fires (Hall, 2001). Under normal conditions, a major fire in an occupied highrise poses excessive demands on the fire department, for occupant evacuation as well as firefighting. An earthquake can only exacerbate this situation. In this section, we briefly review several notable large non-earthquake highrise fires, to determine certain common trends in these events, and then discuss these trends in the light of the post-earthquake environment.

3.7.1   1988 First Interstate Bank Building

The First Interstate Bank building, in 1988 the tallest in the city (and the state of California), was built in 1973, one year before a highrise sprinkler ordinance went into effect, and had sprinkler protection only in the basement, garage, and underground pedestrian tunnel. The 62-story office tower measures 124 feet by 184 feet and has a structural steel frame, protected by a sprayed-on protective coating, with steel floor pans and lightweight concrete The exterior curtain walls are glass and aluminum. A complete automatic sprinkler system installation was approximately 90 percent complete, but valves controlling the sprinklers on completed floors were closed.

The fire was reported at 10:37pm and originated on the 12th floor, with the cause thought to be electrical in origin, and extended to floors above, primarily via the outer walls of the building (windows broke and flames penetrated behind the spandrel panels around the ends of the floor slabs).There were only 40 occupants of the building (normally 4,000 in the daytime).There was heavy exposure of flames to the windows on successive floors as the fire extended upward from the 12th to 16th floors. The flames were estimated to be lapping 30 feet up the face of the building. The curtain walls, including windows, spandrel panels, and mullions, were almost completely destroyed by the fire. There were no "eyebrows" to stop the exterior vertical spread. Minor fire extension also occurred via poke-through penetrations for electricity and communications, via HVAC shafts, and via heat conduction through the floor slabs. The fire extended at a rate estimated at 45 minutes per floor and burned intensely for approximately 90 minutes on each level. This resulted in two floors being heavily involved at any point during the fire. The upward extension was stopped at the 16th floor level, after completely destroying four and one-half floors of the building, Figure 3-13.

Figure 3-13 First Interstate Bank Building burned-out floors

One fatality occurred – a maintenance worker took an elevator to the floor of origin to check on the fire alarm condition, and was confronted with severe fire conditions as soon as the elevator doors opened. Falling glass and other debris created a major problem during this incident. Virtually all of the exterior curtain wall, from the 12th through 16th floors, was destroyed and fell to the ground. The falling glass and debris caused significant damage to pumpers hooked-up to the Fire Department connections. The hose lines were cut several times and had to be replaced, under the constant danger of additional falling materials. The entire perimeter of the building, for over 100 feet out from the walls, was littered with this debris. Fortunately, a tunnel between the building and the parking garage across the street provided a safe path into the building at the basement level for both personnel and equipment. Without this tunnel it would have been very difficult to maintain the necessary logistical supply system and to avoid injuries to personnel from the falling debris.

The Los Angeles City Fire Department described the suppression effort as the most challenging and difficult highrise fire in the city's history. It took a total of 64 fire companies and 383 fire fighters more than 3 1/2 hours to control the fire – this constituted about 40% of the on-duty staffing. The fire is of significance because of the interior and exterior fire spread, the significant internal smoke spread, and the role of modern office environment materials and their arrangement in relation to the fire growth and development. (Abstracted from Klem, 1988, and Routley, 1988).

3.7.2 1991 Meridian Plaza

The One Meridian Plaza building is a 38-story 240-ft. by 92-ft. steel framed office building completed in 1973, in the heart of Philadelphia across the street from City Hall. At the time of the fire, Feb. 23, 1991, the building was in the process of being sprinklered, but the installation was not complete and the building was considered still unsprinklered. Cause of ignition was oil-soaked rags on the 22nd floor of the building, with the fire being reported at 8:30pm on a Saturday evening. The fire department arrived to find a well-developed fire with heavy smoke, which spread primarily via exterior vertical flame impingement due to window failure. Firefighting was severely hampered by failure of the entire building's electrical system, and major problems in the standpipe system due to irregular settings of the pressure reducing valves. These problems contributed to the death of three firefighters on the 28th floor when they became disoriented in the heavy smoke and their SCBAs ran out of air. Although the Philadelphia Fire Department responded with 51 engine companies, 15 ladder companies and over 300 firefighters, firefighting was eventually abandoned, and the fire burned for over 19 hours. The fire was stopped when it reached the 30th floor, which was protected by a functioning automatic sprinkler system – only ten sprinklers were required to be activated.

During the fire, falling debris damaged hose lines. The steel frame of the building was severely damaged, with steel beams sagging as much as three feet. Granite facing of the building was dislodged due to warped steel framing, and posed a subsequent hazard, which resulted in a prolonged evacuation of the entire block and surrounding buildings. (Abstracted from Klem, 1991, and Routley et al, 1991).

3.7.3 1993 World Trade Center

At 12:18pm on Feb. 26, 1993 a van filled with explosives exploded in the second basement area of the World Trade Center (WTC) complex, in lower Manhattan, immediately killing 6 people. In the immediate area of the explosion, the floor slabs for two basement levels collapsed onto vital electrical, communications, and domestic water systems equipment for the complex. Also, masonry firewalls and fire doors separating the buildings within the complex were penetrated by the force of the explosion, which enabled dense, black, super-heated smoke from the explosive materials and the ensuing fire to quickly move into numerous elevator shafts. The initial speed of the smoke spread was influenced most by the explosion, and forces associated with the "stack effect," a natural, ever-present condition in highrise buildings.

Approximately 1 hour and 15 minutes into the incident, all remaining electrical power to the affected buildings within the complex was shut down. Many occupants entered stairways and were confronted by the smoke. Later, due to the electrical power failure, the occupants experienced total darkness in the stairways and resorted to other means of illuminating their exit paths. The unexpected mass evacuation further influenced adequate performance of the exiting system since the stairways soon filled with people who were waiting for an opportunity to enter stairways and held doors open, thus allowing smoke from the respective floors to migrate into the stairways.

The loss of the normal electrical service and of the emergency generators also affected the standpipe and sprinkler systems for most of the buildings. The primary water supply for the standpipe systems and some of the sprinkler systems was municipal water mains and electric fire pumps. The primary water supply for the sprinkler systems in the towers was gravity tanks, which were not affected. With the loss of electrical power, the primary water supply was limited to that provided by the normal pressures in the water distribution system. Furthermore, the loss of electrical power to domestic water pumps limited the capability of the sprinkler systems in the towers to that water in the gravity tanks. Fortunately, the fire did not propagate from the basement levels and thus did not challenge the performance of the remaining fire protection features.

Successful occupant response during a fire emergency in this complex is dependent upon a transfer of information from emergency personnel in the operations control center. In this incident, however, the control center was destroyed by the explosion, leaving occupants without vital information from emergency responders. As a result, the occupants' response to the fire was uncoordinated, underscoring the necessity for all building occupants to understand and be trained in proper fire safety procedures.

The New York City Fire Department responded to the explosion and fire at the World Trade Center with 16 alarms, involving hundreds of fire fighters at the height of activities. This commitment represented approximately 45 percent of the New York City Fire Department on-duty resources and was the largest single response in the history of the New York City Fire Department. Several fire crews were committed to the suppression of cars and other combustible materials burning in the basement. However, the vast majority of the fire fighters were committed to time-consuming tasks of search and rescue in all areas of the seven highrise buildings and assisting in the care of escaping occupants. Over 1,000 people were injured, many from smoke inhalation, but none died from the exposure. This was primarily because there was a limited amount of combustibles that were initially ignited, and because of the basement floor collapse, there was limited fire spread to adjacent materials. Because of the limited burning there was a significant dilution of the products of combustion as they moved through this massive building complex. (Abstracted from Isner and Klem, 1993)

3.7.4   2001 World Trade Center

The events of Sept. 11, 2001 are well known –hijacked airlines crashed into World Trade Center tower 1 (WTC 1) at 8:46 am and WTC 2 at 9:03 am, resulting in immediate major structural damage to both towers, and massive fires which resulted in the collapse of WTC 2 at 9:59 am, and the collapse of WTC 1 at 10:29 am. About 3,000 persons died, including 2,830 building occupants, 157 airliner passengers and crew, and 343 emergency responders (FEMA, 2002).

The 22-story Marriott World Trade Center Hotel (WTC 3) was hit by a substantial amount of debris during both tower collapses, and was substantially destroyed. WTC 4, 5, and 6 (9, 9 and 8 stories, respectively) were set afire by flaming debris, and suffered partial collapses from debris impacts and fire damage to their structural frames. WTC 7 (47-story steel frame) was set afire and burned unattended for 7 hours before collapsing at 5:20 pm. The falling debris also damaged water mains around the WTC site, which significantly reduced the amount of water available for firefighting. However, fireboats responded and drafted water from the nearby Hudson River, to provide emergency firefighting supply.

Relative to the problem of fire following earthquake, the collapse of WTC 1 and 2 are somewhat less significant than the collapse of WTC 7. WTC 1 and 2 were unique structures – the world's tallest buildings when constructed, and of an atypical structural configuration. Relatively light steel truss joists, whose fire protection may have degraded over time, supported the floors. Combined with the massive initial damage from the impact, and the massive fires associated with the plane's fuel, the weakening and collapse of these structures due to fire is not surprising. WTC 7 however, is of more conventional construction, and was not significantly weakened by impact by either the plane, or debris. The performance of WTC 7 is therefore of significant interest because it appears the collapse was due primarily to fire, rather than any impact damage from the collapsing towers. Prior to September 11, 2001, there was little, if any, record of fire-induced collapse of large fire-protected steel buildings.

WTC 7 stored a significant amount of diesel oil and had a structural system with numerous horizontal transfers for gravity and lateral loads. Loss of structural integrity was likely a result of weakening caused by fires on the 5th to 7th floors. The specifics of the fires in WTC 7 and how they caused the building to collapse remain unknown at this time, but the basic point to be learned is that unfought fires can significantly weaken a modern steel-framed building sufficiently to cause collapse.

3.7.5 Highrise fire following earthquake

Major highrise fires are relatively rare events under any circumstances, and almost non-existent under recent large urban earthquakes. The few major non-earthquake highrise fires reviewed above demonstrate that highrise fires under ordinary conditions are a complex subject, characterized by:

- Potential for floor-to-floor, and building-to-building fire spread

- Effectiveness of sprinkler systems

- High priority and time-consuming occupant evacuation

- Difficult and labor-intensive firefighting conditions

Earthquakes simply act to exacerbate these characteristics, by degrading active and passive fire protection features and enhancing fire and smoke spread. That is, strongly shaken highrise buildings can be expected to sustain fairly large inter-story drifts, resulting in some broken windows, cracked walls, buckled fire and elevator doors, broken HVAC ducting, etc. The openings thus created more readily permit the entry of hot gases, and the spread of fire.

Beyond fire spread within one building is the spread of fire from a building to a highrise. The plume of smoke and hot gases rising from a nearby fire, even if in a low-rise building, can cause ignition in upper stories of an adjacent highrise. This does not usually occur under ordinary circumstances, due to active fire department exposure protection, but is likely to happen if fire approaches a highrise building following a strong earthquake, due to the absence of the fire department. Even sprinklering of the highrise may not prevent this, since sprinklers may be set off on several floors, overwhelming the hydraulic capacity of the system and permitting ignition.

Under a worst-case scenario, which is entirely credible in the dense concentrations of highrises in the CBD portions of major cities, this ignition of upper stories can proceed from one highrise to another, resulting in a phenomenon akin to a forest fire "crown fire". This process may be aided by the fall of burning debris from an involved highrise onto the roofs, skylights and other portions of intervening lower buildings, igniting them, their fire then spreading to the next highrise, and so on. This of course is predicated on the basis of the virtual absence of the fire department, which is not unreasonable. On-duty firefighting personnel in San Francisco for example (population 700,000, over 500 highrises in the city) are approximately 300, organized into 41 engine companies and 18 truck companies.

Due to excessive demands on the fire service following an earthquake, active suppression on their part cannot be relied upon. In many cases, the installed detection, alarm and suppression systems may also not be reliable. That is, fire/smoke alarm detectors may or may not function properly, following strong shaking. Dust raised by the shaking will often activate smoke detectors, so that dozens of alarms may simultaneously be received over a number of floors of the building. Alternatively, shaking may dislodge detectors or otherwise damage their circuitry, so that they cannot function. This problem extends to the central fire alarm panels, located in mechanical rooms and/or in the Fire Marshall's room (required in most modern highrises, usually located in or near the lobby). These panels are usually not seismically qualified, or even properly mounted. That is, the panels may be freestanding or inadequately secured, so that they will overturn or fall during strong shaking.

A typical highrise building firewater and sprinkler system is shown schematically in Figure 3-14, and consists of the following elements:

- The water service connection to the building

- A fire pump with (sometimes without) emergency power (and associated fuel supply)
- A fire pump control panel
- A secondary water supply (a tank, often located in the basement)
- Sprinkler risers, mains, branches and sprinklers

Each of these elements can sustain damage, rendering the system inoperative. Since many urban CBD's are located in or adjacent to old waterfronts, the municipal water supply system in the vicinity of the CBD will often sustain a significant number of water main breaks, due to the soil failure in the soft waterfront soils, filled-in areas, etc, resulting in temporary loss of municipal water pressure immediately following the earthquake. Water supply for highrise fire protection thus reverts to the secondary supply. In older highrises, the tanks may not be adequately secured for strong shaking. Even relatively small rocking or sliding of a tank can break pipe connections, eliminating the secondary water supply.

Fire pumps will usually be adequately secured to their foundations, for ordinary mechanical reasons. Depending on piping configuration and anchorage displacement however, pumps may be badly damaged, if the differential displacement between the piping anchorage and the pump location is of more than a few centimeters. Since the municipal electrical supply will often fail towards the end of the ground shaking, emergency power is critical to the functioning of the fire pumps.

Emergency power usually takes the form of a diesel engine and/or generator set. These are spring-mounted, for ordinary vibration isolation purposes. Under strong ground shaking, these spring mounts may resonate, resulting in excessive displacements and tearing of connections. Seismic snubbers are required, to restrain excessive displacements. The diesel sets will be started by power from a battery set. These batteries have been observed to often not be adequately restrained, so that they will be tossed about, breaking their circuitry. The emergency power fuel day tank may also not be adequately restrained, resulting in an ignition hazard itself, beyond the simple problem of loss of fuel. For all these reasons, emergency power, if experienced designers have not detailed it for seismic forces, is not reliable. Modern building codes in high seismicity zones require higher reliability of life-safety related systems, such as the emergency power and firewater system, but the requirement is a general one and experienced designers are necessary for satisfactory detailing and performance. In certain locations building code requirements have only been in place for about the last decade, so that older highrises, built before these requirements, often lack these features. Seismic criteria for sprinkler systems has been in place since 1947.

The last element in the firewater system is the sprinkler piping. Building code requirements for sway bracing (both lateral and longitudinal) at nominal intervals are probably satisfactory. Anchorage of pipe hangers into certain types of floors (precast planking for example) may not be entirely satisfactory - failures occurred in the October 1, 1987 Whittier Narrows earthquake. Simple pounding of sprinkler heads with suspended ceilings may be a problem - breakage of heads due to this were also observed in the Whittier earthquake. The breakage is serious in that it directly results in water damage and, more importantly, if occurring in more than a few heads, will exhaust the hydraulic capacity of the sprinkler piping.

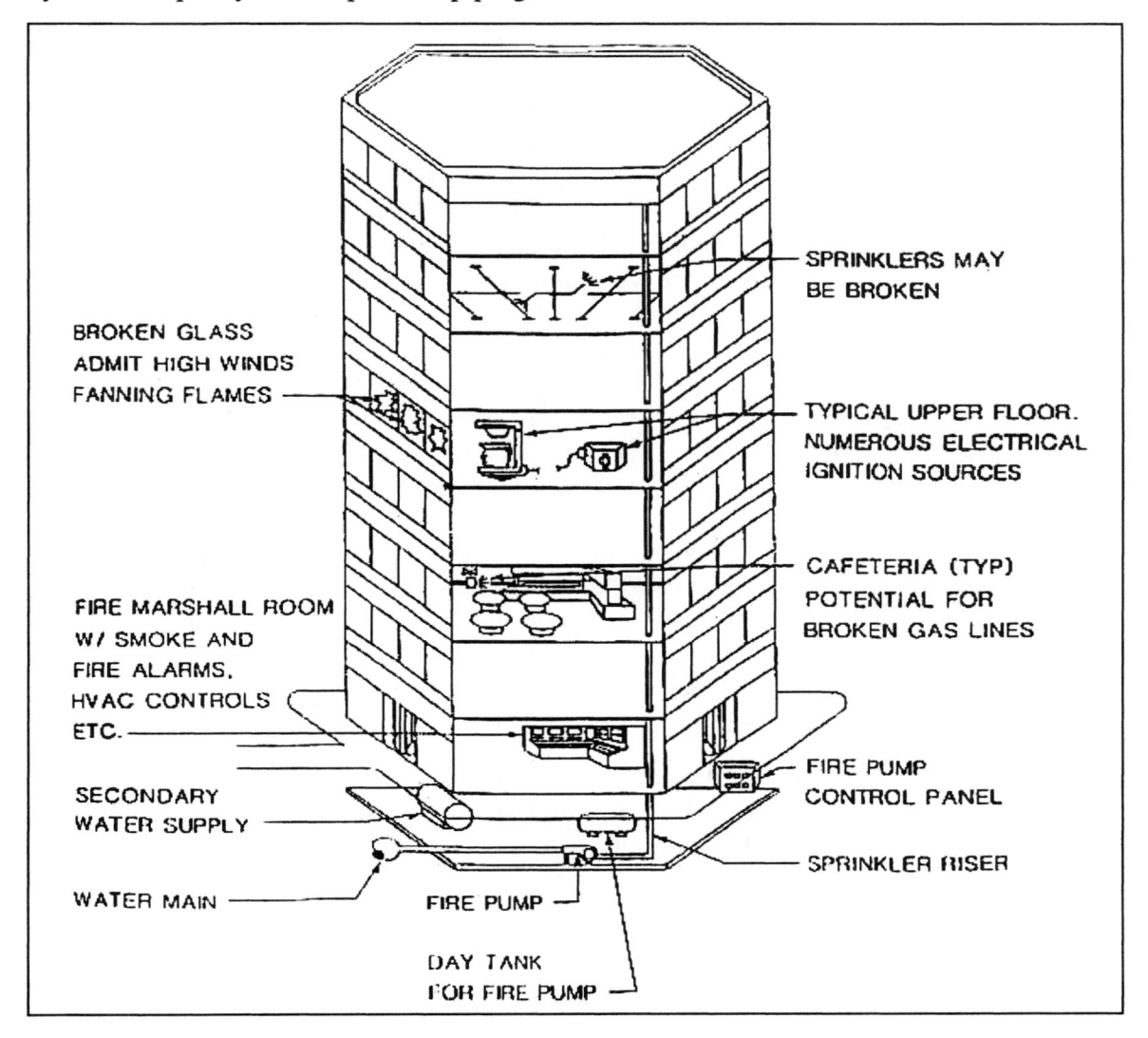

Figure 3-14 Highrise Building Fire Protection System Schematic

3.8 Industrial Experience

That major fires may occur in industrial facilities following a major earthquake, should come as no surprise, as demonstrated by the Tupras refinery experience in the 1999 Turkey earthquake discussed in Chapter 2. An extensive discussion of industrial fire experience and protection is beyond the scope of this Monograph. However, we briefly discuss several aspects, with relevance to the fire following earthquake problem:

Hazardous Materials: Many industrial facilities contain flammable and explosive materials that can be released in an earthquake. Selvaduray for example has documented over 1200 releases of hazardous materials in past earthquakes, Table 3-4. Fires due to hazmat releases have been documented in a number of earthquakes (e.g., Scawthorn and Donelan, 1984, Selvaduray, 2002). One incident of particular note is documented by Selvaduray (2002) in the 1995 Kobe earthquake:

> *A 20,000-ton LPG tank sprung a leak at the flange of the main valve. Since the leaking flange was on the tank-side of the valve, the leak could not be shut off. The leak was detected shortly after the earthquake, and the employees who detected the leak made the first attempts to stop the leak by covering the flange with blankets and pouring water, in an attempt to freeze it, thus stopping the leak. However, gradual settlement of the pipeline leading away from the tank, and aftershocks, resulted in the leak becoming worse. The secondary containment walls had also been damaged, thus increasing the risk posed. The Kobe Fire Department was informed of the leak at around 10:00 a.m. in January 17, 1995 [the earthquake had occurred at 5:46am]. In view of the possibility of a major conflagration, and even explosion, an evacuation warning was issued at 6:00 a.m. on January 18, 1995 (i.e., approximately 24 hours after the earthquake and leak). This affected approximately 70,000 residents who had difficulty finding alternate housing in the post-disaster environment. All fire departments that were capable of responding to this incident did so. This facility had three 20,000-ton LPG tanks, and fortunately only one was damaged. Further, one of the neighboring tanks was practically empty so that the contents of the leaking tank could be transferred, a process which was completed only in January 30, 1999 (i.e., approximately two weeks later).*

The gravity of this situation is demonstrated by non-earthquake LPG releases, such as occurred in 1970 at Port Hudson MO, where an LPG pipeline ruptured. Since LPG is heavier than air, under still wind conditions it collects in a low-lying cloud which can explode in an *unconfined vapor cloud explosion* (UVCE) manner. This occurred at Port Hudson, in a TNT-equivalent of 50 tons of TNT. Even more serious is the *boiling-liquid-expanding-vapor-explosion* (BLEVE), which can result from a fire overheating a vessel containing flammable materials. An example of this was the BLEVE that occurred in 1984 at the San Ixhuatepec LPG facility near Mexico City, when leaking LPG formed a vapor cloud which ignited and which then resulted in a BLEVE, killing 500 people (Pietersen, 1988). Fire departments are well aware and trained for UCVE and BLEVE incidents, but in the confusion of an earthquake, may not be informed (as in Kobe), or have the resources to deal with the incident. BLEVE situations for example, require massive amounts of water deluging a vessel to prevent boiling over – the facilities' systems may be damaged in an earthquake, and the fire department apparatus otherwise committed.

Containment: Most hazmat storage facilities have secondary containment to protect against catastrophic release. However, as occurred in the LPG incident above, the earthquake is a common-cause failure mode, in which the shaking or permanent ground deformations may cause breach of the secondary as well as the primary containment.

Safe-shut-down: Industrial facilities rely on safe-shut-down procedures. An earthquake may impair the power or control required for SSD however.

Fire Protection: Industrial facilities handling or storing flammable materials typically have fire loops and special sprinkler, deluge or other fire suppression systems. All too often, however, the water sources for these systems, and/or the energy sources for the pumps to pressurize these systems, may be very vulnerable to earthquake. As seen at the Tupra_ refinery in the 1999 Turkey earthquake, where the firewater source was from via a pipeline from a lake 45km away even though the refinery was next to the sea, earthquake vulnerability is often not thought through.

In summary, industrial facilities are potentially very vulnerable to earthquakes, and require earthquake-specific analyses to assure their safety.

3.9 Summary Observations

The accumulation of experience based on observations of the above events, and others which space does not permit discussing here, leads to the conclusion that the potential exists for large conflagrations in an urban area, particularly in a region with a large wood building stock. An analysis of 79 conflagrations during the period 1914-1942 indicates that wood shingle roofs, high winds and inadequate water supply are the principal factors leading to conflagrations, one or more of these factors occurring in more than one third of all conflagrations. Under adverse meteorological and other conditions, therefore, conflagrations in US cities are still possible, and may burn for several days, replicating events such as 1871 Chicago, 1904 Baltimore, 1923 Berkeley, and 1991 East Bay Hills, usually in a mass fire or conflagration mode rather than a firestorm. Highrise building fires are quite possible, further contributing to the demand on fire department resources. Industrial facilities are vulnerable, and releases of hazardous materials can result in massive explosions, leading to additional multiple ignitions.

Table 3-3. Principal Factors Contributing to Conflagrations in US and Canada 1914-1942

Factor	No. times contributing	%
wood shingle roofs	38	48%
inadequate water distribution system	30	38%
high winds (> 30 mph)	27	34%
lack of exposure protection	20	25%
inadequate public protection (i.e., fire department inadequacies)	17	22%
unusually dry or hot conditions	11	14%
congested access	10	13%
delay in discovery of fire	8	10%
private fire protection inadequate	7	9%
delay in giving alarm	5	6%

Table 3-4 Hazardous Materials Incidents in Past Earthquakes

Earthquake Name	Date	M (magnitude)	No. Incidents	Earthquake Name	Date	M (magnitude)	No. Incidents
Long Beach	Mar 33	6.3	13	Coalinga	May 83	6.7	26
Fukui	Jun 48	7.3	4	Nihonkai-Chubu	May 83	7.7	67
Kern County	Jul 52	7.7	4	Nemuro-hanto oki	Jun 83	7.4	1
Alaska	Mar 64	8.0	30	Morgan Hill	Apr 84	6.2	1
Niigata	Jun 64	7.5	30	Nagano-ken Seibu	Sep 84	6.8	1
Tokachi-oki	May 68	7.9	17	Chile	Mar 85	7.8	5
Santa Rosa	Oct 69	5.6	1	Mexico City	Sep 85	8.1	2
San Fernando	Feb 71	6.4	29	Palm Springs	Jul 86	5.9	2
Nicaragua	Dec 72	6.3	1	El Salvador	Oct 86	5.4	2
Izuhanto-oki	May 74	6.8	1	New Zealand	Mar 87	6.3	1
Peru	Oct 74	7.5	2	Ecuador	Mar 87	6.1	3
Tangshan, China	Jul 76	7.8	1	Whittier-Narrows	Oct 87	6.1	20
Izuoshima-kinkai	Jan 78	7.0	1	Chibaken Toho oki	Dec 87	6.7	11
Miyagiken-oki	Jun 78	7.4	23	Quebec	Nov 88	6.0	3
Santa Barbara	Aug 78	5.1	2	Loma Prieta	Oct 89	7.1	490
Imperial Valley	Oct 79	6.6	6	Costa Rica	Mar 90	6.9	1
Livermore	Jan 80	5.8	8	Kushiro oki	Jan 93	7.5	5
Chibaken Chubu	Sep 80	6.1	1	Hokkaido Nansei oki	Jul 93	7.8	11
Greece	Feb 81	6.7	1	Northridge	Jan 94	6.7	387
Urakawa-oki	Mar 82	7.9	2				

3.10 Credits

Table 3-1, Scawthorn. Table 3-2, National Interagency Fire Center. Table 3-3, NFPA 1942. Table 3-4, Selvaduray 2002. Figures 3-2, 3-3, 3-5, Library of Congress. Figure 3-4 left, http://users.anet.com/~flannery/fire.html. Figure 3-4 right http://www.chicagohs.org/fire/conflag/pic-firemap.html. Figure 3-6, NFBU. Figures 3-7, 3-8, 3-9, Eidinger. Figure 3-10, California OES. Figures 3-11, 3-13, 3-14, Scawthorn. Figure 3-12, Routley, 1991.

3.11 References

Eidinger, J. 1993. Fire Conflagration and Post-Earthquake Response of Power and Water Lifelines, 4th Department of Energy Natural Phenomena Hazards Mitigation Conference, October.

EBRPD, 1982, Blue Ribbon Urban Interface Fire Prevention Committee, East Bay Regional Park District.

FEMA. 2002. *World Trade Center,* Building Performance Study, FEMA 403, Federal Emergency Management Agency, Washington.

Glasstone, S. and P.J. Dolan, 1977. The Effects of Nuclear Weapons. Dept. of Defense and Dept. of Energy. Washington.

Goltz, J., J. Decker and C. Scawthorn, 1993. The Southern California Wildfires of 1993, EQE Review, Winter 1994, San Francisco (available at http://www.eqe.com/publications/socal_wildfire/scalfire.htm)

Goodsell J.H. and C.M. Goodsell. 1871. History of the Great Chicago Fire, October 8, 9, and 10, 1871. Chicago. (available at http://chicago.about.com/library/blank/blfiregoodsell01.htm)

Hall, J. R. 2001.High Rise Building Fires, National Fire Protection Association, Quincy.

Hanson, N. 2001. The Great Fire of London, Wiley, Hoboken.

Isner, M. and Klem, T.J. 1993. Fire Investigation Report, World Trade Center Explosion and Fire, New York, NY, Feb. 26, 1993, National Fire Protection Association, Quincy.

Klem, T.J. 1988. Fire Investigation Report, First Interstate Bank Building Fire, Los Angeles, California, May 4, 1988, National Fire Protection Association, Quincy.

Klem, T.J. 1988. Fire Investigation Report, One Meridian Plaza Fire, Philadelphia, Pennsylvania, Feb. 23, 1991, National Fire Protection Association, Quincy.

NBFU, 1923, Report on the Berkeley, California Conflagration of September 17, 1923, National Board of Fire Underwriters.

NBFU, 1951, Report on the City of Berkeley, California, National Board of Fire Underwriters, Committee on Fire Prevention and Engineering Standards.

NFPA. 1942. Conflagrations in America Since 1914. National Fire Protection Association, Boston.

Pietersen, C.M. 1988. Analysis of the LPG disaster in Mexico City, J. Haz. Mat., 20:85-108.

Routley, J.G. 1988, Interstate Bank Building Fire (May 4, 1988), Report TR-022, U.S. Fire Administration, Federal Emergency Management Agency, Washington.

Routley, J.G. C. Jennings and M. Chubb. 1991, High Rise Office Building Fire, One Meridian Plaza Fire, Philadelphia, Pennsylvania (Feb. 23, 1991) Report TR-049, U.S. Fire Administration, Federal Emergency Management Agency, Washington.

Routley, J.G., 1991. The East Bay Hills Fire Oakland-Berkeley, California (October 19-22, 1991), Report 060 of the Major Fires Investigation Project conducted by TrlData Corporation under contract EMW-90-C-3338 to the United States Fire Administration, Federal Emergency Management Agency.

Scawthorn, C. 1989. Fire Following Earthquake in Highrise Buildings. Chapter in Topical Volume on Fire Safety in Tall Buildings, Council on Tall Buildings and Urban Habitat, M. Sfintesco, editor, Bethlehem.

Scawthorn, C. and J. Donelan. 1984. "Fire-Related Aspects of the Coalinga Earthquake." In Coalinga, California Earthquake of May 2, 1983: Reconnaissance Report. Report 84-03. Berkeley, CA: EERI.

Selvaduray, G. 2002. "Hazardous Materials - Earthquake Caused Incidents And Mitigation Approaches", Chapter in Earthquake Engineering Handbook, Chen and Scawthorn, editors, CRC Press, Boca Raton.

USSBS. 1945. The United States Strategic Bombing Survey, Summary Report (European War), Washington DC.

USSBS. 1946. The United States Strategic Bombing Survey, Summary Report (Pacific War), Washington DC.

4 Analysis and Modeling

4.1 Introduction

This Chapter presents a methodology for estimating fire losses following earthquake, developed in the 1970s (Scawthorn et al, 1981), and widely employed since that time (AIRAC, 1987; NDC, 2002; HAZUS, 1999; ICLR, 2001). The basic methodology can be seen in Figure 4-1, the Fire Following Earthquake process, which depicts the main aspects of the fire following earthquake problem. Fire following earthquake is a process, which begins with the occurrence of the earthquake. In summary, the steps in the process are:

- *Occurrence of the earthquake* – this will presumably cause damage to buildings and contents, even if the damage is as simple as knockings things (such as candles or lamps) over.

- *Ignition* – whether a structure has been damaged or not, ignitions will occur due to earthquakes. The sources of ignitions are numerous, ranging from overturned heat sources, to abraded and shorted electrical wiring, to spilled chemicals having exothermic reactions, to friction of things rubbing together.

- *Discovery* – at some point, the fire resulting from the ignition will be discovered, if it has not self-extinguished (this aspect is discussed further, below). In the confusion following an earthquake, the discovery may take longer than it might otherwise.

- *Report* – if it is not possible for the person or persons discovering the fire the immediately extinguish it, fire department response will be required. For the fire department to respond, a Report to the fire department has to be made.

- *Response* – the fire department then has to respond.

- *Suppression* – the fire department then has to suppress the fire. If the fire department is successful, they move on to the next incident. If the fire department is not successful, they continue to attempt to control the fire, but it spreads, and becomes a conflagration. The process ends when the fuel is exhausted – that is, when the fire comes to a firebreak.

This process is also shown in Figure 4-2, which is a Fire Department Operations Time Line. Time is of the essence for the fire following earthquake problem. In this figure, the horizontal axis is Time, beginning at the time of the earthquake, while the vertical axis presents a series of horizontal bars of varying width. Each of these bars depicts the development of one fire, from ignition through growth or increasing size (size is indicated by the width or number of bars).

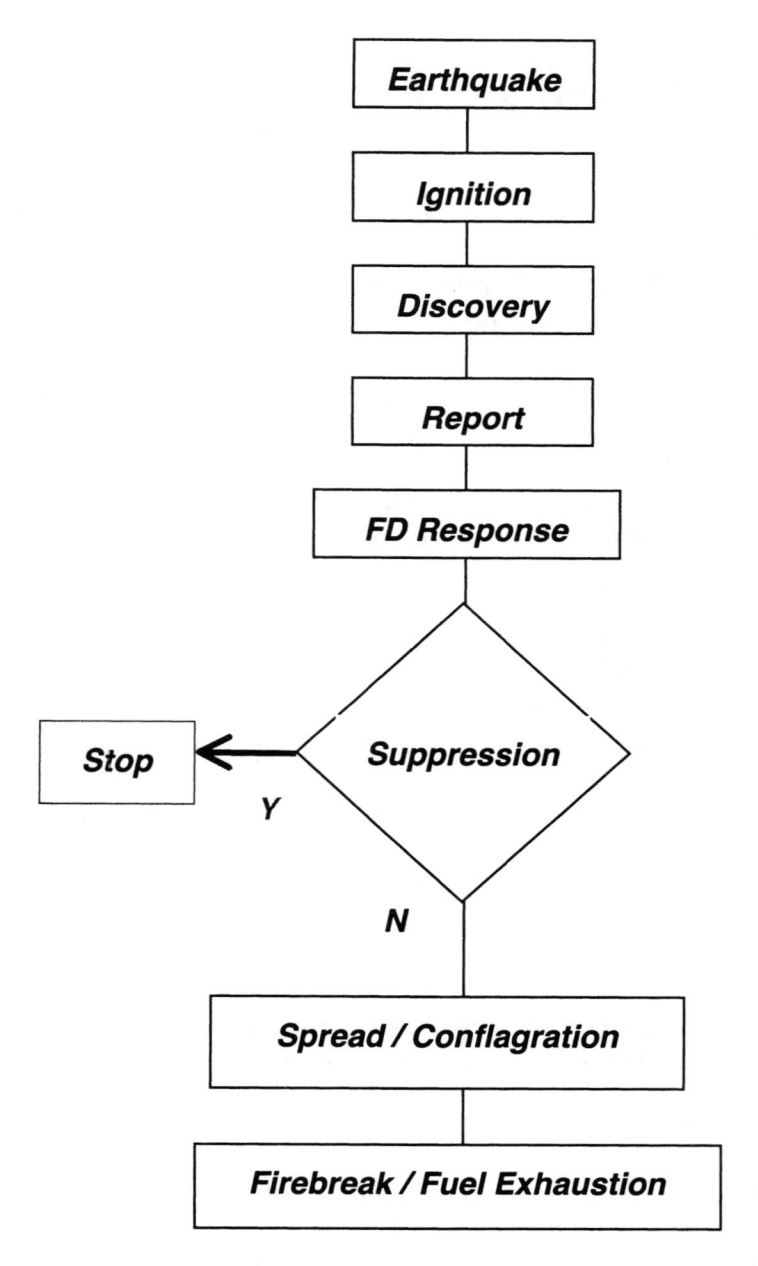

Figure 4-1 Fire following earthquake process

FIRE FOLLOWING EARTHQUAKE

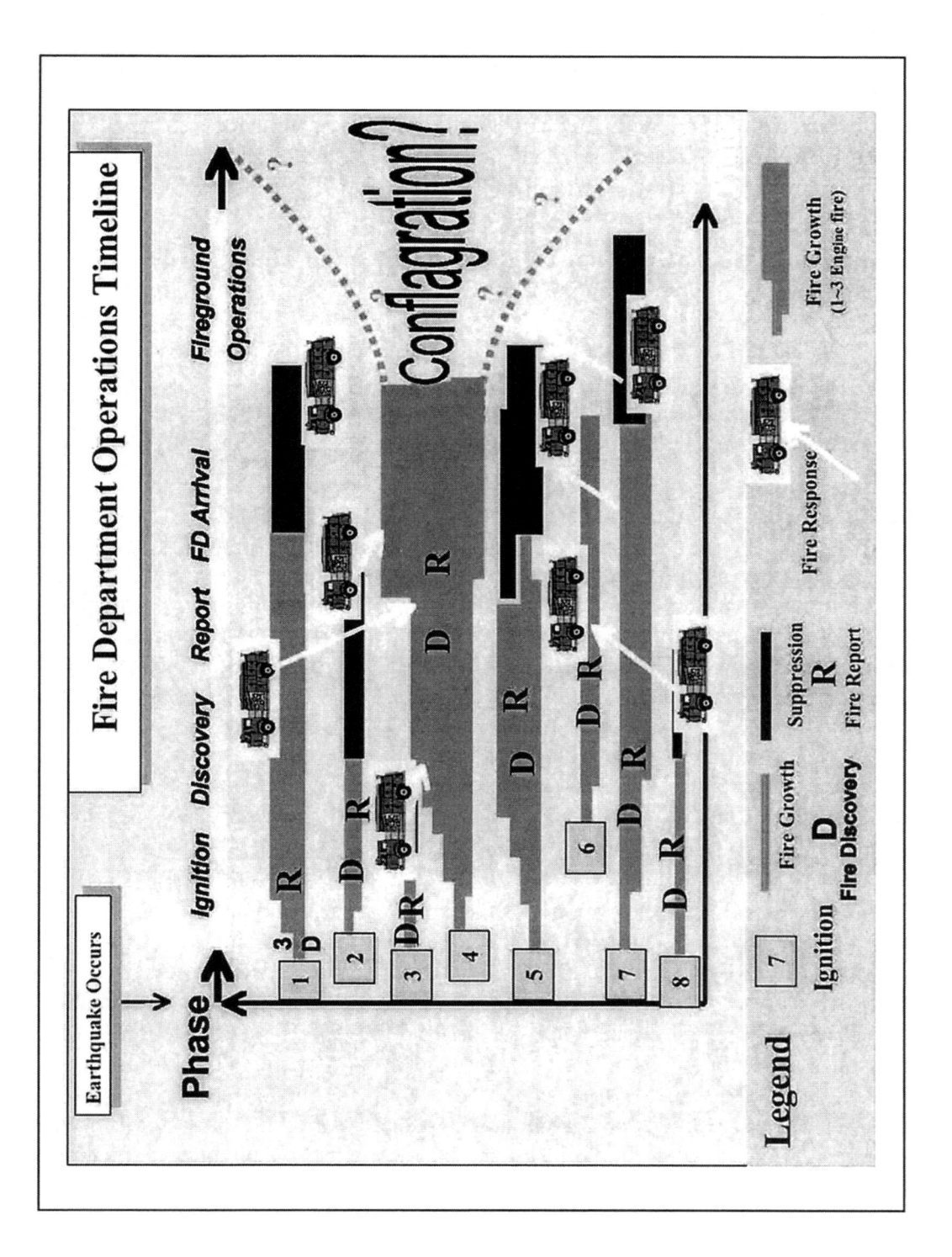

Figure 4-2 Fire department Operations Time Line

Beginning at the left of Figure 4-2 (that is, at the time of the earthquake), is the occurrence of various fires or ignitions (denoted by the number of the fire in a square box, see the legend at the bottom of the figure). Some of these fires occur very soon after the earthquake, while others occur sometime later (due for example to restoration of utilities). The mechanism of these ignitions is no different following an earthquake than at other times, although the earthquake can create unusual circumstances for ignition to take place. The primary difference due to the earthquake is the large number of simultaneous ignitions.

Following this ignition, or Fire Initiation, phase there is a period during which the fire is undiscovered but grows. A typical rule of thumb in the fire service is that the rate of growth of an uncombated fire in this phase will double each seven seconds. In Figure 4-2, the size of the fire is denoted by its number of bars. That is, each bar for a particular fire represents one engine required for control and/or suppression. Thus if, with time, a fire in Figure 4-2 proceeds from one bar to two and then three, this denotes that the fire is growing and now requires three Class A fire engines to control the fire (Class A fire engines have approximately 1200 gpm of pumping capacity).

The letter "D" denotes discovery of the fire. Discovery under the post-earthquake environment is often no different than at other times, although discovery may be impeded due to damaged detectors, or distracted observers. Upon discovery, citizens may themselves attempt to combat the fire, and will sometimes be successful. We are concerned herein only with ignitions that citizens cannot/do not successfully combat, and which require fire department response. The letter "R" denotes receipt of the Fire Report by the fire department. Under normal circumstances, fires are reported to the fire department by one of four methods:

- Telephone
- Fire department street boxes (voice or telegraph)
- Direct travel to a fire station by a citizen (so-called "citizen alarms")
- Automatic detection and reporting equipment, usually maintained by private companies

Under normal circumstances, with the exception of citizen alarms, these methods all communicate the occurrence of a fire within seconds, which is critical in the timely response and suppression of structural fires, and in the size and 'design' of modern fire departments. In the critical minutes following an earthquake, review of earthquake experience has indicated that at present citizen alarms are likely to be the only feasible method for reporting fires, in areas of strong ground shaking. The telephone system may or may not sustain damage, but almost definitely will be incapacitated due to overload. Fire department street boxes are generally no longer in use (in California, only San Francisco maintains street boxes). Automatic detection and reporting equipment account for only a fraction of commercial property. Such equipment may be damaged in an earthquake, and will likely produce many false alarms, leading to lack of response to real fires, due to the inability to discriminate the real from the false alarms. Several other, unconventional, reporting methods may be employed. These include:

- Amateur short-wave radio operators
- Helicopter observation

- Ground reconnaissance by police or fire personnel

With regard to the last of these other methods, present post-earthquake damage reconnaissance planning on the part of several larger California fire departments has been reviewed and, in general, found to be unrealistic and inadequate for identifying fires at a sufficiently early stage to prevent conflagration. A fundamental flaw in most of these plans is the performance of the post-earthquake damage reconnaissance by the fire personnel themselves, employing fire apparatus (i.e., engines and trucks)[12]. The flaw lies in the fact that following an earthquake, these personnel and apparatus will almost immediately be redirected from the reconnaissance to actual fire or other emergency response.

Helicopter observation and amateur radio operators similarly will typically only be able to identify and report fires after they have reached greater alarm status (that is, when it is too late). Further, most fire officers have no special training in aerial observation or command.

Following receipt of the Fire Report by the fire department, apparatus will respond, if available (in Figure 4-2, arrival of apparatus at the fireground is denoted by the engine number within a circle-note that herein we only track fire engines, since only engines can suppress serious fires). Response may be impeded by blocked streets due to collapsed structures, or by traffic jams. Upon arrival, the fire may be combated per normal procedures or, if the general situation is sufficiently serious, minimal tactics may be all that is possible. Minimal tactics may constitute:

- Deluge (so-called "flood and run" tactics)

- Abandonment of the burning structure and protection only of exposures

- Recognition that exposed structures cannot be protected, with fall back to a defensible line (e.g., abandonment of a city block, with attempt to stop the fire at a wide street)

- Total abandonment, that is, recognition that either little or nothing can be done, that the fire will burn itself out at an identified firebreak with or without fire department intervention, or that other situations are more critical and demand the apparatus

Water supply is a critical element, and earthquake damage to the water system may reduce supply, thus altering tactics. Due to the interconnectedness of a water supply system, earthquake damage at some distance from a fireground may still result in reduced supply.

In Figure 4-2, we denote increasing control of a fire by the reduction in the number of bars (i.e., engines required for control). As suppression progresses and control of the fire becomes near total, engines will be released by the incident commander for more pressing emergencies elsewhere. Movement of these released apparatus is denoted by a diagonal arrow showing travel of an engine from one fire to another. As fires are controlled, engines eventually converge on one

---

[12] Basic fire service apparatus consists of two types: engines (or pumpers), that typically carry a pump of about 1200 gpm capacity, 2000 ft. of hose, 500 gallons of water, some tools and small ladders and several firefighters. The second type are trucks (also termed ladders, aerials, hook and ladders, etc.), which carry large ladders of about 75 or 100 ft. length, a much larger variety of tools and ladders, and additional firefighters. A ladder truck by itself cannot pump water so that the pumper is one basic measure of firefighting capacity. For typical non-earthquake structural fires however, the limiting factor is more often manpower, than pumping capacity.

or several large fires, or conflagrations. Growth and spread of conflagrations is a function of building materials, density, street width, wind, water supply and firefighting tactics.

In this methodology, the process depicted in Figure 4-2 can be coded in a computer program. An algorithm determines ignitions, assigns a number to each fire, and tracks fire growth. Algorithms also determine fire reporting time and fire engine arrival. Each fire engine is tracked from location to location. Damage and hydraulic performance to the water supply can be determined using detailed loss models of the water system such that the availability of fire flows at any fire location and any time after the earthquake can be tracked; the effectiveness of fire department apparatus to control a fire at a given location can then be estimated based on the size of the fire, the quantity of fire department apparatus at the fire, and the quantity of water available for fighting the fire. Final burnt area for each ignition is thus calculated as a function of fire growth and applied fire suppression capacity. Each aspect of the problem is next discussed in more detail.

4.2 Ignition

Available twentieth century United States fire following earthquake experience provided in Table 2-1 has been reviewed, so as to identify earthquake-caused ignitions requiring fire department response. A detailed review of this data is beyond the scope of this chapter. Based on this review, it has been determined that post-earthquake ignitions are typically a random event due to:

High inertial ground motions (peak ground accelerations), resulting in overturning or breakage of building contents (e.g., ignitions due to open flames or chemical reaction), or

Excessive structural deflections, resulting in abrasion or other damage to electrical wiring and consequent short-circuiting, or

Breakage of underground utilities (such as natural gas lines), which provides a fuel source for the ignition (more prevalent in areas subject to liquefaction, landslide, or surface faulting).

Building collapse is an extreme example of (i) above, and will greatly increase the probability of ignition. Gas piping may be ruptured by either (i) or (ii). For example, a water heater may overturn, thus breaking the connection, or a building may slide on its foundation and shear connections. Less typical but observed modes of ignition are heat due to friction or sparking due to pounding of structures.

4.2.1 Ignition Rate

Post-earthquake fire ignitions have been normalized by building density, and regressed the data against shaking intensity. Two equivalent measures of density and intensity are used:

For building density:

- Millions of square feet of total building floor area

- SFED (Single Family Equivalent Dwellings = 1500 sq. ft. of floor area), are chosen as a measure which is readily understandable by laymen (they can be thought of as detached "houses"). A large building of 1.5 million sq. ft. for example would be 1000 SFED

For seismic intensity:

- Peak Ground Acceleration (PGA), which is a measure more common in the technical community (see Chapter 4, this volume, for definitions of both MMI and PGA), and

- Modified Mercalli Intensity (MMI), which is a common measure of shaking intensity

Ignition models based on SFED and MMI have often been used for purposes of communications with laypersons and the fire service (who have a different jargon than earthquake engineers).

Figure 4-3 shows normalized fire ignitions as a function of MMI vs thousand SFED, and PGA vs. million sq. ft. of building floor area corresponding to equations 4-1 and 4-2. A clear trend of increasing ignitions with increasing MM1 can be seen. The actual regression is:

$$IGN/\text{thous SFED} = 0.015(MMI)^2 - 0.185(MMI) + 0.61 \qquad r^2 = 0.2, n=59 \quad \text{Equation 4-1}$$

In terms of ignitions per thousand single family equivalent dwellings (SFED) as a function of MMI, and

$$Ign/MMSF = 0.028 \exp(4.16\, PGA) \qquad r^2 = 0.2, \qquad n=59 \quad \text{Equation 4-2}$$

In terms of ignitions per million square feet of building floor area (MMSF) as a function of peak ground acceleration (PGA, g). The relations are based on 59 data points collected by Scawthorn (ICLR, 2001) for events in California since 1971, and have been widely employed in the insurance industry. If all 20th C. California events are employed (i.e., 1906 San Francisco, 1933 Long Beach etc events), the relations are not significantly different.

Another regression of the empirical data set was done by (Eidinger et al, 1995). Figure 4-4 shows the normalized function. This model is based on fire ignitions requiring fire department response, for historical earthquakes in California and Alaska (Table 4-1) and is used in HAZUS 99.

$$Ign/MMSF = -0.025 + (0.59 * PGA) - (0.29 * PGA^2),\ r^2 = 0.34 \quad \text{Equation 4-3}$$

Equations. 4-1 to 4-3 all satisfy a null hypothesis with 95% confidence (ie, there is 95% confidence that a correlation exists). However, there is considerable uncertainty on estimates derived using these relations. More reliable correlation of post-earthquake ignition would take into account whether or not an area has buried natural gas lines; the nature of the building stock and occupancies, the extent of damage to buildings; whether or not the electric power to the area is cut-off within the first few seconds of the earthquake; whether or not the electric power is restored to an area rapidly and before people can isolate / adjust appliances, season and time of day, etc. This is an area for future research.

Table 4-1 Fires Following U.S. Earthquakes - 1906 – 1989, used for Eq (3)

City, Year of Earthquake	PGA (g)	Intensity (MMI)	Ignitions	Ignitions per million sq. ft
Coalinga 1983	0.36	VIII	1	0.30
Daly City 1989	0.12	VI	3	0.05
Anchorage 1964	0.71	X	7	0.24
Berkeley 1906	0.44	VIII-IX	1	0.16
Berkeley 1989	0.07		1	0.013
Burbank 1971	0.21	VII	7	0.16
Glendale 1971	0.15	VI-VII	9	0.13
Los Angeles 1971	0.15	VI-VII	128	0.09
Los Angeles 1933	0.15	VI-VII	3	0.01
Long Beach 1933	0.53	IX	19	0.26
Marin Co. 1989	0.12	VI	2	0.02
Morgan Hill 1984	0.21	VII	4	0.40
Mountain View 1989	0.21	VII	1	0.02
Norwalk 1933	0.28	VII-VIII	1	0.05
Oakland 1906	0.44	VII-IX	2	0.06
Oakland 1989	0.07		0	0.00
Pasadena 1971	0.21	VII	2	0.04
San Francisco 1989	0.21	VII	27	0.08
San Francisco 1906	0.44	VII-X	52	0.26
San Francisco 1957	0.12	VI	0	0.00
San Fernando 1971	0.53	IX	3	0.37
San Jose 1984	0.36	VIII	5	0.02
San Jose 1906	0.36	VIII	1	0.08
Santa Clara 1906	0.44	VIII-IX	1	0.22
Santa Cruz 1989	0.36	VIII	1	0.04
Santa Cruz Co. 1989	0.28	VII-VIII	24	0.03
San Mateo Co. 1906	0.36	VIII	1	0.14
Santa Rosa 1969	0.36	VIII	1	0.06
Santa Rosa 1906	0.71	X	1	0.18
Whittier 1987	0.28	VII-VIII	6	0.1

The general trend of equations 4-1 to 4-3 is shown in Table 4-2, which provides a simple rule of thumb for estimating the number of ignitions that a fire department will have to cope with. Combing this with another rule of thumb, that the total building floor area (in SFED) is about half the population in thousands, allows fire departments and emergency planners to quickly estimate the number of ignitions they may be confronted with. For example, for a town of population 50,000 subjected to MMI IX, the total SFED is then about 25,000, and the number of ignitions about 8.

These rates of ignition do not account for possible intentional ignitions arising out of several motives (that is, arson). It can be argued that arson will be a significant problem, since property owners are in general aware that while they may not be covered for shaking damage, their fire coverage includes fire following earthquake. As will be seen, it takes only a relatively few additional ignitions to overwhelm fire department resources, with possible ensuing conflagration. Countering this is the point that, while past earthquakes have seen examples of arson, these appear to have been relatively few, and did not occur immediately after the shaking but rather days or weeks later. That is, in the immediate post-earthquake period (the first minutes to hours) of injury, emotion, confusion and multiple, simultaneous ignitions potentially leading to conflagration, people will be too overwhelmed by events to consider arson. Later, when this occurs to the relatively few persons who will actually commit this crime, the fires they set will occur in a period when the fire departments are back on a more normal footing, have been reinforced from outside the stricken area, and presumably will be able to handle these fires. We estimate that the latter scenario is more relevant, and hence have not included an allowance for arson in the methodology. Arson ignitions can be included by simply increasing the number of initial ignitions, if one wishes to do so. The author has specifically investigated this aspect (i.e., intentional fires) following major earthquakes during the 1980s and 1990s, and confirmed the above observations.

While only 1971 San Fernando and later data are employed in these relations, the results are about the same had all 20th century events been employed. Interestingly, the trend is also quite similar to that observed in Kobe in the 1995 earthquake.

Post-earthquake ignitions for a particular locality can thus be calculated as a random Poisson process with mean probability determined as a function of MMI or PGA and building inventory (i.e., SFED, or millions of building floor area).

Table 4-2 Approximate No. of SFED per ignition, vs. MMI

MMI	VII	VIII	IX	X
1 Ign. Per SFED	12,000	7,000	3,000	1,000

As will be discussed further below, the potential for conflagration is a skewed process, in the sense that a small increase in the number of ignitions can lead to a very large increase in the likelihood of one or more large fires. Therefore, one of the potentially best mitigation measures would be to lower the ignition rate.

FIRE FOLLOWING EARTHQUAKE

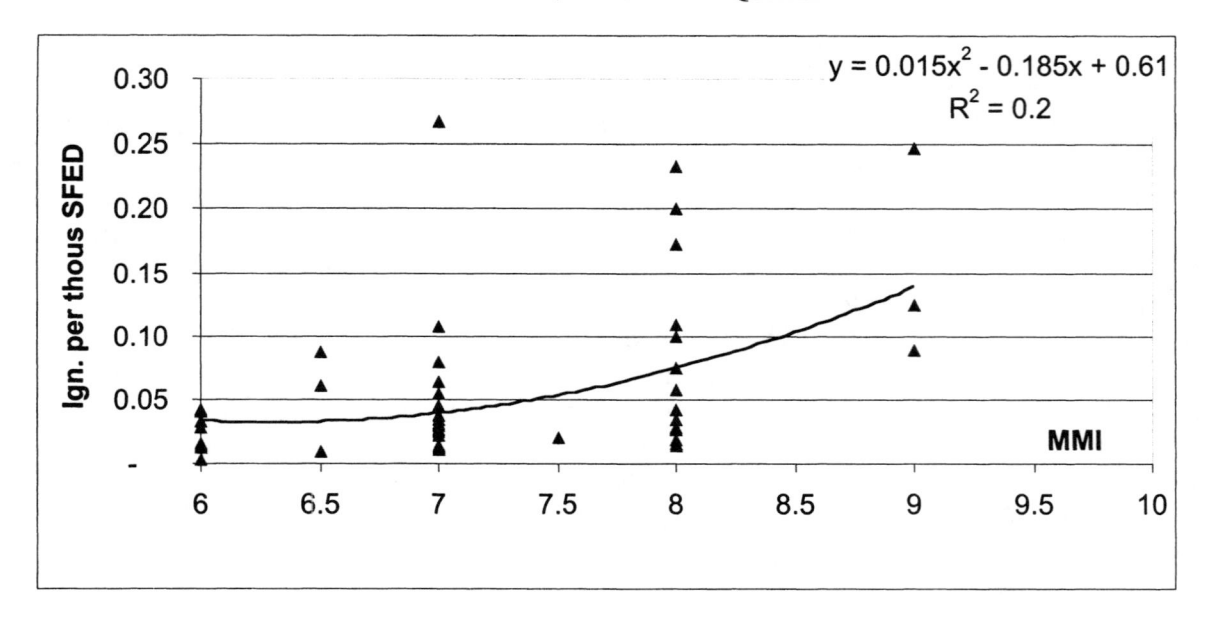

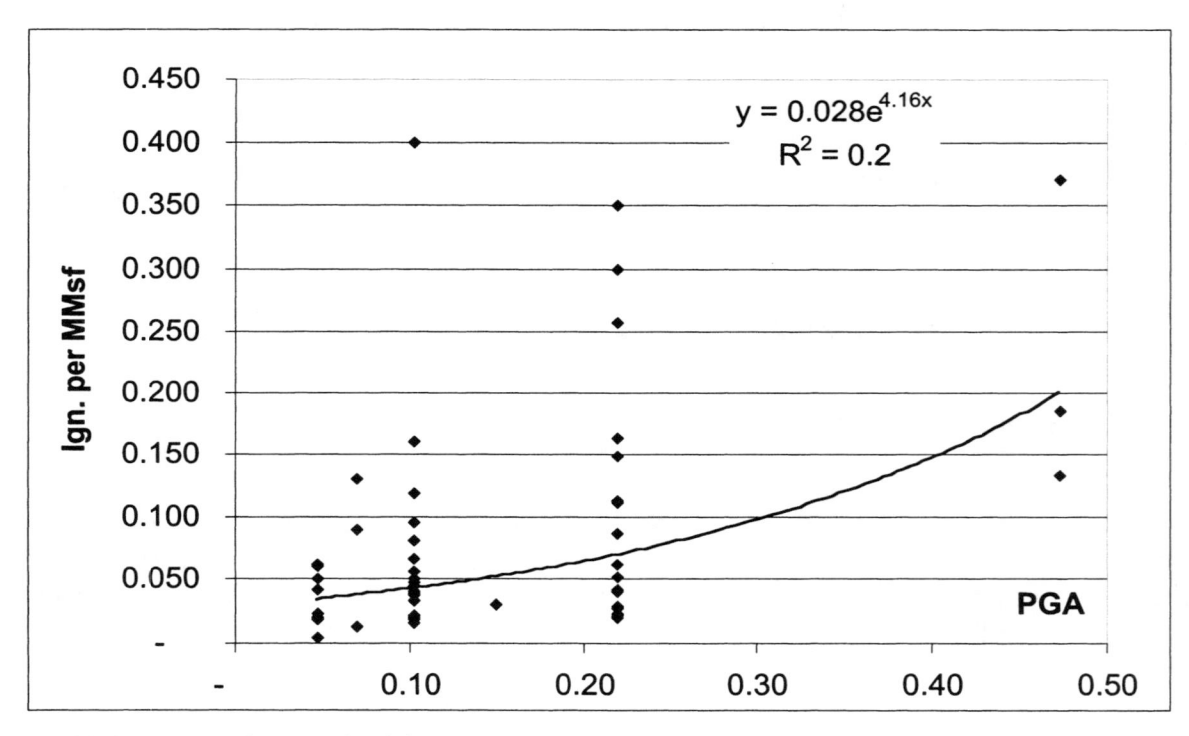

Figure 4-3 Post-earthquake ignition rate, based on San Fernando and later data – (t) Ignitions per thousand SFED vs. MMI, and (b) Ignitions per million sq. ft. floor area vs. PGA (No. data = 59)

FIRE FOLLOWING EARTHQUAKE

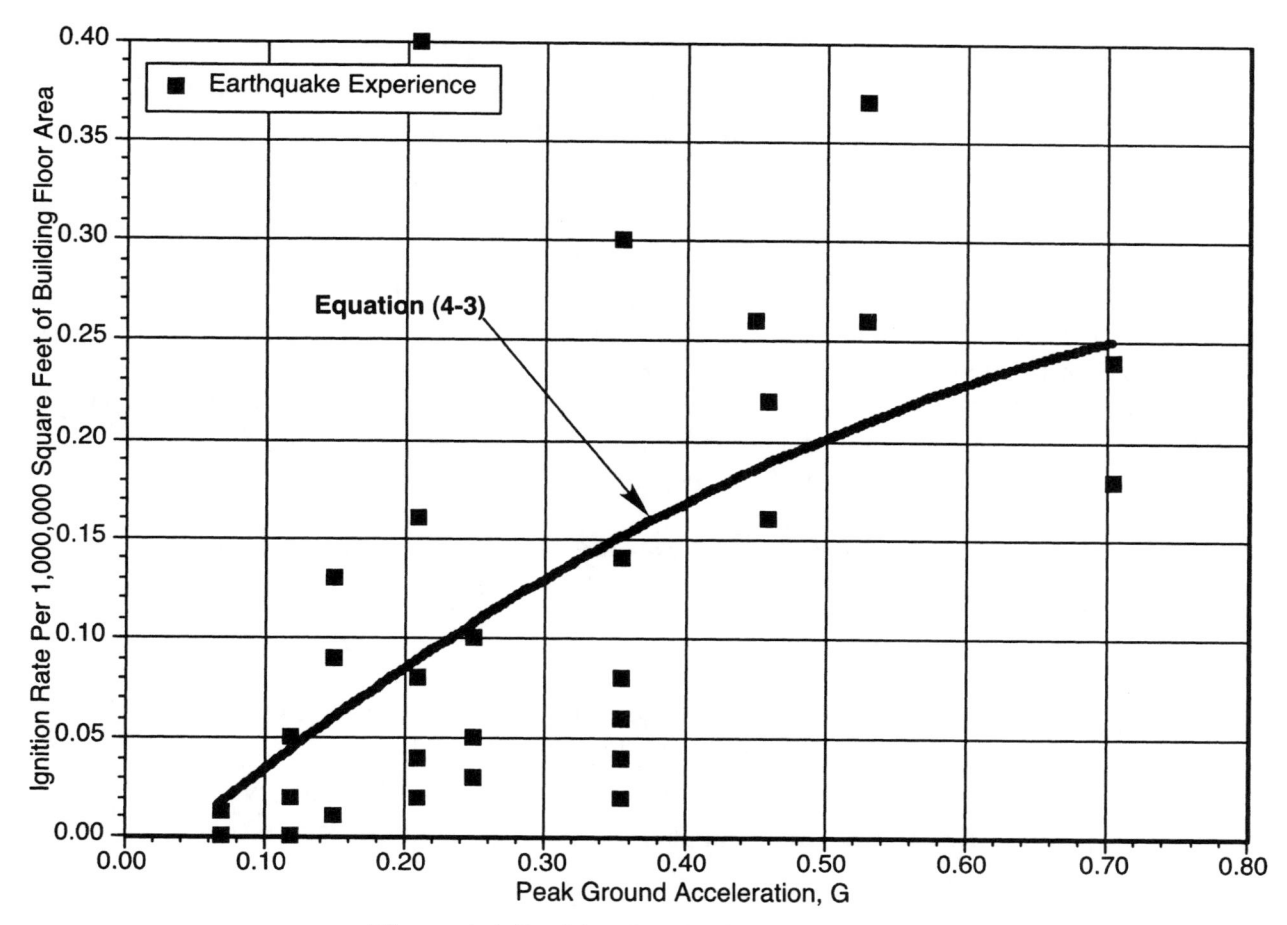

Figure 4-4 Ignition Model for Equation (3)

4.2.2 Locations and Causes of Ignitions

It is of value to understand the locations and cause of ignitions, both for estimation and mitigation purposes.

Table 4-4 lists the general property use for 77 Los Angeles Fire Department earthquake related fires. More than 70% (66) of the earthquake-related fires occurred in single- or multiple-family residences, as might be expected from the building stock that is typical in the San Fernando Valley. The sources of heat types of spark, heat or flame that started the fires are described in Table 4-5. The major cause of ignition was electric arcing as the result of a short circuit, although gas flame from an appliance is also a recurring source of ignition. This data can be grouped into three simple categories of ignition, as shown in Table 4-3, which tends to confirm a general trend observed in many earthquakes, which is that about a third of the ignitions will be electrical in nature, a third gas, and a third other.

Table 4-3 General Sources of Ignition, LAFD Data, Northridge Earthquake

Source	Fraction
Electrical	56%
Gas-related	26%
Other	18%

This is also confirmed by a recent report on Improving Natural Gas Safety in Earthquakes by the ASCE and the California Seismic Safety Commission, (SSC 2002) which observed that *"the number of post-earthquake fire ignitions related to natural gas can be expected to be 20% to 50% of the total post-earthquake fire ignitions."*.

The material that first ignites is identified in Table 4-6. Where identification could be made, escaping natural gas (presumably from a broken gas line is the single most common ignition material.

Table 4-4 Property Use for 77 LAFD Earthquake-Related Fires
4:31 TO 24:00 hrs, January 17, 1994

General Property Use	Frequency
One or Two Family Residential	35
Multi-Family Residential	20
Public Roadway	6
Office	4
Primary / Secondary School	2
Vacant Property	2
Restaurant	1
Commercial	1
Power Production/Distribution	1
Other	4
Unknown	1

Table 4-5 Forms Of Heat Ignition, 77 LAFD Earthquake-Related Fires
4:31 TO 24:00 hrs, January 17, 1994

Form of Heat Ignition	Frequency
Gas Appliance Flame	13
Short Circuit, Mechanical Damage	6
Short Circuit, Insulation	6
Short Circuit, Other	5
Normal Electrical Equipment Heat	5
Spark from Equipment	4
Direct Spread	4
Heat from Gas Appliance	2
Escaping Wood/Paper Ember	2
Overloaded Electrical Equipment	2
Faulty Electrical Contact	2
Electric Lamp	2
Spontaneous	2
Liquid Fuel Appliance Heat	2
Discarded Hot Ember	2
Rekindle	1
Catalytic Converter	1
Match/Lighter	1
Unknown	17

Table 4-6 Material First Ignited for 77 LAFD Earthquake-Related Fires
4:31 To 24:00 hrs, January 17, 1994

Material First Ignited	Frequency
Natural Gas	13
Sawn Wood	5
Man-Made Fabric	5
Wood	4
Cotton Fabric	4
Flexible Plastics (i.e., wire insulation)	3
Rubbish	2
Tree/Brush	1
L.P.-Gas	1
Gasoline	1
Class II Combustible Liquid	1
Rigid Plastics	1
Fiber Board, Wood Pulp	1
Other Wood	1
Plastic/Vinyl Fabric	1
Other	3
Unknown	29

4.2.3 Further Examination of Causes of Ignitions

In addition to the observations made in Section 4.2.2, the ignition rate is probably also a function of:

The occupancy type of the building: residential, commercial, or industrial. The occupancy type determines the building contents, type of use, and usage hours. Commercial buildings for example are used more during the day than during the night. Residential buildings on the other hand are used more at night.

The structure material, in particular: wood structures vs. non-wood structures. A spark from a short circuit more likely turns into a fire in a wood building than in a non-wood building.

Time of day: during meal times, more electrical and gas appliances are in use. This increases the potential for ignitions as compared to nighttime. Similarly, time of year is important in that gas or oil appliances are used in the winter for home heating.

The Tokyo Fire Department in Japan has developed a set of curves considering the above characteristics (Tokyo Fire Department, 1997). A typical set of curves is presented in Figure 4-5. Six similar sets of curves can be used to estimate the number of ignitions as a function of the following occupancies and building materials: residential wood and non-wood buildings, commercial wood and non-wood buildings and industrial wood and non-wood buildings. After calibration against recent US earthquakes such as San Fernando (1971), Morgan Hill (1984), Whittier (1987), Loma Prieta (1989) and Northridge (1994), these curves can be used in the US.

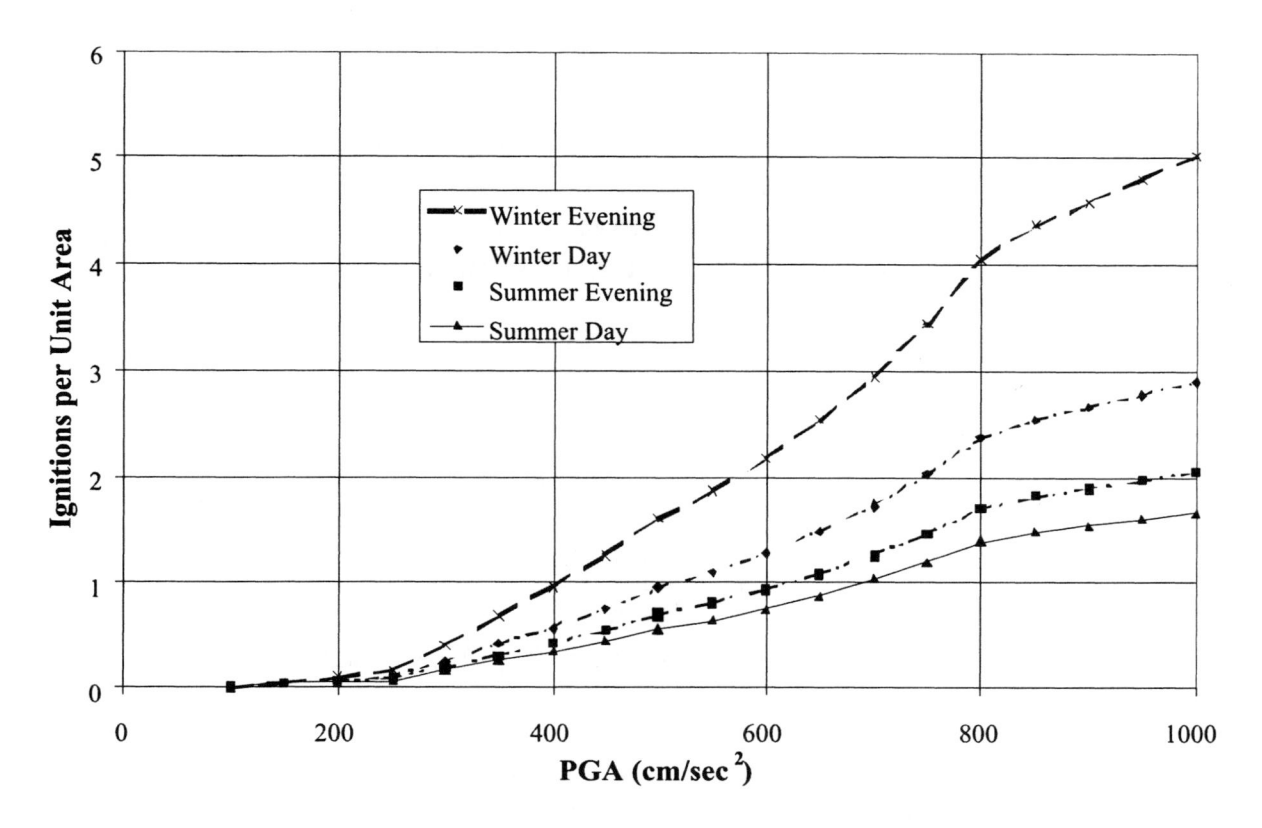

Figure 4-5 Fire Ignition rate as a Function of Season and Time of Day

The four ignition curves in Figure 4-5 represent the variation in the number of ignitions for earthquakes occurring at different times of day and different seasons. The four ignition curves correspond to earthquakes occurring in summer day time, summer evening time, winter day time, and winter evening time. Summer, winter, day, and evening are all considered as fuzzy variables with fuzzy weights shown in Figure 4-6 and Figure 4-7. Combining the membership functions in these figures, the membership surfaces for summer day, summer evening, winter day and winter evening is derived. Typical summer evening fuzzy weights are shown in Figure 4-8.

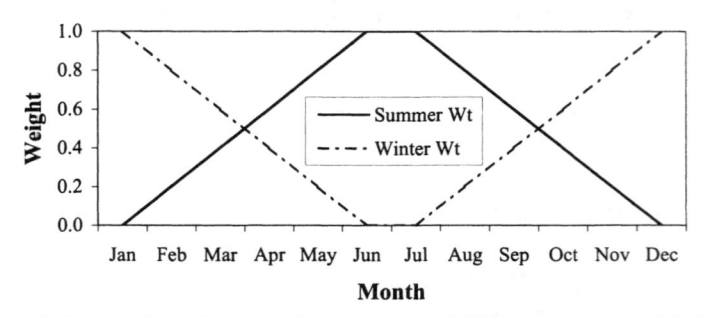

Figure 4-6 Ignition Rates – Summer and Winter Fuzzy Weights

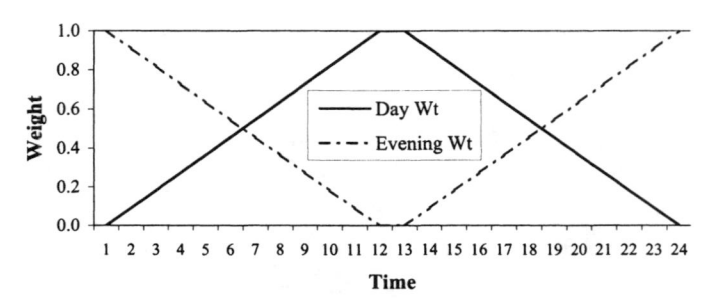

Figure 4-7 Ignition Rates – Time of Day Fuzzy Weights

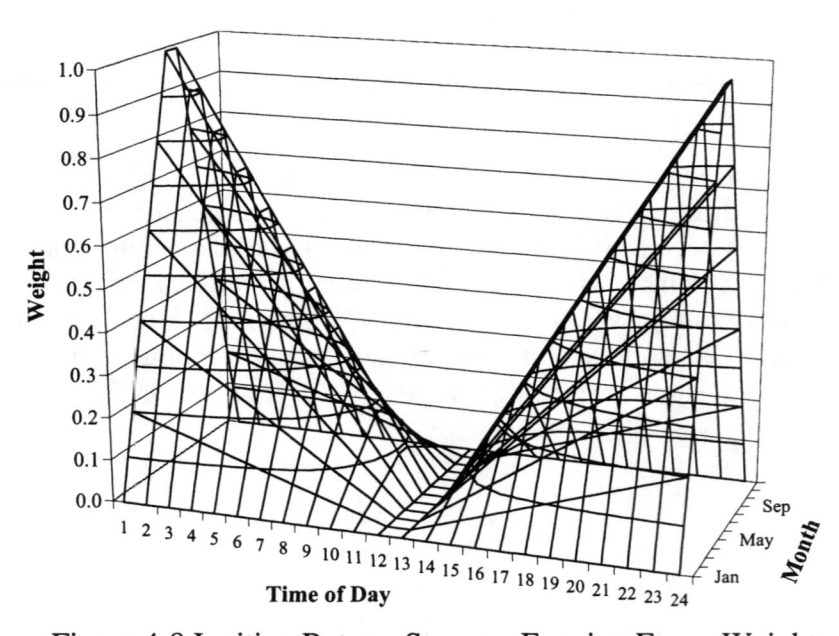

Figure 4-8 Ignition Rates – Summer Evening Fuzzy Weights

The number of ignitions for an earthquake occurring on a certain date and time of day is obtained by first calculating the membership value of the time and season of the earthquake from each of the summer day, summer evening (Figure 4-8), winter day, and winter evening surfaces. These four weights are used to obtain a weighted average of the number of ignitions from the four curves corresponding to the occupancy and structure type of the building.

The number of ignitions estimated using the above ignition model includes both fires starting immediately after the earthquake and those starting some time after the earthquake. A typical cause of these later ignitions is the restoration of electric power. When power is restored, short circuits that occurred due to the earthquake become energized and can ignite fires.

4.3 Fire Report and Response

Some data is available on time of fire occurrence relative to the earthquake. Figures 2-19 and 4-9 for example shows data from the 1994 Northridge earthquake, which indicate that about 50% of earthquake-related fires are reported within several hours of the earthquake.

How fires are reported can vary. Citizen alarms are likely to be the dominant or only method of reporting fires in the minutes following a major earthquake. This means that, following discovery, a citizen running or driving to the nearest fire station will report a fire. Thus, time of reporting involves a period of delay (herein termed Earthquake Delay, accounting for initial confusion, traffic difficulties, etc.) plus travel time from the location of the fire to the nearest fire station. Travel time can be determined based on either direct or right angle travel, and vehicle speed is a variable.

Time of fire engine travel to the fire following receipt of report is similarly based on distance and vehicle speed. Under normal conditions, fire engines average between 15 and 20 mph in responding to a fire (i.e., if distance traveled is divided by total elapsed "roll-time," the result is in the range of 15 to 20 mph-of course, higher speeds are attained during certain portions of the travel). Delays are possible, due to detours or traffic jams. Debris blockage of streets is a potential issue, and was observed to be a significant factor in Kobe in 1995. For typical US urban situations, this has been considered with the conclusion that typically this should not be a major impediment although it may occur in selected districts. Depending on time of day, traffic jams may be a more critical factor. Based on a review of post-earthquake traffic conditions in U.S. earthquakes, typical delays can be accounted for by the above Earthquake Delay factor.

4.4 Fire Growth and Spread

4.4.1 The Hamada Model

Fire growth is a particularly complex phenomenon. Most research in the United States has concentrated on fire growth within one room (so-called "compartment fires"), and only very limited research has been performed on U.S. inter-building urban fire growth (Takata, 1968; Woycheese et al, 1999). More work has been performed on wild lands fire spread (Rothermel, 1983), but this work is of limited applicability to the urban situation. Of available urban fire spread equations, the Hamada equations are most useful (Hamada, 1951, 1975; Horiuch, n.d.; Scawthorn et al., 1981; Terada, 1984). These equations are based on Japanese experience in twentieth century conflagrations, both in peacetime (e.g., following the 1923 Kanto earthquake)

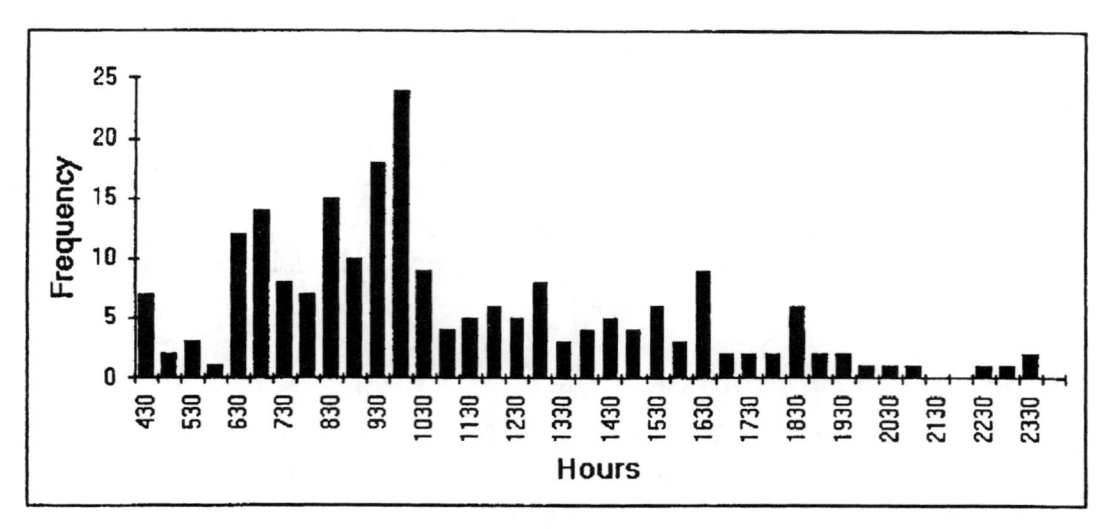

Figure 4-9 Santa Monica Fire Department Incident Reports
January 17, 1994 Northridge Earthquake, 0431-2400

and wartime. More recently, work has been initiated on fire spread based on the physics of fire assuming building-to-building fire spread is due to thermal radiation and fire-induced plume (Himoto and Tanaka, 2002). However, currently, fire spread modeling can only be based on empirical fire spread models, such as the Hamada (Section 4.4.1) or TOSHO (Section 4.4.2) models.

The Hamada equations provide an elaborate model of fire spreading for urban Japan, accounting for build-up to a fully involved fire, etc. From his model, conservatively using parameters for only a fully involved fire, a result which can be used (Scawthorn, 1981) is that[13]:

$$N(x_t, V) = [(1.5 \delta)/a_o^2] K_s [K_d + K_u] \qquad \text{Equation 4-4}$$

where

$N(x_t, V)$ is the number of low-rise buildings destroyed by fire spread, per fire outbreak;

x_t is time in minutes;

V is wind speed, m/sec;

a_o, is average building plan dimension, m;

K_d = length of fire in downwind direction (see Figure 4-11), from the initial ignition location, in meters, defined in Eq-4-5:

$$K_d = [a_o + d)T_4]x_t \qquad \text{Equation 4-5}$$

K_s = width of fire from flank to flank, in meters, defined in Eq 4-6:

---

[13] The Hamada equations are used in the fire spread model in HAZUS (2002).

$$K_s = [(a_o/2) + d] + [(a_o + d)/T_s] (x_t-T_s) \qquad \text{Equation 4-6}$$

K_u = length of fire upwind, from the initial ignition location, in meters, defined in Eq 4-7:

$$K_u = [(a_o/2) + d] + [(a_o + d)/T_u] (x_t-T_u) \qquad \text{Equation 4-7}$$

δ = total plan area of wooden (non fire-resistant) buildings divided by (total area minus water area) of the subject regional sub-section;

$$\delta = \frac{\sum_{i=1}^{n} a_{in}^2}{(TractArea - TractOpenArea)} \qquad \text{Equation 4-8}$$

where

a_{in} = plan dimension of building i, which is not fire resistant (n)

Tract Open Area = non-urban fire resistant area (lakes) within the tract

$$T_i = (1-f_b)[3 + 0.375a_o + (8d/(c_{4i} + c_{5i}V))] + f_b[5 + 0.625a_o + 16d/(c_{4I} + c_{5i}V)]/C(V)$$
$$\ldots \text{Equation 4-9}$$

$$C(V) = c_{1I} (1 + c_{2i}V + c_{4i}V^2) \qquad \text{Equation 4-10}$$

where d is the average building separation, m; T_4, is the time in minutes the fully developed fire requires to advance to the next building in the downwind direction; T_s is similar in the side-wind and T_u is similar in the up-wind directions; f_b is the ratio of fire-resistant buildings; $i = 4,s,u$, and c_{ji} (j=1 to 5) given in Table 4-7.

Table 4-7 Constants for Eqns. 4-1 to 4-6

i	c_{1i}	c_{2i}	c_{3i}	c_{4i}	c_{5i}
4	1.6	0.1	0.007	25.0	2.5
s	1.0	0.0	0.005	5.0	0.25
u	1.0	0.0	0.002	5.0	0.2

The Hamada equations are based on the hypothesis of an urban area being a series of equal blocks of structures, with equal spacing between structures. The plan dimension of the average structure is denoted "a", with average separation distance between structures "d". The "built-upness", or building density ratio δ is defined by equation Equation 4-8.

Figure 4-10 shows fire spread as an irregular oval, which is typical of fires burning through a relatively uniformly distributed fuel load, with constant wind velocity. In actual urban conflagrations, patterns of fire spreading are more variable, and are a function of wind speed and changes in direction, varying fuel availability in differing directions, and of fire suppression tactics. Burnt area is approximately $(K_d + K_u) * (K_s)$.

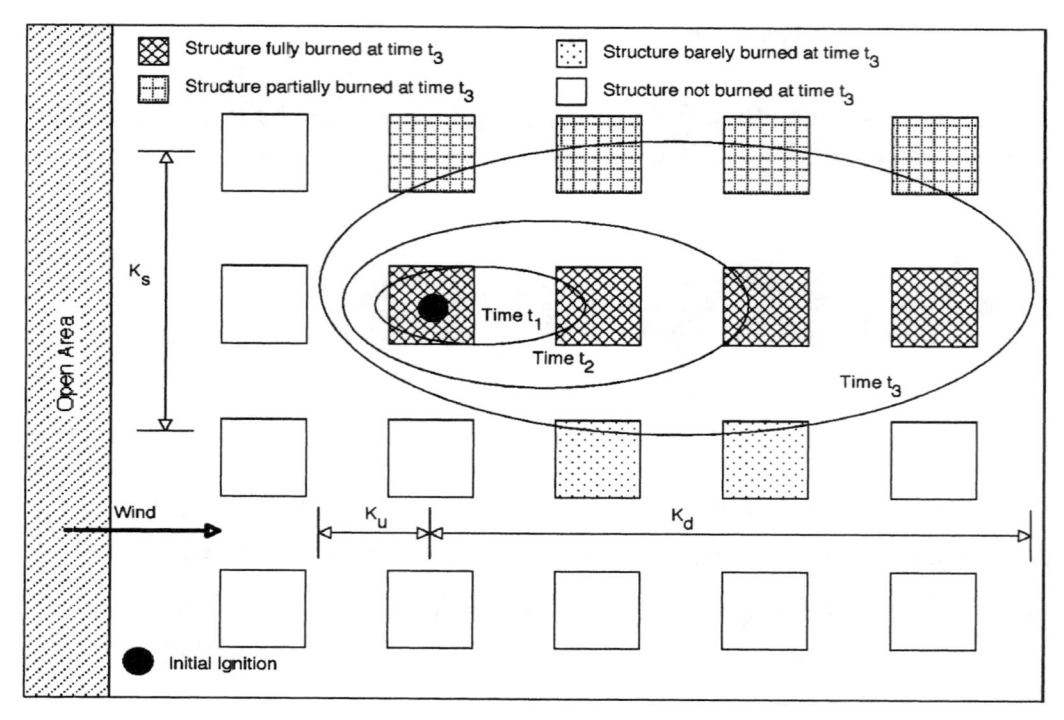

Figure 4-10 Typical Fire Spread

In order to explore the question of utility of these equations in the U.S. context, observed fire spread in various U.S. twentieth century fires was compared with the fire spread predicted using these equations (AIRAC, 1987), Figure 4-11, finding that, with the exception of spread by branding among wood buildings under high winds, estimation using the Hamada equations agreed well with observation.

FIRE SPREAD: ESTIMATION VS OBSERVATION

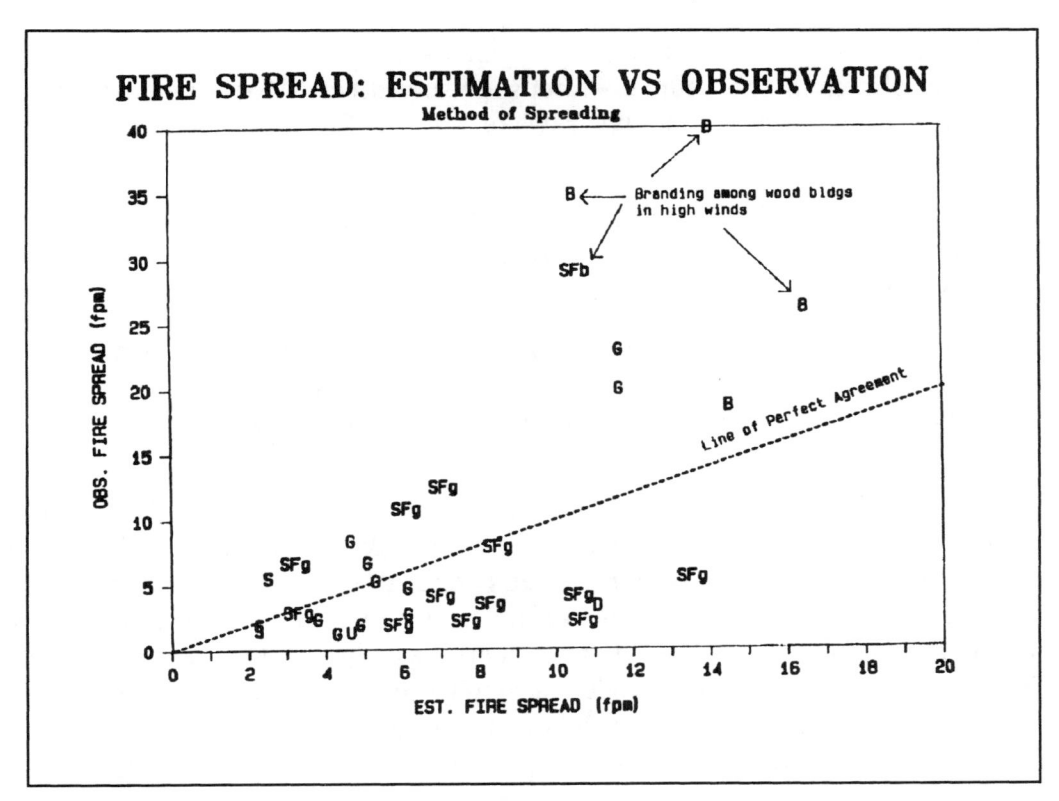

Figure 4-11 Fire spread estimated with Hamada equations, vs. fire spread observed in U.S. non-earthquake conflagrations

A discussion of some of the fire spread variables follows:

Variable a. It is assumed that an urban area is represented as a series of equal square size (plan area) structures, with equal spacing between structures. The plan dimension of the average structure is denoted "a", and hence its plan area is a^2. For example, say a subdivision in an urban area has 500 houses, varying in size from 2,000 square feet to 3,000 square feet (floor area). Also assume that 80% of these structures are 2 stories, and the remainder are 1 story. For modeling purposes, this subdivision is represented as follows:

The average floor area is 2,500 square feet.

20% of the structures have plan area of 2,500 square feet.

80% of the structures have plan area of 1,250 square feet (half of 2,500 square feet total floor area, divided by 2 stories).

The average plan structure dimension is as follows:

$a = (0.20 * \sqrt{2500}) + (0.80 * \sqrt{1250}) = 38.3$ feet = 11.7 meters

Variable d. It is assumed that the spaces between structures in a subdivision can be represented by an average separation distance, d. For purposes of this model, the separation distance, d represents the typical distance between structures within a single block. This distance accounts for side yards, backyards and front yards, but does not include streets and sidewalks. Figure 4-11 shows a hypothetical urban area with four rows of houses in one direction, and five rows of houses in the other direction. The figure does not show city streets. Most subdivisions have city streets spaced such that a block has only two rows of houses. The modeling of city streets is covered later on.

Variable δ. The "built-upness," or building density ratio is defined by equation (4-8). To put the "built-upness" value in context, a value of 0.35 represents a densely built area, and a value of 0.10 represents an area that is not very densely built.

Floor-to-floor fire spread in modern mid and highrise construction typically is slower however, and we have modeled this by using the above equations, reduced by a factor. Note that mid and highrise fire resistiveness varies substantially by jurisdiction, depending on local fire codes and enforcement.

The estimates of both the Hamada and TOSHO (next Section) models are for fire spread within one city block or a built-up district, and do not account for fire spread across firebreaks, such as streets. World War II and other data were reviewed (Bond, 1946), to develop estimates of the probability of a typical fully developed building fire crossing a firebreak, under various wind speeds and with and without active fire suppression efforts (Figure 4-12, see AIRAC, 1987 for details). Fire growth and spread then is modeled using these equations, taking into account probability of crossing a firebreak (with or without active fire suppression capability being present). Fires are tracked and merged where they meet (i.e., areas are not "burnt twice").

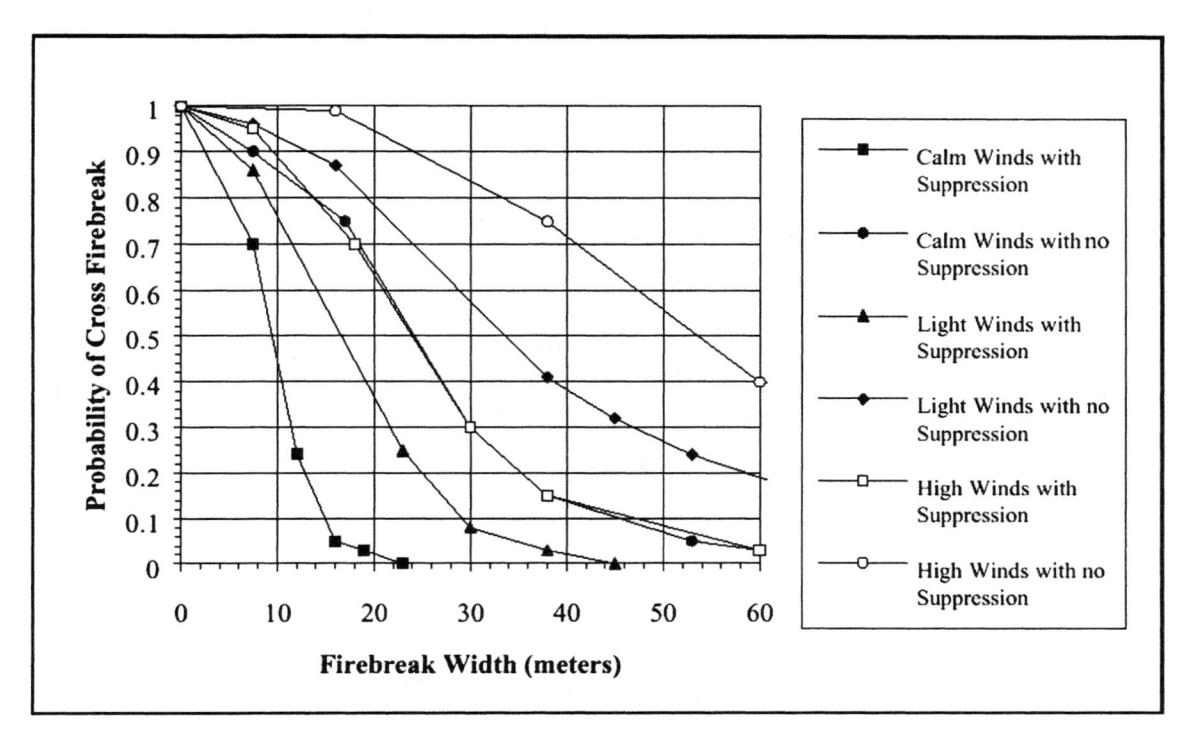

Figure 4-12 Probability of Crossing Firebreak

4.4.2 The TOSHO Model

The Tokyo Fire Department developed in 1997 the so-called TOSHO model. Like the Hamada model, the TOSHO model determines the fire spreading speed in four directions as a function of time: downwind, upwind and the two directions across wind. The TOSHO model requires a specific description of the exposure at risk and is therefore location independent. For the sake of completeness, it is presented in some detail.

Based on the Tosho Model, the fire spreading speed, $V(t)$, at any time, t, after the initial ignition is determined from:

$$V(t) = \frac{V_f}{1 + \{1.3 - 0.3\exp(-0.3t)\}\{(V_f/V_0 - 1\}\exp[-\{0.5V_f/(V_f - V_0)\}t]} \qquad \text{(Equation 4-11)}$$

where:

V_o is the initial fire spreading speed in (m/h), at the startup time of the ignition

V_f is the final fire spreading speed in (m/h),

The initial fire spreading speed, V_o, is computed from:

$$V_0 = \delta \cdot g(h) \cdot (1 - c') \qquad \text{(Equation 4-12)}$$

where:

$$\delta = \frac{\frac{\{r(U)a(a'V_w + b'V_m) + (a+2.6)d'Vc\}}{(a'+b'+d')} + \frac{r(U)\{d(a'+b')^2 V_{nn} + (d-1.3)(a'+b')d'(V_{nc}+V_{cn}) + (d-2.6)d'^2 V_{cc}\}}{(a'+b'+d')^2}}{a+d} \quad \text{(Eqn 4-13)}$$

where:

- a : length of one side of a building in meters
- d : spacing between buildings (door to door) in meters
- a' : the ratio of non-damaged bare wood structures
- b' : the ratio of non-damaged fire resistive structures
- c' : the ratio of fire resistive structures
- d' : the ratio of totally damaged wooden structure
- U : wind speed in m/sec
- h : humidity in %
- $r(U)$: factor depending on wind speed, U, and is equal to
- $r(U) = 0.048U + 0.822$
- $g(h)$: factor depending on humidity, h, and is equal to
- $g(h) = -0.005h + 1.371$
- V_w : fire speed inside wood buildings
- V_m : fire speed inside fire resistive buildings
- V_c : fire speed inside collapsed buildings and is equal to
- $V_c = 98./(1.+3.9 \exp(-.094\ U^2))$ (m/h)
- V_{nn} : fire speed from non-collapsed to non-collapsed buildings
- V_{nc} : fire speed from non-collapsed to collapsed buildings
- V_{cn} : fire speed from collapsed to non-collapsed buildings
- V_{cc} : fire speed from collapsed to collapsed buildings

The final fire spreading speed, V_f, is computed from:

$$V_f = \frac{V_u + \exp\{-50(k-0.14)\}V_1}{1+\exp(\{-50(k-0.14)\})} \quad \text{(Equation 4-14)}$$

where:

k : coefficient that performs a smooth transition for V_f from V_l for small values of k (< 0.14) to V_u for high values of k ($\gg 0.14$) and is calculated from

$$k = (1-c')(a''+0.85b'') \{m(1-x)-0.1\}^{1.2}(U-4.9-8x)^{0.33}$$

m : Built-upness, building density ratio, or building coverage ratio

x : total damaged ratio of wooden structures

$$x = \frac{d'}{a'+b'+d'} = \frac{0.54}{1+680\exp(-0.10\alpha)} - 0.0024$$

α : acceleration in gal

$a'' = a'+0.0018b'\alpha$ if $b'-0.0018b'\alpha < 0$ 	then $a'' = a'+b'$

$b'' = b'-0.0018b'\alpha$ if $b'-0.0018b'\alpha < 0$ 	then $b'' = 0$

however, if $m(1-x)-0.1 <0$ or $U-4.9-8x <0$ then $k = 0$

V_l : fire spreading speed influenced by radiation

$$V_1 = (1-x)^2[6a_uV_o(m^{1.5}-m^2)+b_1](1-c')(a''+.85b'')(0.1U+0.1)^{0.5} + V_o$$

$$a_1 = 0.31/m + 0.52$$

$$b_1 = (-0.1U - 1.8)/m + 2.7$$

V_u : fire spreading speed influenced by wind speed and temperature

$$V_u = 0.46(1-x)^2[a_uV_o\{(1-c')(a''+.85b'')+1.6((1-c')(a''+.85b''))^{-0.5}(U+0.1)^{-0.4}\}+b_u]m^{0.2}+V_o$$

$$a_u = \{1.4(U+1.0)^{-0.61}+0.47\}/m+(4.4U^{0.19})-5.6$$

$$b_u = \{-8.9U^{0.75}-8.6\}/m+(0.041U^{3.1})+49$$

It is assumed that an urban area is represented by a series of equal square (plan area) structures of average dimension a_o and with equal spacing between structures d. It is also assumed that the structures are arranged into city blocks. Each city block has an average dimension b_w and an average block separation (also referred to as fire break width) f_b.

The built-upness, m, differentiates between densely and sparsely populated areas. A value of built-upness equal to 0.35 represents a densely built area, while a value of 0.10 represents an area which is not very densely built. The built-upness, m, is computed from the

$$m = \left[\frac{b_w^2}{(b_w+f_b)^2}\right]\left[\frac{a_o^2}{(a_o+d)^2}\right]$$

where:

  a_o : average building dimension

  d : average distance between buildings

  b_w : average block width

f_b : firebreak width

The fire is assumed to spread in an elliptical shape, through an evenly distributed fuel load, with the longer axis in the wind direction. In an actual urban conflagration, fires exhibit this trend initially, but the final shape of the fire spread differs, as different fuel loads are experienced, fires merge, and different fire suppression actions take place. At each time step the fire dimensions in the downwind direction K_d, the upwind direction K_u and the side wind direction K_s are updated by using the present fire spreading speeds.

The fire spreading speed differs in the downwind direction, from both the upwind and side wind directions due to the difference in wind speed. The actual wind speed is used in computing the fire speed in the downwind direction, while zero wind speed is used in computing the fire speed in both the upwind and sidewind directions.

The fire burnt area at any time step is computed by calculating the area of the fire ellipse as follows:

$$\text{Fire Area} = \pi \left(\frac{K_d + K_u}{2} \right) K_s$$

Before the TOSHO model was introduced, the Hamada Model was the most commonly used model for fire spread. The TOSHO spreading speeds are generally lower than the Hamada fire spreading speeds. The decrease reflects a more realistic processing of more recent data such as the Sakata conflagration (1976) and the 1995 Kobe earthquake.

The TOSHO model has a factor to account for damage to the building inventory, whereas the Hamada model does not. In the three earthquakes with most fire spread (1906 San Francisco, 1923 Kanto, 1995 Kobe), damage to buildings had significant influence on fire spread, as follows:

- 1906. Widespread collapse of masonry (brick) buildings led to damage to many service connections. The loss of water pressure and water inventory through leaking service connections hampered firefighting activities. The fallen masonry created piles of debris that prevented the water and fire departments from rapidly closing the meters to the service connections.

- 1906, 1923, 1995. The collapsed buildings resulted in great masses of wood that was much like kindling. The wood debris ignited more readily and allowed more rapid spread of the fires.

This suggests that future enhancements to fire spread models might incorporate inventory-specific factors such as building collapse rate and style of water and natural gas service line installations, more of which will be discussed in Chapters 6 and 7.

4.5 Fire Response and Suppression

As discussed above, fire growth and fire engine location are tracked concurrently. Under normal conditions, urban fire department response to a structural fire is usually a minimum of two fire engines and one ladder truck (additional apparatus responds in high value or extra hazard areas). These normal responses will not be possible following a large earthquake, since fires may outnumber fire engines. Based on review of actual operations following earthquakes, and discussions with senior fire department officials, it is likely that following an earthquake, initially only one engine will respond to reported fires, to suppress the fire and/or size-up the situation. Thus, initial response should be modeled in this manner, initially allocating one engine to each fire, and additional engines as available. Where fires outnumber first line engines, the difference is termed "excess fires".

A question arises whether fire department resources will initially be totally and primarily devoted to fire suppression, since it should be recognized that other demands (search and rescue, hazardous material response, emergency medical treatment) will also be placed on these resources. For most analysis, it should be assumed that the fire department's first priority will be fire suppression. Note that this assumption is optimistic, especially since building collapses and hazardous materials releases may involve large numbers of victims. This aspect has been reviewed with senior officials of several fire departments, and their opinion is that some fire department resources will have to be diverted from firefighting to these other services. However, experience has shown that serious fires typically receive first priority, for the following reasons:

- Fire service training and tradition

- Fires are dynamic while building collapses are relatively static-that is, a fire situation will worsen if neglected, while the building collapse and rescue situation can often wait several hours (indeed often must await the arrival of heavy equipment)

- Ability of other services (police and others) to assist in building collapses, emergency medical treatment and hazardous materials management (via isolation and evacuation), while only the fire service is equipped to handle serious fires

The impact of this assumption is to decrease total expected losses due to fire following earthquake.

In addition to each jurisdiction's fire suppression resources (i.e., the department's first line and reserve engines, other equipment and personnel), auto and mutual aid need to be considered. These resources of course arrive somewhat later, from more distant locations.

The size of fire at first engine arrival time (in terms of actively burning SFED's, and water required for suppression) is the next parameter in the model, and should be compared with that engine's suppression capability. Engine suppression capability should be modeled using guidelines appropriate for typical structural fires under non-earthquake conditions, and modified to take into account reduced available water (due to earthquake damage to the water supply) and likely fire department use of minimal tactics (discussed above).

That is, given a burning area, required fire flow under normal conditions can be computed on the basis of 4 gallons per minute (gpm) for each 100 cubic feet (cf) of occupancy directly involved in the fire or immediately exposed (Kimball, 1966). For larger fires, this volumetric calculation is based on only perimeter defense. Depending on construction and available manpower, a fire engine typically can apply up to 1500 gpm (using the monitor and or handlines, note however if 1500 gpm is to be applied entirely by handlines, additional personnel are needed). This may be reduced depending on the post-earthquake condition of the water supply system. Thus one engine can typically attack a maximum burning volume of 37,500 cf (typically, 3,000 to 4,000 square feet of floor area) if the monitor can be efficiently used, or half of this (about one SFED) if hand lines are used and additional personnel are not available. If minimal tactics are employed (i.e., no attacks, perimeter protection only), then the capacity of one engine can be considered to be increased (e.g., up to three or four hundred linear feet of perimeter).

Damage to the water supply was of course of prime importance in the 1906 San Francisco, 1989 Loma Prieta, 1994 Northridge and 1995 Kobe earthquakes, and will likely be critical in future earthquakes. As discussed above, although the earthquake severed the main water transmission lines into San Francisco, this was not critical – there existed sufficient water in reservoirs, etc., within the city itself, for initial firefighting needs. Rather, hundreds of breaks in the local distribution network, due to large ground displacements and liquefaction, resulted in the system hemorrhaging water and losing pressure. In order to account for damage to the water supply, two approaches can be employed:

- A detailed hydraulic model of the system with estimated damage due to shaking, permanent ground displacements, etc., is preferred. This approach however may be beyond the scope of some investigations
- An alternative method is to review the regional water supply systems, and likely areas of liquefaction and permanent ground displacement. Based on this review, for areas where permanent ground displacements are determined to severely impact the water supply system, remaining water supply functionality and reduction in fire suppression capability can be estimated judgmentally. This approach is described in detail in ATC (1992)

At the time of first arrival, if the fire (in terms of SFED's) is larger than the suppression capability (in terms of SFED's) of the allocated (one or more) engines, this is termed a "large fire." The number of engines required to suppress or protect a fire depends on the size of the fire involved – for a single family dwelling, it is one to two engines, depending on exposures, while for larger fires, it depends on the exposed perimeter. That is, under post-earthquake conditions, it is assumed that firefighters will not engage in aggressive attack, but rather simply try to protect exposures and prevent the fire from progressing. For conflagrations in low-rise construction, the number of engines required is approximately one engine per 300 feet of perimeter of the fire, plus some additional engines for down-wind 'brand patrol' – that is, suppressing fires at an early stage which are caused by burning brands being carried downwind.

Large fires are assumed fought by available engines in a downwind perimeter defense, and to initially spread elliptically at down/up/side wind rates through a uniformly spaced gridwork of buildings, This assumption of a uniform gridwork of buildings has been used in all urban fire modeling to date, and is reasonable. The spacing, story height, etc., of this gridwork are a function of building density and built-upness (the latter is the ratio of property devoted to buildings contrasted to the total area, including streets, parks, etc.). Fire spreading rates are decreased, and the initial elliptical shape altered, as the available firefighting capability increases. Eventually, for each fire, one of three things happens:

- Fires are suppressed. That is, firefighting capability exceeds needs (either initially, or with build-up of engines as other fires are suppressed and their engines redirected, and/or mutual aid engines arrive) and the fire is surrounded, controlled and suppressed
- The fire is too large and capability is exceeded by need. The fire burns relatively freely within a city block. At each street or other (down/up/side wind) firebreak, the fire crosses with crossing probability as discussed above. This probability is a function of firebreak width, wind direction and, especially, suppression capability. If sufficient engines and

water are available, many fires will typically be stopped at the first wide street. Note that strong winds have an important effect on downwind firebreak crossing probabilities. In this context, branding (windborne transmission of flaming debris) is sometimes an important factor, especially where wood roofs are prevalent (these are banned in San Francisco, but are common in some other jurisdictions). In most cases, a fire is not expected to cross more than a few typical streets, so that most large fires will be stopped, or burn themselves out, within a few blocks. Again, moderate to strong winds will extend this stopping distance significantly

- The fire reaches an "ultimate" firebreak-that is, a large expanse of water, the edge of the urbanized area, etc. Available engines are sufficient to defend exposures along the remaining perimeter, and the fire is controlled and suppressed

4.6 Final Burnt Area

The above method, termed a *simulation model*, is followed for each fire. Fires need to be tracked and merged where they meet (i.e., areas should not be "burnt twice"). Final burnt areas are summed, to arrive at total final burnt area. The implementation of this simulation model for a large urban area is clearly a data- and computationally intensive matter, so that fires following earthquakes are usually estimated using computer programs, termed a simulation code. Several such codes exist including EQEFIRE^{sc}, HAZUS^{sc}, URAMP^{sc}, SERA and RiskLink. The following sections discuss examples from the HAZUS^{sc}, URAMP^{sc}, SERA and RiskLink codes.

4.6.1 HAZUS

HAZUS is a program developed by the National Institute for Building Sciences, for the Federal Emergency Management Agency (HAZUS, 1999). As noted in the HAZUS Technical Manual (HAZUS, 1999):

HAZUS is designed to produce loss estimates for use by federal, state, regional and local governments in planning for earthquake risk mitigation, emergency preparedness, response and recovery. The methodology deals with nearly all aspects of the built environment, and a wide range of different types of losses. Extensive national databases are embedded within HAZUS, containing information such as demographic aspects of the population in a study region, square footage for different occupancies of buildings, and numbers and locations of bridges. Embedded parameters have been included as needed. Using this information, users can carry out general loss estimates for a region. The HAZUS methodology and software are flexible enough so that locally developed inventories and other data that more accurately reflect the local environment can be substituted, resulting in increased accuracy (Figure 4-13).

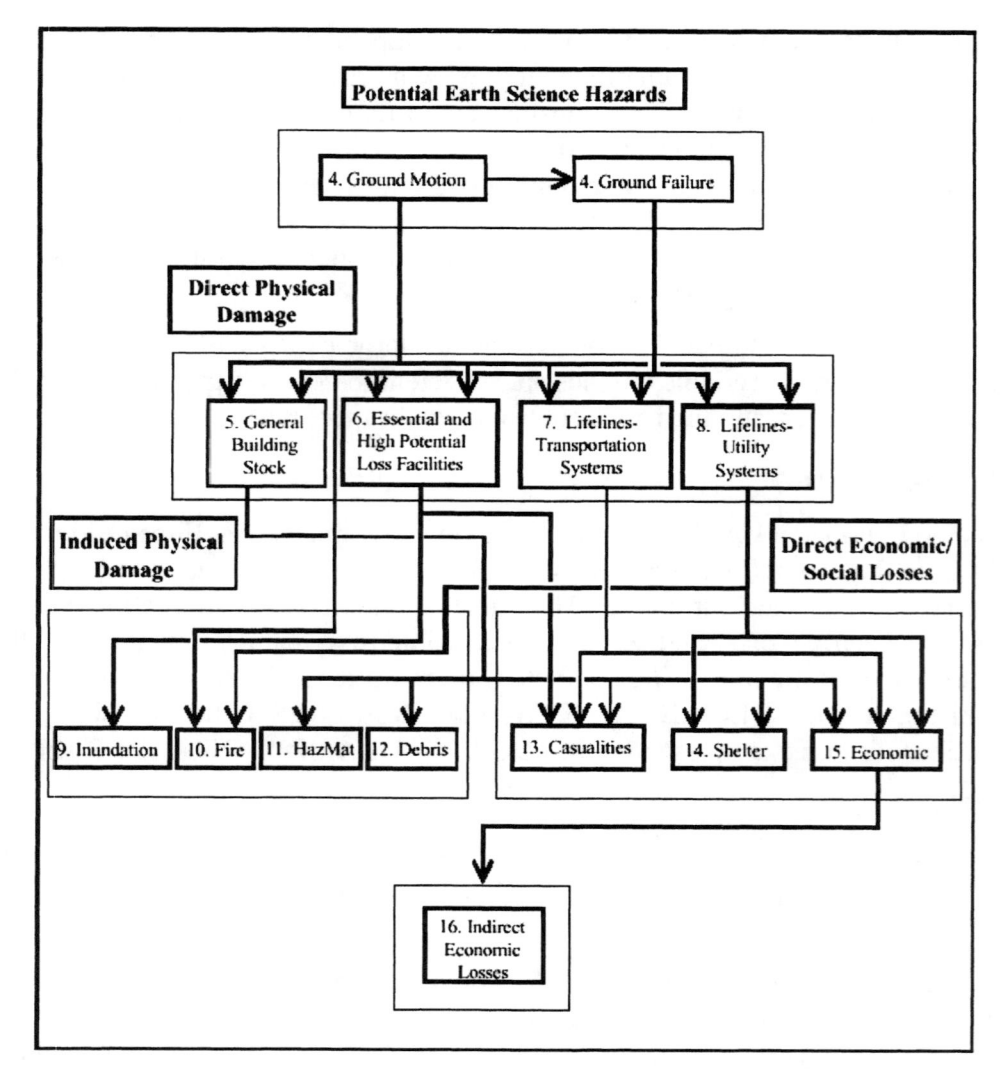

Figure 4-13 Flowchart of the HAZUS Earthquake Loss Estimation Methodology

Among other aspects of earthquake damage dealt with within HAZUS is a model for fire following earthquake. The HAZUS' fire following earthquake model uses the ignition model (Equation 4-3), the Hamada fire spread model (Equation 4-4) and a fire suppression model that varies according to the availability of water (for complete details the reader is referred to HAZUS, 1999). The HAZUS model does not include data from more recent events such as ignition rates from the 1994 Northridge earthquake, or fire spread rates from the 1995 Kobe earthquake, etc. The HAZUS model assumes the water system inventory solely based on population density, and does not require the user to provide the actual characteristic of the water system – a major simplification for ease-of-use, but also a major oversimplification. Another limitation of HAZUS is its deterministic nature – that is, it provides only point estimates, not a range or lower and upper bounds or other measure of uncertainty on its results. Given the highly stochastic nature of the fire following earthquake problem, understanding the uncertainty in any estimates is very important.

HAZUS has been used in many regional studies of earthquake losses, in which estimates of fire following earthquake have been made. In a study of the State of South Carolina Bouabid J. et al

(2002) report that a M6.3 event in the state would cause about 42 fires, while a M7.3 event would cause about 255 fires. In a study of potential earthquake effects in Manhattan (New York City), Tantala et al (NYCEM, 2000) report results for several events, shown in Table 4-8.

Table 4-8 HAZUS Fire following earthquake estimates, New York City

Scenario	Magn.	Avg. Ret. Period (yrs)	No. Ignitions	Exposure ($ mills)	Exposure People	GPM Demand
A	5.0	2,475	10	216	362	10,218
B	6.0	19,500	111	632	24,620	153,764
C	7.0	160,000	169	15,200	68,638	213,518

These results agree well with previous estimates, where Scawthorn and Harris (1989) estimated 130 ignitions for the Scenario B event (M 6.0), versus the NYCEM estimate of 111.

The HAZUS methodology has been examined in a validation study (FEMA, 2001), which compared 'predicted' numbers of ignitions for recent earthquake, against those observed, Table 4-9. As the validation study observed:

- HAZUS did not consistently predict the number of fire ignitions region wide. It may be difficult to determine what should be considered a single ignition. The more important statistic, however, is that HAZUS predicts a high potential dollar loss caused by fires. These predictions are much larger than documented losses. It should be noted that HAZUS predicts only the potential for loss, assuming a wind speed, fire engine response time, etc., and that all the buildings in the burned area are a total loss. A more detailed study might explore the actual conditions at the time of the earthquakes with the predicted parameters.

The editors note the HAZUS validation study might have been problematic. The editors caution that any user of HAZUS should consider its predictions as being only as good as the inventory and the models used in the analysis. With refinements in the inventory, and with ongoing improvements in models, better predictions can be made.

Table 4-9 HAZUS Validation Study – Fire Following Earthquake

	Ignitions / Dollar Loss	
Earthquake	Predicted	Documented
Coalinga	1 / < $1 million	3/NA
Whittier	33-43 / $40-70 million	90 / $426 thousand
Loma Prieta	14-38 / $30-100 million	58 / NA
Northridge	72-101 / $70-1,340 million	110 / $13.6 million

NA: not available

Table 4-10 Ignition Rates by Occupancy

MMI (PGA, in g)	Number of Ignitions per million ft²		
	Residential	Commercial	Industrial
VI (0.09 - 0.18)	0.024	0.007	0.002
VII (0.18 - 0.34)	0.071	0.019	0.005
VIII (0.34 - 0.65)	0.177	0.047	0.012
IX (0.65 – 1.24)	0.397	0.106	0.026
X (>1.24)	0.819	0.218	0.055

4.6.2 URAMP^{SC}

The URAMP^{SC} software (Utilities Regional Assessment of Mitigation Priorities) estimates losses, and incorporates avoided losses as benefits in a benefit-cost framework, to determine the most effective seismic risk reduction program for a water utility. URAMP^{SC} was developed for the California Governor's Office of Emergency Services (OES), and incorporates the functionality of EPANET, a hydraulic modeling software package distributed by the EPA, to assess the impact of earthquake damage on water network performance. The URAMP^{SC} software allows the user to assess the benefits of individual mitigation strategies, and to determine which combination of mitigation measures best enhances water network performance in the aftermath of an earthquake.

The URAMP^{SC} analysis modules include modern seismic hazard models to facilitate regional network analysis, and consider both deterministic (e.g., scenario earthquakes) and probabilistic seismic risk assessments, Figure 4-14. The damage estimation modules utilize engineering-based damage and mitigation models, based, whenever possible, on published methodologies. In addition to direct damage to utility facilities, URAMP^{SC} estimates potential damage and economic losses due to fire-following earthquake, and other economic impacts on the provider utility and community, such as lost revenue, business impacts of fire, and cost of sewage clean-up. Many of the economic models are based on earlier "Seismic Reliability Assessment Studies" performed by the U.S. Army Corps of Engineers (Seligson et al, 2003).

The fire following earthquake module of URAMP^{SC} was developed by Scawthorn, and follows the general methodology outlined above. Ignition rates are based on the same data and methods described above, and are also provided in the format shown in Table 4-10. To simplify the data needs (i.e., reduce the burden of collecting detailed building inventory data), URAMP^{SC} employs a simplified methodology, in which neighborhoods can be classified as High (H), Medium (M) or Low (L) density Residential (R), Commercial (C) or Industrial (I) occupancies, Table 4-11.

Table 4-11 Building Densities, by Occupancy

Occupancy	Building Floor Area Range (mills sq ft) per sq. mile
R-H	15+
R-M	6~15
R-L	2~6
C-H	40+
C-M	8~40
C-L	2~8
I-H	25+
I-M	6~25
I-L	2~6

Note: High (H), Medium (M) or Low (L) density;
Residential (R), Commercial (C) or Industrial (I) occupancies

Using this scheme, final burnt areas were determined using the above methodology in a Monte Carlo simulation methodology, with results shown in Table 4-12. In this Table, results are provided for two extremes: Water Supply Reliability factor = 0, or 1.

A Water Supply Reliability factor is employed because, in a detailed analysis of potential losses due to fire following earthquake, and the effect of a water system on these losses (under post-earthquake, possibly damaged conditions), it is normal to employ the results of a water system vulnerability analysis. Such an analysis will indicate portions of the service area likely to not have water, and thus be more prone to fire spread (since the fire department will have reduced suppression capabilities, having to resort to alternative water sources – a time consuming process during which the fire continues to spread). However, the scope of URAMP^{SC} was not based on such a detailed analysis of system vulnerability. Therefore, URAMP^{SC} employed a Water Supply Reliability Factor (WSR factor): **$0 \leq WSR \leq 1$,**

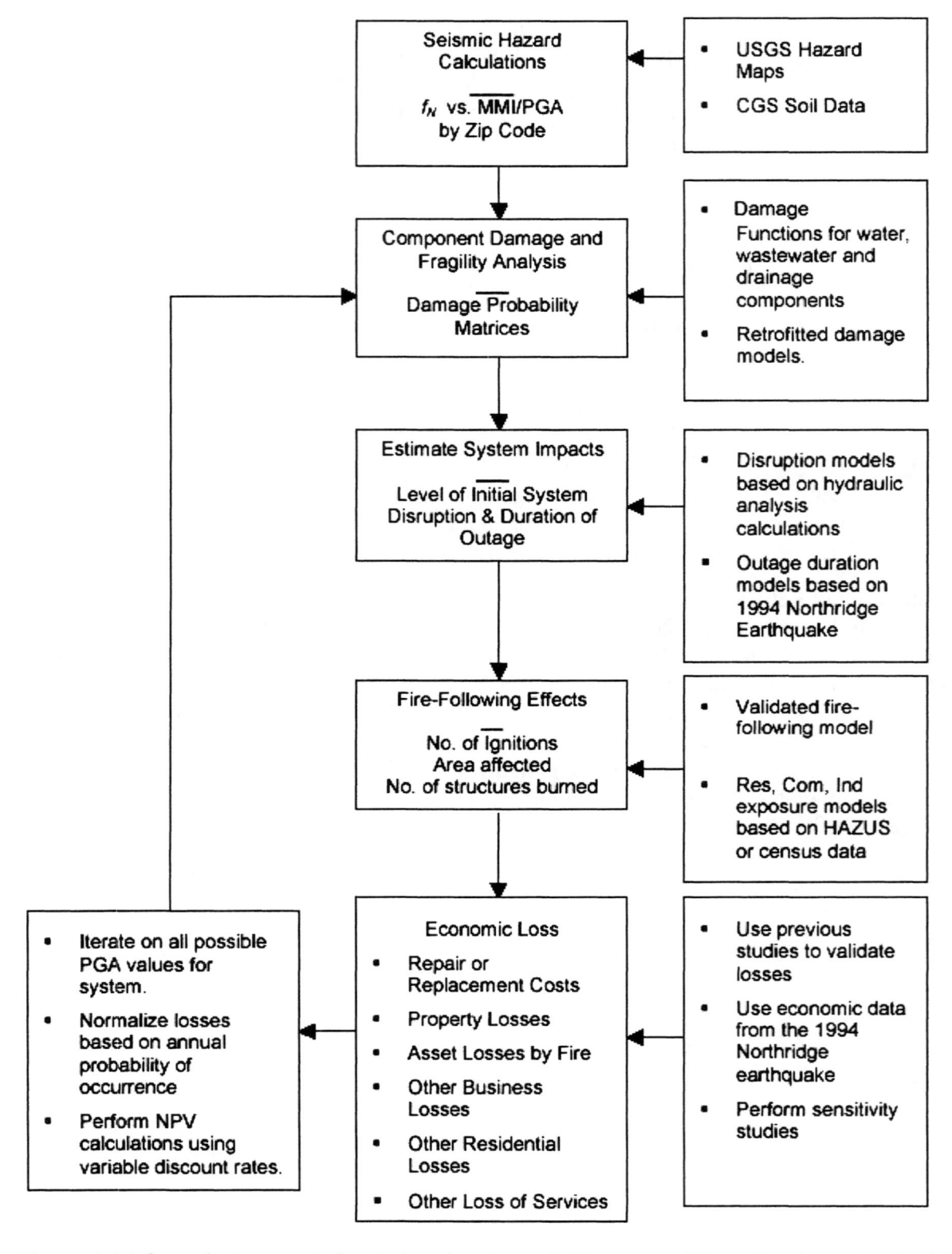

```
                    ┌─────────────────────┐        ┌──────────────────────┐
                    │ Seismic Hazard      │        │ • USGS Hazard        │
                    │ Calculations        │◄───────│   Maps               │
                    │                     │        │ • CGS Soil Data      │
                    │ f_N vs. MMI/PGA     │        │                      │
                    │ by Zip Code         │        │                      │
                    └─────────┬───────────┘        └──────────────────────┘
                              │
                              ▼
                    ┌─────────────────────┐        ┌──────────────────────┐
                    │ Component Damage and│        │ • Damage Functions   │
                    │ Fragility Analysis  │◄───────│   for water,         │
  ┌────────────────►│                     │        │   wastewater and     │
  │                 │ Damage Probability  │        │   drainage           │
  │                 │ Matrices            │        │   components         │
  │                 └─────────┬───────────┘        │ • Retrofitted damage │
  │                           │                    │   models.            │
  │                           ▼                    └──────────────────────┘
  │                 ┌─────────────────────┐        ┌──────────────────────┐
  │                 │ Estimate System     │        │ • Disruption models  │
  │                 │ Impacts             │◄───────│   based on hydraulic │
  │                 │                     │        │   analysis           │
  │                 │ Level of Initial    │        │   calculations       │
  │                 │ System Disruption & │        │ • Outage duration    │
  │                 │ Duration of Outage  │        │   models based on    │
  │                 └─────────┬───────────┘        │   1994 Northridge    │
  │                           │                    │   Earthquake         │
  │                           ▼                    └──────────────────────┘
  │                 ┌─────────────────────┐        ┌──────────────────────┐
  │                 │ Fire-Following      │        │ • Validated fire-    │
  │                 │ Effects             │◄───────│   following model    │
  │                 │                     │        │ • Res, Com, Ind      │
  │                 │ No. of Ignitions    │        │   exposure models    │
  │                 │ Area affected       │        │   based on HAZUS     │
  │                 │ No. of structures   │        │   or census data     │
  │                 │ burned              │        │                      │
  │                 └─────────┬───────────┘        └──────────────────────┘
  │                           │
  │                           ▼
  │                 ┌─────────────────────┐        ┌──────────────────────┐
┌─┴──────────────┐  │ Economic Loss       │        │ • Use previous       │
│• Iterate on all│  │ • Repair or         │        │   studies to         │
│  possible PGA  │  │   Replacement Costs │        │   validate losses    │
│  values for    │◄─│ • Property Losses   │◄───────│ • Use economic data  │
│  system.       │  │ • Asset Losses by   │        │   from the 1994      │
│• Normalize     │  │   Fire              │        │   Northridge         │
│  losses based  │  │ • Other Business    │        │   earthquake         │
│  on annual     │  │   Losses            │        │ • Perform            │
│  probability of│  │ • Other Residential │        │   sensitivity        │
│  occurrence    │  │   Losses            │        │   studies            │
│• Perform NPV   │  │ • Other Loss of     │        │                      │
│  calculations  │  │   Services          │        │                      │
│  using variable│  │                     │        │                      │
│  discount rates│  │                     │        │                      │
└────────────────┘  └─────────────────────┘        └──────────────────────┘
```

Figure 4-14 Overall Approach for Estimating Potential Losses and Benefits Associated with Mitigation Measures

where

- WSR =1 is perfect reliability, meaning that the system's capability to furnish volume and pressure of water following the earthquake is the same as prior to the earthquake[14], and
- WSR = 0 is perfect failure, meaning that the system cannot furnish any volume of water, due to damage by the earthquake. Note that the damage resulting in complete loss of service at a point may have occurred at some distance, but still may be capable of draining the system.

Note that even with an undamaged water system, fires may still spread if there are high winds (such as occurred in the 1991 East Bay Hills fire), or due to there being more fires than the fire department has apparatus or personnel to respond to.

Another feature of URAMP^{SC} was its consideration of brush zones. The western US, especially California, are subject to wildland fires which, at the urban-wildland interface, can result in major losses to structures. This problem exists under normal conditions, and also is a factor to be considered in the fire following earthquake problem. Examples of these are very numerous, and include the 1991 East Bay Hills fire, the series of major fires in southern California in 1992 and other events described in Chapter 3 (see Table 3-1). Substantial numbers of residential properties are in heavily vegetated suburban areas, in which under certain circumstances (esp. hot dry windy conditions) firespread is greatly enhanced due to the vegetation serving as a continuous fuel path between buildings. In recognition of this, the California Department of Forestry (CDF) has mapped these areas, which are informally termed brush zones, Figure 4-15. Firespread in these areas is much more severe, for certain seasons. Analysis for brush zone areas under average wind conditions were performed, and Table 4-13 shows the final burnt area as a percentage of the total brush zone area, for residential properties and densities, for WSR factors of 0 and 1.0[15].

Additional details of the URAMP^{SC} model are available in URAMP^{SC} (2002).

---

[14] Note that this assumes the system is a well-designed system, providing adequate fire protection under normal circumstances.

[15] Generally, significant numbers of commercial and industrial properties are not located in brush zone areas, so that analyses were only performed for residential properties.

Table 4-12 Mean Area Burned as a function of Building Density

WSR =1	MMI				
	OCCUP	7	8	9	10
	R-H	3.23%	7.89%	16.88%	31.72%
	R-M	1.21%	2.98%	6.54%	13.01%
	R-L	0.13%	0.75%	2.01%	4.37%
	C-H	0.00%	8.89%	18.85%	35.00%
	C-M	0.00%	2.34%	5.18%	10.39%
	C-L	0.00%	0.00%	0.00%	0.00%
	I-H	0.00%	0.00%	0.00%	5.32%
	I-M	0.00%	0.00%	0.00%	1.27%
	I-L	0.00%	0.00%	0.00%	0.00%

WSR = 0	MMI				
	OCCUP	7	8	9	10
	R-H	5.34%	12.83%	26.57%	47.13%
	R-M	1.68%	4.14%	9.02%	17.71%
	R-L	0.16%	0.91%	2.43%	5.27%
	C-H	7.59%	17.93%	35.80%	59.91%
	C-M	1.52%	3.80%	8.36%	16.45%
	C-L	0.00%	0.00%	0.00%	0.00%
	I-H	0.89%	2.23%	4.89%	9.84%
	I-M	0.17%	0.40%	0.89%	1.85%
	I-L	0.00%	0.00%	0.00%	0.00%

Table 4-13 Mean Area Burned as a function of Building Density
BRUSH ZONE CONDITIONS

WSR =1	MMI				
	OCCUP	7	8	9	10
	R-H	14%	31%	57%	83%
	R-M	3%	7%	14%	27%
	R-L	0%	1%	3%	6%

WSR = 0	MMI				
	OCCUP	7	8	9	10
	R-H	37%	68%	92%	100%
	R-M	7%	16%	32%	55%
	R-L	0%	2%	6%	12%

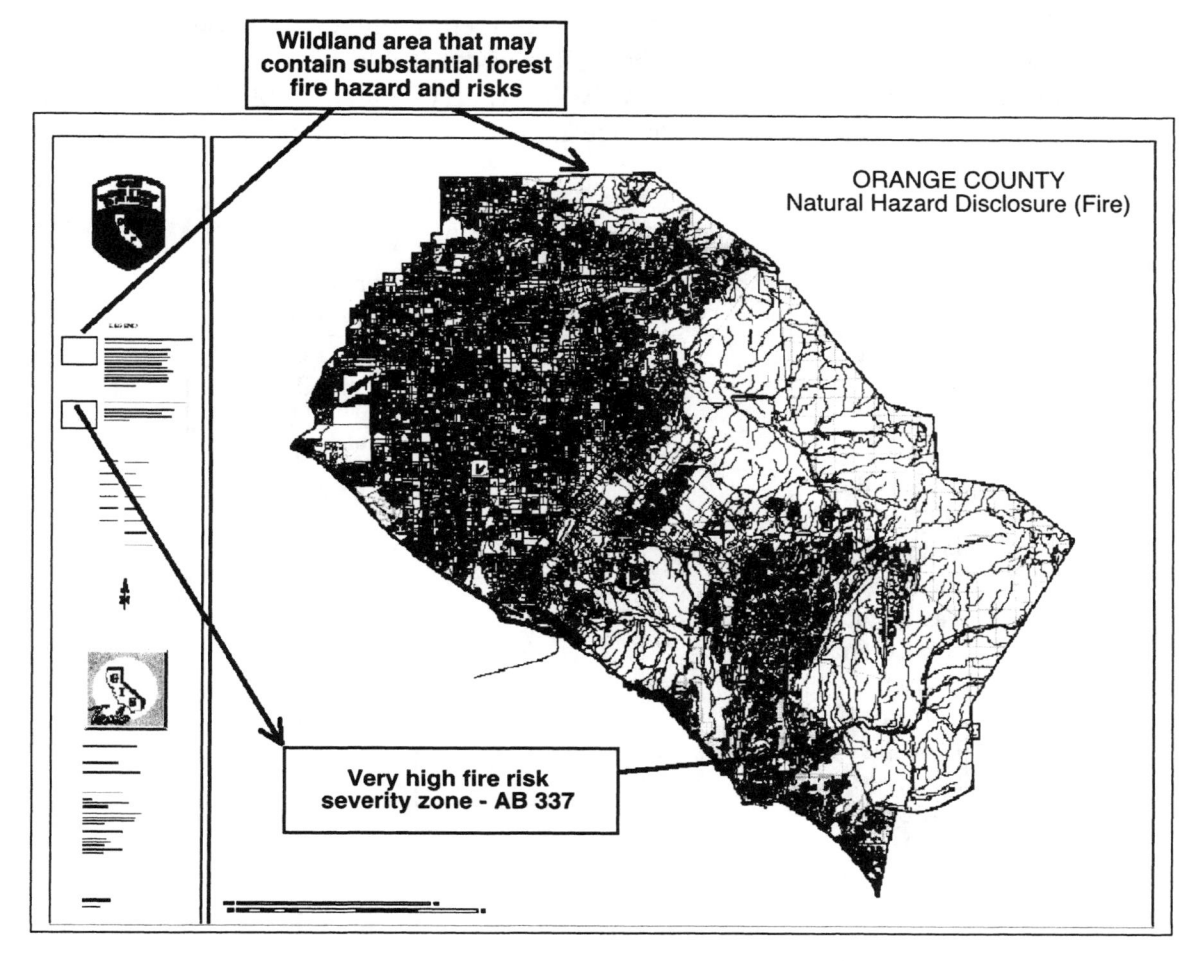

Figure 4-15 Example Natural Hazard Disclosure (Fire) Map, Orange County CA

4.6.3 SERA

SERA (System Earthquake Risk Assessment) is a GIS-based simulation model for lifeline systems. It has been adopted by water utilities: EBMUD (Diemer et al 1995), San Diego Water Department (Eidinger et al 2001), and transit systems (Bay Area Rapid Transit District) (Eidinger et al 2003). The SERA fire model incorporates detailed analysis of the seismic vulnerability of the water system, time dependent hydraulic performance of the water system after an earthquake, and a detailed fire ignition, fire spread and fire suppression model, similar to the ones described in Chapter 4. Using the WSR nomenclature (Section 4.6.2), the model has WSR that varies by location, as leaks depressurize the water system, tanks run dry, and pumps are used to restore supply, as calculated every minute for 24 hours after the earthquake.

The SERA model was applied to the East Bay Municipal Utility District's water system as it was built in 1992. This water system then included about 3,700 miles of pipeline, 175 reservoirs, 139 pump stations, 6 water treatment plants, and served 22 cities in the east bay area of San Francisco. The SERA model was run to simulate fire losses within the EBMUD service area for five scenario earthquakes, and under five possible conditions of the water systems: as-is, and with four possible levels of seismic upgrade costing from $13 million to $202 million to implement. One of these five scenario earthquakes was a repeat of the 1989 Loma Prieta earthquake; that being useful to benchmark the SERA model. Table 4-14 shows the number of fire stations and engines then located within the EBMUD service area. Table 4-15 shows the number of fire stations and engines available to response as mutual aid, but at a distance from the EBMUD service area.

Table 4-14 Number of Fire Engines Within Service Area

City	Fire Stns	No. Engines
Alameda	4	6
Albany	1	3
Berkeley	8	13
Castro Valley	3	6
Emeryville	2	5
Hayward	4	8
Oakland	33	44
Piedmont	1	3
San Leandro	4	14
San Lorenzo	4	11
Danville	4	8
El Cerrito	2	4
Hercules	1	1
Lafayette	1	3
Moraga	2	2
Pinole	1	6
Pleasant Hill	18	47
Richmond	7	14
Rodeo - Crockett	4	12
San Pablo	1	3
San Ramon	2	3
Walnut Creek	4	8

Table 4-15 Number of Fire Engines Outside of Service Area Available for Mutual Aid

COUNTY	NUMBER OF STATIONS	NUMBER OF ENGINES	DISTANCE TO DISTRICT(MILES)
Alameda County Mutual Aid	32	55	8
Contra Costa County Mutual Aid	30	75	10
San Francisco County Mutual Aid	41	41	10
Napa County Mutual Aid	5	5	15
San Mateo County Mutual Aid	53	115	20
Marin County Mutual Aid	33	74	20
Solano County Mutual Aid	7	6	30
Santa Clara County Mutual Aid	69	127	35
Sonoma County Mutual Aid	22	56	50
Santa Cruz County Mutual Aid	4	12	90
TOTAL	296	566	

FIRE FOLLOWING EARTHQUAKE

Over 500 simulations of the fire following earthquake model were run. Each simulation allowed for variation in the level of damage to the water system, the number of fire ignitions, the probability that an individual fire will spread, etc. Average results are presented in Table 4-16. The value of structures burned are presented in year 1992 dollars. The wind speeds are assumed to be 0 m/s (calm), 3 m/s (light) or 12 m/s (high).

Table 4-16 Value of Structures Burned, x $1,000,000

Scenario		Level of Seismic Upgrades				
Earthquake	Wind Speed	As Is	$13	$16	$84	$202
Hayward M 7	Calm	$136	$125	$122	$101	$57
	Light	$332	$282	$265	$165	$122
	High	$1,856	$1,841	$1,836	$1,805	$1,633
Hayward M 6	Calm	51	44	42	28	11
	Light	156	123	113	48	20
	High	687	685	685	680	584
Calaveras M 6.75	Calm	10	10	10	9	8
	Light	19	19	19	18	14
	High	262	242	235	194	177
Concord M 6.5	Calm	4	4	4	4	3
	Light	7	7	7	6	6
	High	52	52	52	52	48
Loma Prieta M 7	Calm	0.3	0.3	0.3	0.3	0.3
	Light	0.5	0.5	0.5	0.5	0.5
	High	4.1	4.1	4.1	4.1	4.1

Some observations are made as to the trends seen in Table 4-16.

- The benchmark analysis using the actual 1989 Loma Prieta M 7 earthquake shows that the forecasted losses due to fire were $0.3 million. Actual losses in the District service area were well under $1 million, and winds were calm at the time of the earthquake. The model appears to be reasonably calibrated, at least for this event.

- Had the water system been seismically upgraded at the time of the 1989 Loma Prieta earthquake, losses from fire would not have been different than what was actually experienced. This makes sense, in that there was rather modest damage to the water observed in that earthquake. In the 1989 Loma Prieta earthquake, the EBMUD water system suffered 135 pipe repairs, some minor damage to a very few tanks and pump stations, about 3% of all customers lost service, with last customer restored in less than 3 days. The actual fires that occurred in the District were not at the shoreline locations that sustained the worst pipe damage.

- Five possible levels of seismic upgrade of the water system were considered, four of which are listed in Table 4-16. (Two other levels of seismic upgraded, costing from $365 million to $1.1 billion, were not seriously considered for implementation due to low benefits versus their cost). Ultimately, the $202,000,000 upgrade program was adopted by the water utility. This $202,000,000 program was forecasted to reduce fire losses by about 2/3 for large scenario earthquakes that could affect the EBMUD service area, such as the Hayward M 7 earthquake, but with less reduction in losses for earthquakes on the Concord or Calaveras faults. The relative fire "ineffectivensss" of seismic upgrade for the Concord and Calaveras fault scenarios is in part because these earthquakes do not strike through the center of the EBMUD service area, as would a Hayward M 7 event; and also in part because the built environment near the Calaveras and Concord faults reflects more modern suburban areas, with wide streets, more fire resistant construction, etc. than what is prevalent in the older parts of the service area nearer the Hayward fault; there are also far fewer liquefaction zones in these areas, and the existing water system infrastructure in those areas was more seismically rugged than the older infrastructure near the Hayward fault in Berkeley and Oakland.

- It is interesting to note that under a high wind scenario that the level of seismic upgrade to the water system is forecasted to have quite modest net reductions in fire losses. This reflects that even an undamaged water system would not prevent multiple conflagrations should there be dozens or hundreds of nearly simultaneous fire ignitions in the EBMUD service area; the available fire services capability (Table 4-14 and Table 4-15) coupled with other post earthquake problems (longer times to report fires, slower speed to respond to fires) is simply not sufficient to deal with this scenario. This suggests that given a large scenario earthquake at a time of high wind speed, society might be best served with alternate strategies like:

    o Try to reduce the number of ignitions (rapid power outages, delayed restoration of electricity)

    o Try to reduce the availability of natural gas as a fuel supply via gas leaks (smart meters, more seismically-rugged gas pipelines)

    o Enforce building set back requirements for new construction

    o Require non combustible roof coverings for new construction

It is useful to examine the time history of the fire spread process using the SERA model. Figure 4-16 provides the results for one simulation of the Hayward M 7 earthquake, assuming light winds and for the water system in its "as-is" condition:

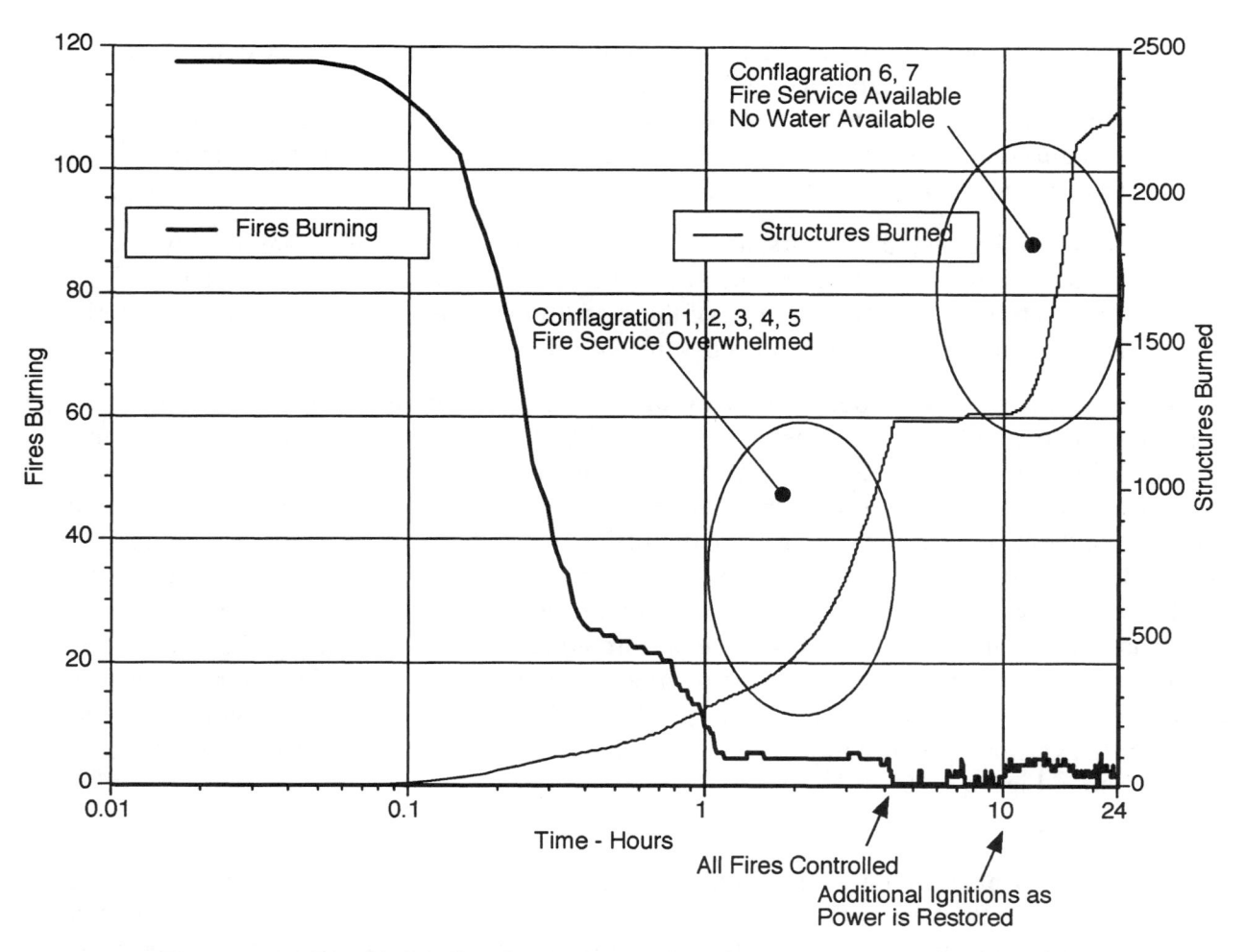

Figure 4-16 Fire Following Earthquake Time History Analysis - EBMUD

- The left axis (fires Burning) shows the number of fires ongoing after the earthquake. For example, at 1 minute after the earthquake, there are 117 fire ignitions. The thick line shows that none of these initial 117 fires come under control for the first few minutes (the curve is flat). There are a total of 169 ignitions in the first 24 hours for the simulation presented in Figure 4-16.

- By 0.1 hours (6 minutes) after the earthquake, some of the fires have been detected and controlled. Within 0.3 hours after the earthquake, about 80% of the 117 initial ignitions have been controlled and about 25 are left burning. By this time, about 100 "equivalent" structures have been burned (thin line, right hand axis). However, the remaining 25 fires are not easily controlled due to a combination of excessive demand for fire service apparatus / personnel, and at some locations, weak water supply being insufficient to control the fire even with adequate apparatus / personnel at the scene.

- Between 0.3 and 0.8 hours, a few more of the remaining 25 fires are controlled, leaving about 20 fires uncontrolled at 0.8 hours. This reflects that some of the more stubborn fires are taking considerably more fire department resources, which take time to mobilize.

- At about 1.2 hours after the earthquake, the number of fires burning is reduced to 5, then 4, until 4 hours after the earthquake. These few remaining fires spread into conflagrations. New ignitions are occasionally occurring, but these are controlled quickly. At 4 hours after the earthquake, all the fires are controlled, and the number of burned structures (right axis) reaches about 1,250. Between 4 and 10 hours after the earthquake, a number of additional ignitions start, but these are quickly controlled, and few additional structures are burned.

- Between time 4 and 24 hours, about 50 more ignitions occur. Most are quickly controlled. In this particular simulation, at time 12 hours due to lack of water at the fire site, high fuel load, etc., a couple of these additional ignitions spread into conflagrations, burning another 1,000 structures by time 24 hours.

In running the SERA Fire module for many simulations, a pattern is observed that suggests that a significant percentage of all fire losses (perhaps 1/3 to 1/2) would occur at a time after all the initial ignitions were controlled, but after the time that hillside reservoirs were emptied (mostly due to leaking pipes). Thus, one of the mitigation strategies for water utilities to consider is to ensure that there is at least some water in all hillside pumped zone areas subject to high risk of fire, at least for about 18 hours (preferably 24 hours) after a large earthquake. Another trend seen in the analyses was that for simulations of a single earthquake with only about 120 ignitions (50 fewer than in Figure 4-16), the fire service performed quite well, with each initial ignition generally leading to just 1 or 2 burned structures, and low chance of conflagration beyond 10 structures. However, if the earthquake caused 220 ignitions (50 more than in Figure 4-16), then many conflagrations would likely occur, and total structures burned could be easily 4 times worse than in Figure 4-16.

4.6.4 RiskLink

Figure 4-17 presents the simulation status some time after the occurrence of a repeat of the 1906 earthquake in San Francisco. The small buildings indicate the location of the fire stations, the dots are the locations of the fires, and the lines with the truck show how the fire engines have been dispatched from the station to the fire. It can be noted that the engines move in a straight line from the station to the fires (city streets are not modeled to calculate engines travel time). The wind speed is constant over the duration of the simulation and is 10 km/h in this simulation. Figure 4-18 presents the status of the fires at the same simulation time as Figure 4-19. The ellipses indicate the burnt area, the gray ellipses represent fires that have been brought under control while the black ones are still burning uncontrolled. The simulation ends when all the fires are either brought under control or are stopped by firebreaks and run out of fuel. Figure 4-19 presents a measure of the ground shaking for a repeat of the 1906 event. A grid has been overlaid over the city and all the parameters are assumed constant within a cell. For this study the size of the cell is 0.01 x 0.01 degree (about 1km x 1km). The level of shaking is quantified in terms of peak ground acceleration (gals or cm/sec^2). The PGA does not always show a uniform decay as the distance from the fault increases because the acceleration is affected by the local soil condition. Figure 4-20 presents the mean burnt area within each cell at the end of a large number of simulations. By overlaying the exposure in each cell with the burned area one can determine the expected fire loss in each cell and for the whole area.

When comparing the burnt area between simulations, it is not uncommon to find the worst-case scenario to be twenty times larger than the most optimistic scenario. Of course, these extreme scenarios have a very low probability of occurrence as compared to the average scenarios. The most sensitive parameters are the number of fires, the wind speed and the location of fires.

FIRE FOLLOWING EARTHQUAKE

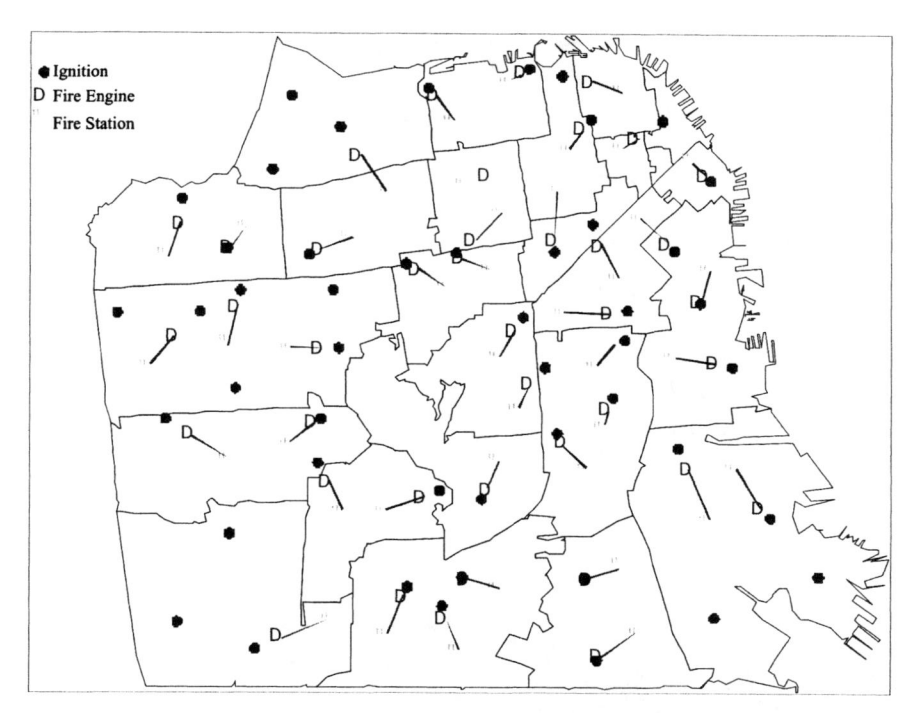

Figure 4-17 Ignitions and deployment of fire engines for San Francisco - RiskLink Simulation

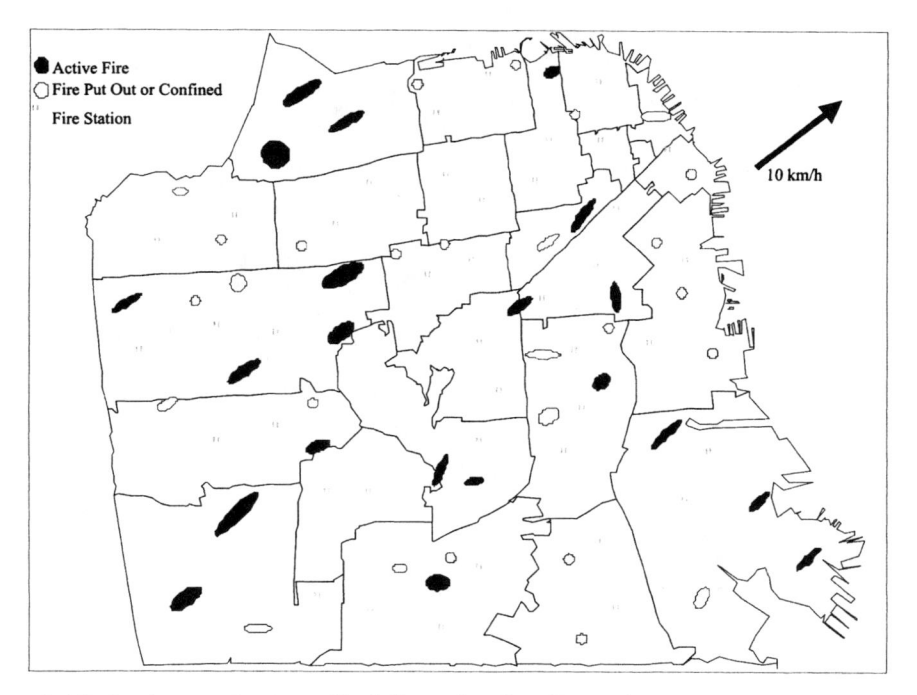

Figure 4-18 Active and controlled fires for San Francisco - RiskLink Simulation

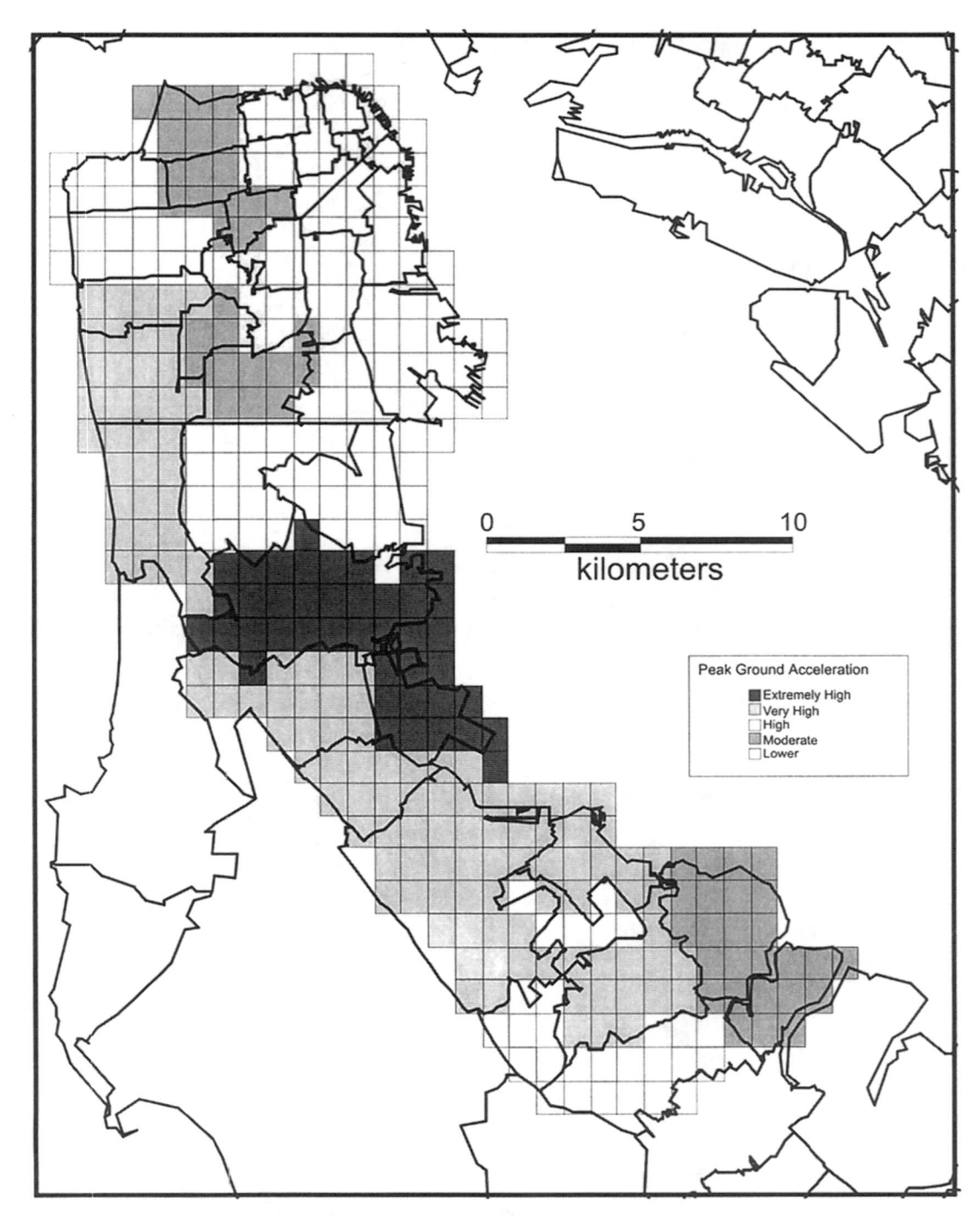

Figure 4-19 Geographical Distribution of Peak Ground Acceleration for the 1906 San Francisco Earthquake (RiskLink Simulation)

FIRE FOLLOWING EARTHQUAKE

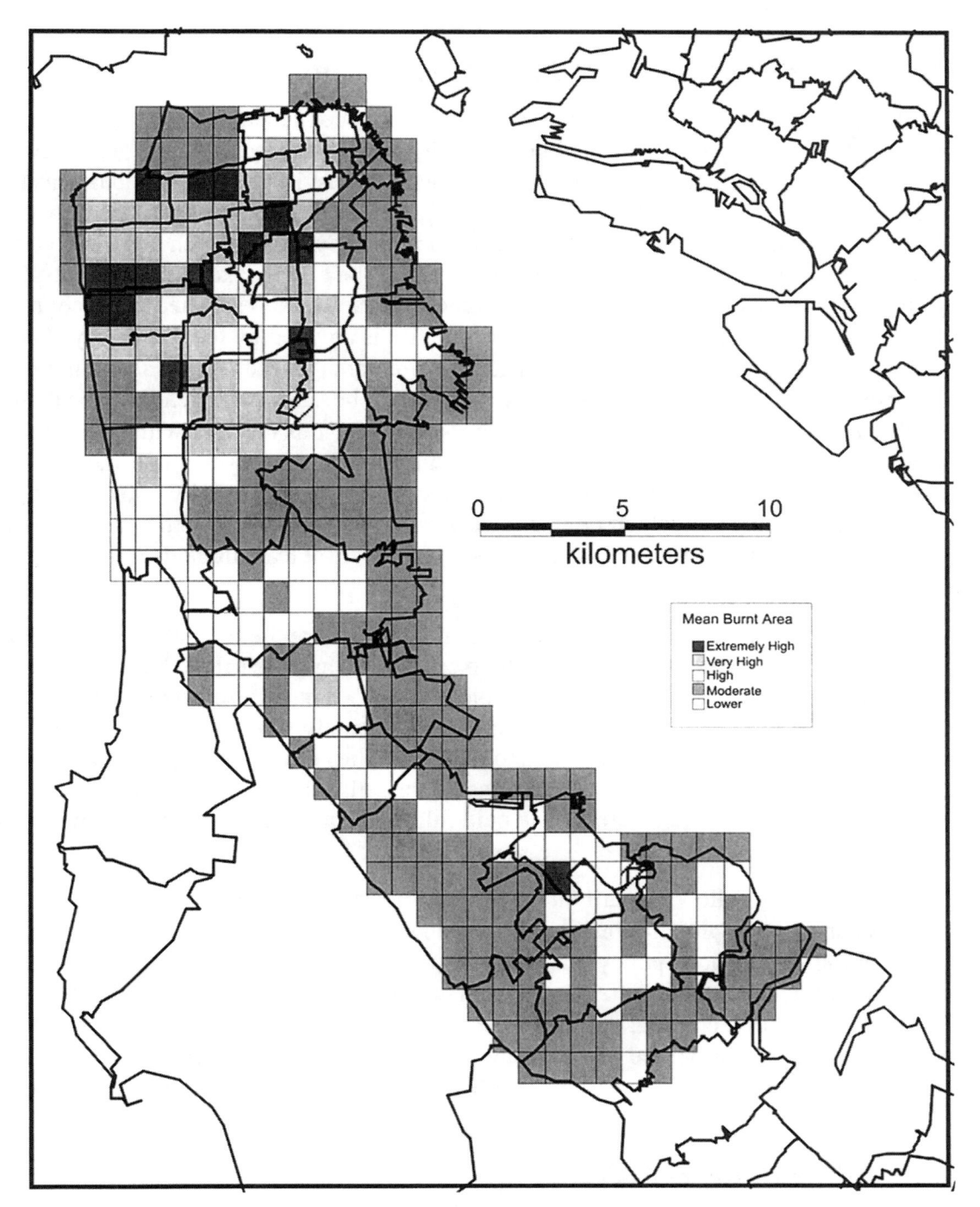

Figure 4-20 Geographical Distribution of Mean Burnt Area for the 1906 San Francisco Earthquake (RiskLink Simulation)

4.7 Insurance Aspects of Fire Following Earthquake

Fire following earthquake is of especial interest to the insurance industry for a very simple reason – earthquake shaking damage is paid for by insurers only when they have explicitly included earthquakes as a peril in the insurance contract – that is, issued an earthquake insurance policy. This is a relatively infrequent coverage. In California, for example, perhaps only 20% to 30% of homeowners purchase earthquake insurance, and the coverage is even a lower fraction in commercial and industrial policies. In Japan in the 1995 Kobe earthquake, probably only 5% of property damage was covered by earthquake insurance. In the US and selected other countries, however, fire damage no matter what the origin of the fire is covered by the fire policy. This means that even if the fire is due to an earthquake, the fire damage (but not the shaking damage) is paid for, from the fire insurance policy. This is important because, while earthquake damage may be widespread, only a small fraction of the property damage is insured, with quite high deductibles (typically 15% deductible in California policies). Thus, only a small fraction of earthquake shaking damage will be paid for by insurance. However, virtually 100% of all properties cover fire insurance, which typically has a very low deductible, so that fire insurers will pay for virtually 100% of fire following earthquake losses, at least in the US and certain other countries[16].

Technically, the fire policy covers only the value of the property at the time of the ignition of the fire – if the building collapsed due to shaking, then caught fire and the rubble burned, then the fire policy would only cover the value of the rubble. After 1906 and in selected other earthquakes, this lead to careful scrutiny of photographs taken after the earthquake, but prior to the advance of the flame front. However, due to the difficulties of differentiating damage due to shaking from damage due to fire, given a pile of ashes, and also due to the emotions in the aftermath of a disaster, most fire insurers have paid full value in the event of fire following earthquake damage.

Thus, the analysis of fire following earthquake is very important to the property insurance industry. This is because a property insurance company can carefully control its risk due to earthquake shaking by judicious policy management, but has much more difficulty controlling its risk due to fire following earthquake (since its mainstream business is the fire insurance business). In the event of a conflagration, a property insurer may be simultaneously confronted with paying for thousands of burned structures, which its cash flow management and re-insurance contracts were not designed to deal with. This topic is discussed further in Scawthorn, Kunreuther and Roth (2002).

4.8 Example: Vancouver, B.C.

The fire following earthquake methodology presented above has been applied in a number of localities, for various purposes, including:

---

[16] Generally speaking, fire following earthquake is insured under the fire policy in English-speaking countries, as a legacy of the Great Fire of London of 1666. Fire following earthquake is insured under earthquake policies in non-English speaking countries, due to a different tradition.

- Assessing the adequacy of the fire service, in terms of the amount and location of resources, such as fire stations, firefighters and equipment, as well as their earthquake planning;

- Examining the reliability of water supply systems to provide firefighting water following a major earthquake;

- In the planning of a metropolitan region, to examine whether major thoroughfares could serve as firebreaks, and how building codes might be employed to reduce the fire following earthquake problem, and

- Examining the potential financial impacts a major earthquake might have on the insurance and finance industry. Note that in the US and many other countries, fire following earthquake is covered under the fire insurance policy, not under the earthquake policy. Therefore, a major conflagration following an earthquake is virtually fully insured, as opposed to major building damage to shaking, which might have little or no impact on the insurance industry. The fire following earthquake is a major concern to the insurance industry, which has supported research and studies in this area for several decades.

An example of analysis of fire following earthquake potential is a recent study performed by the author and colleagues, for the Vancouver, B.C. (Canada) metropolitan region (ICLR, 2001). In summary:

The analysis of potential property loss due to post-earthquake fires in the urbanized portions of the Lower Mainland of British Columbia, included the municipalities and districts of Burnaby, Coquitlam, Delta, New Westminster, City of North Vancouver, District of North Vancouver, Port Coquitlam, Port Moody, Richmond, Surrey, Vancouver, and West Vancouver, which have a total population of 1.76 million, representing 96% of the Greater Vancouver Regional District (GVRD). The Lower Mainland of British Columbia is one of the highest earthquake risk regions in Canada. Seismological research indicates that there are several active seismic sources in the region that have caused destructive earthquakes in the past and are potential sources of future destructive shocks. The region is most affected by earthquakes originating (1) in the shallow crust, (2) on the Cascadia subduction zone, and (3) at depth under Vancouver Island and the Strait of Georgia. Based on the seismotectonics of the region, several *scenario earthquakes* were analyzed, Figure 4-21, including:

- A magnitude 9.0 *megathrust* event on the Cascadia subduction zone,

- Three events of magnitude 7.5 to 7.8 occurring at various locations in the Strait of Georgia, and

- Magnitude 6.0 and 6.5 events randomly placed central to the region of interest (i.e., located beneath the City of New Westminster)[17].

Shaking intensity due to the scenario events varies significantly across the region, ranging from MMI V in areas of good soil to MMI IX in areas of soft soils (e.g., around False Creek, in the

---

[17] Magnitudes are measured on the moment magnitude scale.

City of Richmond, etc)[18]. Because each of these scenario earthquakes has a finite probability of occurrence, and numerous other earthquakes with associated probabilities of occurrence can occur at other locations, all possible earthquakes were also analyzed in a probabilistic analysis, based on the seismotectonics of the region.

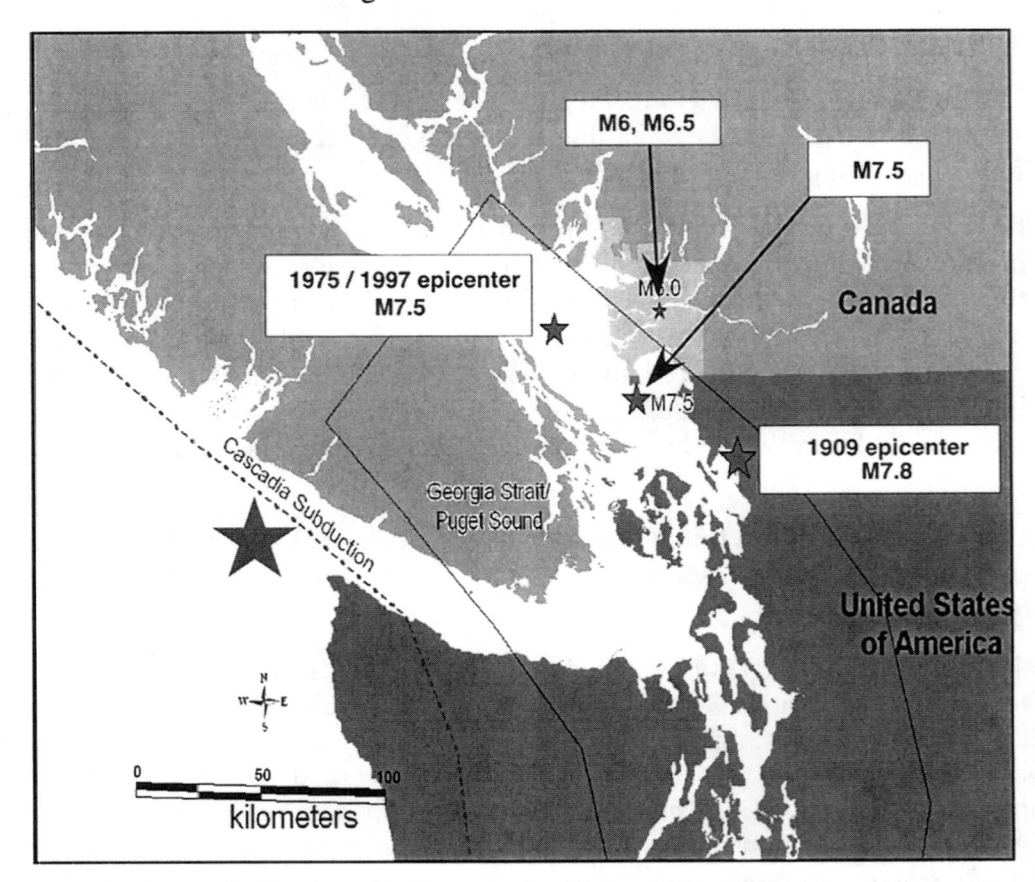

Figure 4-21 Scenario Earthquake Events, Analysis of Fire Following Earthquake, Lower Mainland, British Columbia (Source: ICLR, 2001)

The Lower Mainland of British Columbia is also one of the larger urban regions of Canada, with a high concentration of population and property. An estimate of total property values for buildings, vehicles and infrastructure is approximately $260 billion[19]. Wood buildings represent approximately 58% of the building stock by value, and 65% by total floor area. There are about 1,200 highrise[20] buildings in the region, only about 20% of which are sprinklered. Of significant concern however, is that very few of the highrise buildings in the region have on-site firefighting water supply, in contrast to highrise practice in the US, where highrise buildings in high

---

[18] MMI is the Modified Mercalli Intensity scale, where VI is the initiation of damage to poorly built buildings, and IX is collapse of such buildings, with significant damage to average to good buildings. Corresponding peak ground accelerations range from about 0.02g to 0.45g.

[19] Unless otherwise noted, all monetary amounts are 1999 Canadian dollars.

[20] Definition of a 'highrise' building varies – in this analysis highrise was defined as 7 stories or greater, which is a definition widely employed in the fire service. This estimate is conservative, and there is quite possibly more than 1,200 highrises in the study region.

seismicity areas like Vancouver typically have a 15,000 gallon or larger on-site dedicated water cistern.

The analysis of potential losses due to fires following earthquakes in this example followed the methodology discussed above, in which the process of (a) earthquake occurrence, (b) resulting intensity of shaking at each location in the region, (c) structural damage and resulting probability of ignition, (d) fire discovery and reporting, (e) fire service response and effectiveness given damage to water supplies as well as other demands on the fire service, and (f) fire spread, is treated as a stochastic process, with selected key variables treated as random variables. These random variables' uncertainty is quantified on the basis of data from earthquake and non-earthquake events. The geographic unit of analysis for the region was 3-character postal code (e.g., V6E xxx) – that is, the analysis treated each of the eighty (80) 3 character postal code in the Greater Vancouver region as having uniform shaking intensity, building inventory and other characteristics within that postal code area.

To determine the availability and effectiveness of firefighting and water resources, data was collected from a variety of sources including interviews with key fire service and water department personnel, Fire Underwriters Survey municipal grading summaries, and engineering analyses of water department seismic vulnerabilities, finding that:

- **Firefighting**: The Greater Vancouver region is well protected under normal circumstances by a well-trained and equipped fire service, largely professional but in part volunteer, with a total firefighting staffing of about 2700, of whom about 400 are on duty at any one time. The total number of fire engines in the region is 120, which are housed in 77 fire halls or stations, a significant fraction of which may be seismically vulnerable. In addition to fire engines, additional pumping capacity is furnished by five 'fast attack' fireboats sited around Burrard Inlet (equivalent to five fire engines) and, in the City of Vancouver, the recently constructed Dedicated Fire Protection System (DFPS – discussed further below) and associated hose tenders and special pumps (from a pumping capacity perspective, DFPS can be equated to about 8 fire engines, although its effectiveness is much greater).

- **Water**: the region is almost entirely served by the Greater Vancouver Water District (GVWD, part of the GVRD), which has three supplies – Capilano, Seymour, and Coquitlam. The Capilano and Seymour supplies are north of the Burrard Inlet – they supply Vancouver, Burnaby and other areas via transmission lines that cross the Inlet via submarine crossings that, however, pass through areas subject to liquefaction. The Coquitlam supply is northeast of Vancouver, and no major marine crossings are required to reach the City of Vancouver or a number of other areas. Since the early 1990's, GVWD has been implementing a program of seismic improvements, and is currently conducting a major review of that program to assure appropriate goals are being met. Within each municipality, water is distributed via a gridded network of smaller pipelines, which are anticipated to break in a large earthquake, such that in a number of locations (especially in areas of softer soils), fire hydrants will initially be without adequate water pressure or supply. To prepare for such contingencies, the fire service typically identifies alternative water supplies, and carries special *hard suction* hose and other equipment to access these supplies. Examples of alternative water supplies include water reservoirs, creeks, waterfront access to bays or lakes, and swimming pools. Between Burrard Inlet, the Fraser River and other sources, the study region is generously supplied with opportunities for alternative supply. All fire departments interviewed had equipment for accessing alternative water supplies, but only five out of eight departments appeared to have adequately planned for such contingencies. As noted above, fireboats and Vancouver's DFPS significantly enhance emergency water supply in Vancouver and around Burrard Inlet. However, there are no fireboats located on the Fraser River.

For each scenario, the number of initial fires were estimated based on shaking intensity, and fire department response estimated taking into account effects of the earthquake on transportation, communications, water supply and related factors, Figure 4-22, and Table 4-17. Fire growth was estimated for each fire taking into account fire suppression (or the lack thereof), to arrive at an estimate of final burnt area, for each fire and the sum of all fires. Loss is estimated as the monetary amount corresponding to this final burnt area[21].Based on this data and analyses for the scenario events, it was estimated that the Greater Vancouver region will sustain fires and losses as indicated in Table 4-17.

Uncertainty associated with these estimates is substantial, with coefficients of variation on the order of 0.5 to 1.0. Taking into account all possible earthquakes together with their associated probabilities of occurrence, it was found that the annualized loss for the study region was $ 88.3 million, representing 0.04 % of the value at risk. These losses are of similar relative magnitude to previous estimates for other North American urban areas (NDU, 1992).

With regard to the conditions found in the Greater Vancouver region, it was found that reduction of post-earthquake fire risk could be accomplished via a number of measures, including:

- Development of a regional earthquake plan with a detailed post-earthquake fire element, which would include procedures for rapid location and reporting of post-earthquake fires, mutual aid response to larger fires, identification of alternative water supplies, and related measures.

- Improved provision for alternative water sources, including extension of the DFPS on the other side of False Creek, installation of salt water hydrants along shorelines in higher density areas, provision of fireboats on the Fraser River, and acquisition of additional portable pumps by several of the region's fire departments, similar to the HydroSub recently acquired by the City of Vancouver.

- Installation of water cisterns in sprinklered highrises, and retrofitting of unsprinklered highrises with sprinklers.

- A regional effort at reducing post-earthquake fire ignitions – for example:

    o Vancouver's central business district (CBD) is an enormous concentration of value, yet it is the only major North American city with overhead electric transmission lines. Wood pole mounted transformers abound in the CBD, in many cases only inches from commercial buildings (this is discussed further below).In

---

[21] Note that shaking damage was not quantified in this analysis, and that shaking loss and post-earthquake fire losses cannot be directly summed, as there may be some 'burning of the rubble'.

past earthquakes, pole mounted transformers arced and exploded. Because these were typically in residential areas, few ignitions occurred, but it is expected that many ignitions would result in Vancouver's CBD due to these electrical sources. Undergrounding of the electrical transmission in the CBD would significantly reduce post-earthquake fires.

o   Fire departments could institute public education as to post-earthquake fires, the need for their rapid suppression and, specifically, the need to brace residential water heaters.

Table 4-17 Scenario Events, Fires and Losses
(Loss in Year 2000 C$millions, % of total value at risk)

Georgia Strait M=7.5		1909 Epicenter M=7.8		1975 / 1997 Epicenter, M=7.5		New Westminster M=6.0		New Westminster, M=6.5		Subduction Zone M=9.0	
113 fires		31 fires		101 fires		66 fires		157 fires		155 fires	
Loss	%	Loss	%	Loss	%	Loss	%	Loss	%	Loss	%
$ 2,795	1.04%	$ 1,172	0.43%	$ 2,685	1.00%	$ 1,752	0.65%	$ 8,529	3.12%	$ 4,700	1.74%

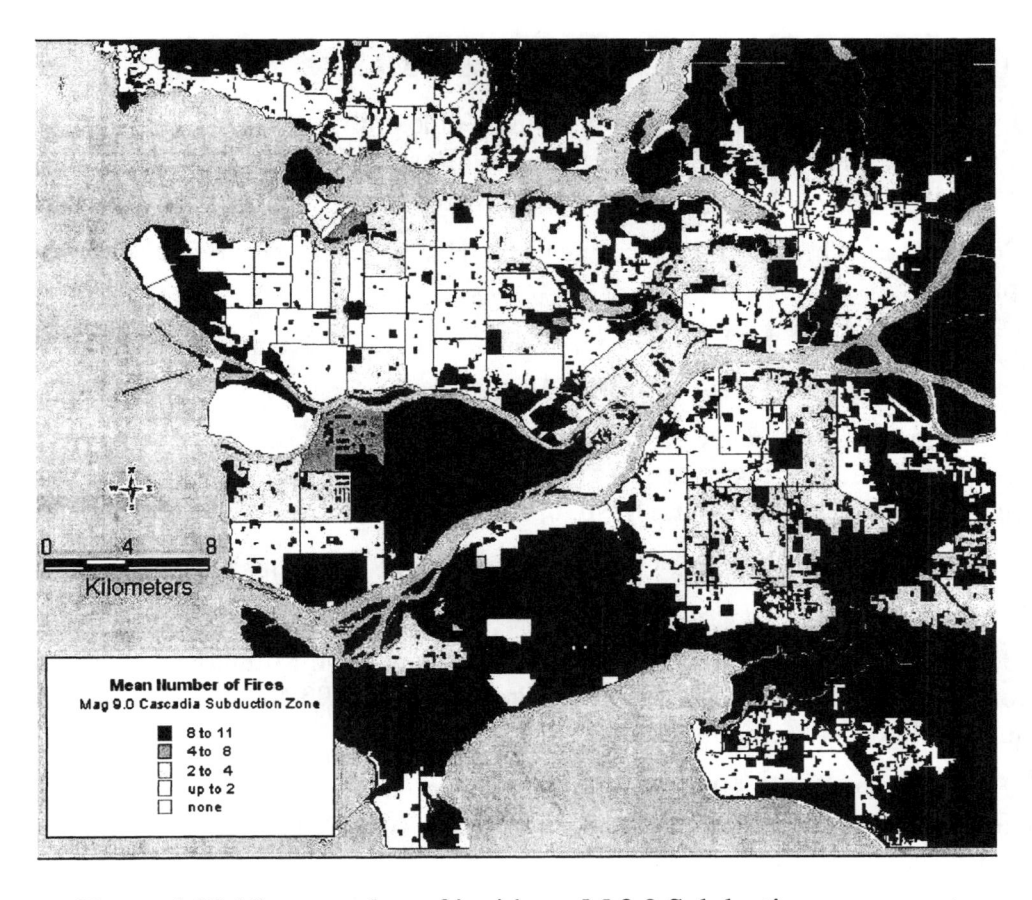

Figure 4-22 Mean number of ignitions, M 9.0 Subduction zone event

4.9 Credits

Figures 4-1, 4-2, 4-3, 4-8, 4-9, Tables 4-4, 4-5, 4-6 Scawthorn et al (1997). Figures 4-4, 4-10, 4-16, Table 4-1 Eidinger. Figures 4-5, 4-6, 4-7, 4-8, 4-20, 4-21. Chris Mortgat. Figures 4-12, 4-13, HAZUS 1999. Figure 4-14, URAMP^{SC} (2002). Figure 4-15, California Department of Forestry. Figure 4-22, ICLR (2001). Table 4-8. NYCEM (2000). Table 4-9. FEMA (2001). Tables 4-10, 4-11, 4-12, 4-13. URAMP^{SC} (2002).

4.10 References

AIRAC. 1987. Fire Following Earthquake, Estimates of the Conflagration Risk to Insured Property in Greater Los Angeles and San Francisco, prepared for the All-Industry Research and Advisory Council, Oak Brook, IL by C. Scawthorn, Dames & Moore, San Francisco.

ATC. 1992. A Model Methodology for Assessment of Seismic Vulnerability and Impact of Disruption of Water Supply Systems. Report ATC-25-1, funded by the Federal Emergency Management Agency. Prepared for Applied Technology Council by EQE International, C. Scawthorn and M. Khater, Redwood City.

Ballantyne, D.B., and C.B. Crouse. 1997. Reliability and Restoration of Water Supply Systems for Fire Suppression and Drinking Following Earthquakes, Report NIST GCR 97-730, Building and Fire Research Laboratory, National Institute of Standards and Technology, Gaithersburg, Maryland 20899

Bond, H. 1946. Fire and the Air War, N.F.P.A., Boston, MA.

Bouabid J. et al. 2002. A Comprehensive Seismic Vulnerability and Loss Evaluation of the State of South Carolina Using Hazus: Part I Overview and Results, Proc. 7th National Conf. of Earthquake Engineering, Earthquake Engineering Research Institute, Oakland.

Bowlen, Fred J., Batt. Chf., S.F.F.D., Outline of the History of the San Francisco Fire, Original Manuscript

Chandler, C.C. et al., 1963. Prediction of Fire Spread Following Nuclear Explosions, PSW Forest and Range Expt. Stn. USFS, Berkeley, CA.

Diemer, D., Miller, M., Pratt, D., Lee, D., 1995. Seismic Evaluation and Improvement Program for a Major Water Utility, Proceedings of the 4th US Conference on Lifeline Earthquake engineering, ASCE TCLEE Monograph No. 6, August.

Du Ree, A.C. 1941. Fire Department Operations During the Long Beach Earthquake of 1933, Bull. Seism. Soc. Am., v. 31, n. 1.

Eidinger, J, and Young, J., 1993. Preparedness, Performance and Mitigation for EBMUD Water Distribution System for Scenario Earthquakes, Proceedings, 1993 National Earthquake Engineering Conference, Memphis, TN.

Eidinger, J. M., and Dong, W, 1995. Fire Following Earthquake, in Development of a Standardized Earthquake Loss Estimation Methodology, prepared for the National Institute of Building Sciences by RMS Inc., February.

Eidinger, J., Collins, F., Conner, M., 2000. Seismic Assessment of the San Diego Water Systems, Proceedings of the 6th International Conference on Seismic Zonation, EERI, Palm Springs, CA.

Eidinger, J., Matsuda, Ed, 2003. Seismic Risk Analysis of the Bay Area Rapid transit Systems, Proceedings of the 6th U.S. Conference on Lifeline Earthquake Engineering, ASCE TCLEE, Long Beach, CA, August.

Eidinger, J., Goettel, K., and Lee, D., 1995. Fire and Economic Impacts of Earthquakes, in Lifeline Earthquake Engineering, Proceedings of the Fourth US Conference, TCLEE Monograph 6, ASCE, August.

EQE Engineering. 1990. Emergency Planning Considerations, City of Vancouver Water System Master Plan, prepared under subcontract to CH2M-Hill, for City Engineering Department, City of Vancouver, B.C.

FEMA 2001. HAZUS99 - SR1, 2001. Validation Study, Developed by: Federal Emergency Management Agency Washington, D.C. Through a contract with: National Institute of Building Sciences Washington, D.C. June 1.

Fleming, R. 1998. Analysis of Fire Sprinkler System Performance in the Northridge Earthquake, NIST-GCR-98-736, Building and Fire Research Laboratory, National Institute of Standards and Technology, Gaithersburg, Maryland 20899

Freeman, J.R., 1932. Earthquake Damage and Earthquake Insurance, McGraw-Hill, New York.

Hamada, M. 1975. Architectural fire resistant themes, No. 21 in Kenchiku-gaku Taikei, Shokoku sha, Tokyo (in Japanese).

Hamada, M. 1951. On Fire Spreading Velocity in Disasters, Sagami Shobo, Tokyo (in Japanese).

Hansen, G., and E. Condon, 1989, Denial of Disaster: The Untold Story and Photographs of the San Francisco Earthquake and Fire of 1906. San Francisco, CA: Cameron and Company.

HAZUS. 1999. Earthquake Loss Estimation Methodology, HAZUS ® 99, Service Release 2 (SR2) Technical Manual, Developed by: Federal Emergency Management Agency, Washington, D.C. Through agreements with: National Institute of Building Sciences, Washington, D.C.

Himoto, K. and Tanaka, T. 2002. A Physically-Based Model for Urban Fire Spread, Proc. 7th international symposium on fire safety science, Worcester, June.

Horiuchi, S. (n.d.) Research on Estimation of Fire Disasters in Earthquakes, Disaster Prevention Section, Comprehensive Planning Bureau, Osaka Municipal Government, Osaka (in Japanese).

Horiuchi, S. (nd.) Research on Estimation of Fire Disasters in Earthquakes, Disaster Prevention Section, Comprehensive Planning Bureau, Osaka Municipal Government (in Japanese).

ICLR. 2001. Assessment of Risk due to Fire Following Earthquake Lower Mainland, British Columbia, report prepared for the Institute for Catastrophic Loss Reduction, Toronto, by C. Scawthorn, and F. Waisman, EQE International, Oakland, CA.

Kimball, W.Y. 1966. Fire Attack! Command Decisions and Company Operations, National Fire Protection Assoc., Boston MA.

McCann, M., Sauter, F., and Shah, H. C., 1980. "A Technical Note on PGA-Intensity Relationship with Applications to Damage Estimation," BSSA, Vol. 7, No. 2, pp 631-637.

Manson, M., 1908. Report on an Auxiliary Water Supply System for Fire Protection for San Francisco, California, Board of Public Works, San Francisco, CA.

National Fire Research Institute, 1995. Report on the Hyogo Ken Nanbu Earthquake and Urban Fires in Kobe City.

NDC 1992. Fire Following Earthquake - Conflagration Potential in the Greater Los Angeles, San Francisco, Seattle and Memphis Areas, prepared for the Natural Disaster Coalition by C. Scawthorn, and M. Khater, EQE International, San Francisco, CA.

NYCEM 2000, Earthquake Loss Estimation Study for the New York City Area, New York City Area Consortium for Earthquake Loss Mitigation Year Two Technical Report 1999-2000, by Michael W. Tantala, Guy J. P. Nordenson, George Deodatis, Department Of Civil Engineering & Environmental Engineering, Princeton University Prepared for: MCEER Multi-Disciplinary Center for Earthquake Engineering Research Red Jacket Quadrangle SUNY at Buffalo NY 14061.

Rothermel, R.C. 1983. How to Predict the Spread and Intensity of Forest and Range Fires, Gen. Tech. Rept. INT-143, USFS, Ogden, UT 84401.

Scawthorn, C. Kunreuther, H. and Roth, R. 2002. Insurance and Financial Risk Transfer, chapter in Earthquake Engineering Handbook, Chen, W.F. and Scawthorn, C., editors, CRC Press, Boca Raton.

Scawthorn, C. 1984. Simulation Modeling of Fire Following Earthquake, Proceedings, 3rd U.S. National Conference on Earthquake Engineering, Charleston, SC, Earthquake Engineering Research Institute, El Cerrito, CA.

Scawthorn, C, A.D. Cowell and F. Borden, 1997. Fire-Related Aspects of the Northridge Earthquake. Report prepared for the Building and Fire Research Laboratory, National Institute of Standards and Technology, NIST-GCR-98-743, Gaithersburg MD 20899.

Scawthorn, C. and S. K. Harris. 1989. Estimation of Earthquake Loss for a Large Eastern Urban Center: Scenario Events for New York City. In Earthquake Hazards and the Design of Constructed Facilities in the Eastern United States, edited by K. H. Jacob and C. J. Turkstra, vol. 558, Annals of the New York Academy of Sciences, New York.

Scawthorn, C., Yamada, Y. and Iemura, H. 1981. A Model for Urban Post-earthquake Fire Hazard, DISASTERS, The International Journal of Disaster Studies and Practice, Foxcombe Publ., London, v. 5, n. 2.

Seligson, H.A. et al. 2003. URAMP^{SC} (Utilities Regional Assessment of Mitigation Priorities) – A Benefit-Cost Analysis Tool for Water, Wastewater and Drainage Utilities: Methodology Development, 7th U. S. Conference on Lifeline Earthquake Engineering, Long Beach CA., Am. Soc. Civil Engineers, Reston.

Steinbrugge, Karl V., 1982. <u>Earthquakes, Volcanoes and Tsunamis – An Anatomy of Hazards</u>, Skandia America Group, 280 Park Avenue, New York NY 10017.

SSC 2002. Improving Natural Gas Safety in Earthquakes, Adopted July 11, 2002, Prepared by: ASCE-25 Task Committee on Earthquake Safety Issues for Gas Systems, for the California Seismic Safety Commission, Report No. SSC-02-03, Sacramento.

Takata, A. and Saizberg, F. 1968. Development and Application of a Complete Fire-Spread Model, vols. I to IV, O.C.D., IITRI, Chicago.

Terada, M. 1984. Application of Fire Prevention Engineering for the Disaster Prevention City Plan, Comprehensive Planning Bureau, Osaka Municipal Government, Osaka.

Tokyo Fire Department. 1997. Determinations and Measures on the Causes of New Fire Occurrence and Properties of Fire Spreading on an Earthquake with a Vertical Shock, Fire Prevention Deliberation Council Report, Page 42, March.

URAMP^{sc}. 2002. URAMP^{sc} Technical Manual, prepared for the California Governor's Office of Emergency Services, by ABS Consulting and ImageCat, Inc., August.

Woycheese, J.P., Pagni, P.J., and Liepmann, D. 1999. Brand Propagation from Large-Scale Fires, 1999. J. Fire Protection Engineering, v. 10, n. 2, pp. 32-44.

5 Fire Department Operations Following Earthquake

5.1 Introduction

This Chapter discusses selected aspects of fire department operations following an earthquake, particularly a large earthquake.

Fire department operations can be divided into emergency and non-emergency operations. Emergency operations include responding to structural and other fires, hazmat incidents, building collapses, transportation and vehicular accidents, rescues and generally almost any non-law enforcement emergency. Many, indeed most, fire departments today also provide emergency medical services (EMS), either as first responder prior to the arrival of an ambulance, or as operators of the municipal ambulance service. Non-emergency fire department operations include fire prevention, inspections, investigations and perhaps, code enforcement.

This Chapter does not discuss 'normal' fire department emergency operations, firefighting or other day-to-day aspects of fire department operations and administration. Rather, we assume some knowledge of these aspects on the part of the reader. For more detail on fire department operations and administration, the reader is referred to standard references (e.g., Bryan and Picard, 1979; ICMA, 1967) and/or individual departmental standard operating procedures (e.g., FDNY, 1997), many of which are available online (e.g., CFD, n.d.). Instead, this Chapter focuses on the special operations and demands on a fire department, typically a large urban fire department, in fighting fires following a major earthquake, particularly a large urban conflagration.

5.2 Nature of the Problem

Following a major earthquake, the demands on a fire department are in many ways no different than at other times – the fire department will still be called on to fight fires, respond to hazmat incidents, perform search and rescue at building and other collapses, and provide EMS. Each of these individual emergency operations may in many ways be no different following an earthquake than at other times – for example, fire departments operations normally include fighting fire in one building and occasionally in multiple buildings. Training and Operations procedures are set up for all these incidents.

However, an earthquake is a qualitatively different experience for a fire department, in several fundamental ways:

- Most importantly, an earthquake results in multiple simultaneous major incidents. Fire departments are accustomed to dealing with major incidents, even several at the same time. Large urban fire departments may even have experience with multiple simultaneous major incidents, where multiple means more than a few. An earthquake, however, can result in dozens if not hundreds of major incidents, all at the same time, to the extent that the safety not of some citizens, or a localized area, but of the entire metropolitan region is at risk.

- At the same time (simultaneously) as the fire department is called on to respond, its capability is under attack and degrading, via lost water supplies, impaired communications, impeded access and other factors which are largely unknown. Compounding this is the element of surprise that is intrinsic to earthquakes, and also the threat of aftershocks. Most if not all of this is unique – it is a very rare experience which most fire service commanders have not had, although their years of experience responding to 'normal emergencies' are the best available training for this situation.

As a result, following a major earthquake, fire department operations must substantially change. Under normal conditions, a fire department's emergency response to a major incident may include:

- Multiple alarms – that is, availability and application of significant resources which can be focused on the incident

- Mutual aid – that is, support from neighboring jurisdictions

- Plentiful supply of water, including the response of water department to increase pressures as needed

- Unimpaired communications

- Response of gas and electric utility personnel, to shut off services to buildings, thus cutting off ignition sources and fuel, and increasing personnel safety

- Emergency right of way on city streets

In a major earthquake, these conditions no longer hold:

- There may be more incidents than fire companies, so that some incidents can't be immediately responded to

- Mutual aid will break down, at least in the initial emergency period. Any earthquake will impact neighboring jurisdictions, who will have their own incidents and won't be able to provide mutual aid. Even if they have spare resources, concerns over possible additional shocks may cause neighboring fire departments to retain their assets

- Water supply will invariably be damaged – this damage will often be greatest in the same neighborhoods where building damage is greatest and the most fires occur, so that no water will be available just where its needed the most

- Communications will break down, so that the department's internal communications, and also with other city agencies and utilities, will be disrupted. Utility personnel won't be able to respond, and gas will fuel fires, and electricity may make firefighting more hazardous

- Traffic on city streets will be chaotic, with signals inoperative and everyone trying to get home while at the same time bridge damage or collapsed buildings will block selected routes.

Because the earthquake threatens the overall safety of the entire City, the above elements in fire department operations, which are usually taken for granted, must now be considered. All of these factors dramatically alter the way a fire department approaches the crisis it faces.

5.3 Departmental Command Considerations

As soon as an earthquake is felt, a number of issues immediately confront the senior management of a fire department. These issues and the necessary response include:

What has happened? That is, the senior management must confirm that it was an earthquake, and find out as quickly as possible the approximate extent of the affected area. The quickest and most effective way to do this via a **radio check** – part of the departments Standard Operating Procedure as soon as an earthquake is felt, should be:

- All Companies should move their apparatus out of the fire stations, and deploy to a pre-determined safe haven (e.g., station driveway or down the street, next to a park, school yard, or other open area)

- Each Company should then perform a radio check – that is, each fire department Company calls in to the department's Dispatch or Call Center, to:

    o Confirm communications are functional and, as concisely as possible,

    o Report it felt what the officer in charge believe was an earthquake, and its approximate intensity, according to a standard terminology (e.g., weak, moderate, strong, very strong – not '*wow, it really shook*')

    o Report the Company's status and availability for service

    o Report any immediately observable signs of damage – this will typically either be buildings in the immediate vicinity, or more distant clouds of smoke (i.e., fire) or dust (i.e., collapsed structure)

- A common mistake that should not be made is for fire Companies to undertake a patrol of their first-due area, to try to find damage. The fire department should not be tasked in the Emergency Plan with patrol or damage survey responsibilities following a major earthquake, since (a) the fire department's resources will be committed to emergency response very quickly following the earthquake, so that they should not be distracted from their response, and (b) because of their rapid commitment, the survey will not be completed. A rapid damage survey of the City is an important task, but it should not be assigned to the fire department. Rather, it should be assigned to the police department, traffic department, building department, or a combination of these and other municipal agencies

- While each earthquake is unique, movement of apparatus to safe havens and a Company radio check can most likely be completed with 2~3 minutes following the cessation of shaking

Emergency Activation and Size-Up. In a major earthquake, the fire department and its City or Jurisdiction's Emergency Plans will be invoked, and the Emergency Operations Center (EOC) activated. A designated senior chief will report to the City EOC, and other chiefs and personnel will report to the department's EOC, which is typically the dispatch or call center (unless that is damaged). By this time, radio checks will have commenced or been completed, and the Incident Commander (IC) will need to consider the following:

- Size-up the Incident - what is the fire and rescue situation in the City? That is, what Companies (i.e., parts of the City) have reported major shaking, and major fires, building collapses or other incidents? What Companies (parts of the City) have not reported yet? Why not?

- Status of the fire department - what units are available, what units are already at incidents, and what stations or other department facilities are damaged?

- Communications – what is the status of the department radios – are all frequencies functioning, are any repeaters or other equipment damaged, what parts of the City are out of touch? Is the dispatch center receiving 911 calls, to confirm the POTS (plain old telephone system) is operating?

- Media – is the television functioning, and what is it reporting? Often, television helicopters will be able to provide very rapid aerial reconnaissance, although it may exaggerate the situation due to its focus on the dramatic. Nevertheless, TV does provide a visual image of selected incidents. Also, within a few minutes, even if the local stations are disrupted, the networks will be in touch with USGS (in Colorado) or other seismological stations, and be broadcasting the earthquake's magnitude, epicenter, and approximate affected area. This is useful information for the IC, since it will inform him as to the potential for mutual aid, and how soon it might be arriving

- While each earthquake is unique, the majority of radio checks should have been completed, and an initial size-up of what has happened, the department's status, and communications functionality can most likely be completed with 5~10 minutes following the cessation of shaking. Invocation of the Emergency Plan and activation of the City's and department's EOC should also occur during this period, although it may take somewhat more time for the EOC to be fully functional

Prioritized Response Mode. At some point, and very early, the IC will be faced with a critical decision, which is to start committing the department's resources, or to hold some or all of the department's resources in reserve until the situation is better understood. In actuality, this decision is not clear-cut – no one will ask the IC whether to commit, or hold in reserve. However, the IC must understand that this decision does need to be made. To begin with, some Companies will be 'free lancing' – that is, self-dispatching to the first injury, fire or damaged building they see. Additionally, dispatchers will be doing what they have been trained to do, and been doing for years, which is to send available units to fires and other emergencies as they are reported. As a result, within a few minutes and unless the IC sizes up the incident and changes the normal response mode, the entire department's complement of apparatus and personnel will be committed. This is very dangerous, because the department can be entirely committed within a few minutes following a major earthquake, and then discover that one or more fires has evolved into a major conflagration. Therefore, IC needs to recognize that the earthquake has become a major incident, and must change the department's normal response mode, and commence a new response mode. This new, prioritized response mode, should be based on the following priorities:

- **Life Safety** – First priority for response is to assure the safety and well being of persons involved in the emergency. Involved persons include victims, bystanders and emergency personnel
- **Control** - Next priority for response to stop the spread or growth of an emergency incident, and bring about its final termination
- **Property Conservation** – Least priority is for those operations or activities required to stop or reduce additional loss to property

The application of these priorities is not as simple as may at first appear. A simple example of this is the choice between sending fire department units to a building collapse with entrapped victims, or to a spreading fire. Since life safety is the highest priority, it would appear that the units should respond to the building collapse, to free the trapped victims. However, if the fire has the potential for spreading to the collapsed building, then it may be better to dispatch the units to the spreading fire, to control the fire so that it doesn't spread to the collapsed building, thereby killing the trapped victims.

Therefore, the prioritized response mode involves:

- Dispatchers taking reports of incidents, but dispatching only to those incidents where life safety is clearly at risk
- When dispatching units, only dispatching the minimum resources required for an initial response. That is, typically, one fire engine or ladder company will be dispatched, to size up an incident, and search for and perhaps extricate victims.
- Dispatch supervisors and supervising fire chiefs monitoring reports of fires and hazmat releases, to identify the potential for a large spreading fire (i.e., conflagration) or toxic cloud

The fundamental goal of the prioritized response mode is the conservation of fire department resources – that is, the avoidance of their dispersal or too early commitment to non-essential tasks.

Other tasks for the IC in this initial period are:

- Initiate a recall of all off duty personnel
- Request mutual aid, to the extent it is available
- Mobilize alternative resources. In a major earthquake, the IC should be resourceful and creative in identifying and mobilizing non-traditional resources. Examples of these include alternative water carriers (e.g., milk company and other tanker trucks can be used to carry water), heavy construction equipment (e.g., large front end loaders can be used to create firebreaks, through residential neighborhoods if necessary[22]).
- Contact county, regional or state Office of Emergency Services or equivalent, if possible, to request additional aid

---

[22] see Scawthorn and Donelan, 1984, where a front-end loader was used to remove a building and create a firebreak, following the 1983 Coalinga earthquake.

- Assure that situation reports continue to be made and logged, so as to maintain a current understanding of the evolving incident:
    - What structures have collapsed, where are they, how many victims might be trapped, what resources are on-scene?
    - What fires are there, where are they, what resources are on-scene, do the fires have the potential for conflagration?
    - What hazmat releases are there, where are they, are they spreading, of what nature, what resources are on-scene?
- Returning to the questions of *What Companies (parts of the City) have not reported yet?* and *Why not?*, the IC should make an effort to confirm that at least some understanding of the situation exists for all parts of the department's jurisdiction. To the extent that radio checks haven't provided this intelligence, the IC needs to employ alternative means. Dispatching fire department Companies to non-reporting areas should not be done. In the absence of other means, the fire department should request police department or other city resources to undertake reconnaissance in non-reporting areas. For example, the City of San Francisco's emergency plan utilizes Traffic Department and Parking Meter Enforcement personnel for this purpose. Parking Meter Enforcement have radios, mobility via their small 'carts', and local knowledge and expertise, so that they are an excellent resource for patrolling.

While each earthquake is unique, this period of prioritized response mode will most likely continue for a period of from several hours, to several days. Several hours have been the experience in California in recent events (e.g., 1989 Loma Prieta and 1994 Northridge, both of which however have been judged to have been 'moderate' events). In a very large earthquake, the period of prioritized response mode could extend for 24 to 72 hours. Kobe in the 1995 earthquake was probably in this operational mode for about 24 hours, and Mexico City in 1985 was in this mode for perhaps 48 hours.

Major Fire Response. In certain situations, one or more fires may grow, or have the potential to grow, into a conflagration. As discussed in previous Chapters, this will most likely be in areas of high building density, often of wood buildings, perhaps under high winds and/or dry conditions. The IC must be alert to the potential for a conflagration. As soon as the potential for a conflagration is perceived, the IC must consider the following actions:

- Identify where the fire currently is, its size, in what directions is it moving, and how fast?
- Identify available resources for combating the fire. In most jurisdictions these resources will be fire engine Companies, and perhaps hose tenders and tank trucks. In selected jurisdictions, other resources might include helicopter or fixed wing aircraft capable of water drops. The IC must identify how many fire engine Companies and other resources are on-scene and otherwise available. For those resources available but not on-scene, where are they, what is their mobility and access, and where might they meet up with the advancing fire?
- Identify available water supplies for use by the fire engine Companies. In most jurisdictions, the normal water supply for fire suppression is the municipal water supply

system. The IC must ascertain the reliability of the water system following the earthquake. Typically, there will be two sources for this information – (a) reports from fire department Companies on-scene, reporting that the fire hydrants are either functioning satisfactorily, or are dry, and (b) reports from the water department, as to the condition of their system. Alternatively, are there bays, rivers, lakes, reservoirs, swimming pools or other alternative water supplies that can be drafted from?

Based on these considerations, the IC must identify **where** the conflagration can be fought, and deploy resources accordingly. A key factor in this is, what broad avenues or other linear firebreaks are in the path of the fire, particularly downwind, where a stand might be made? Responding fire engine Companies need to be able to arrive along this defensive line prior to the fire, sufficiently before so as to be able to connect to hydrants or other water supplies. Prior to committing, an adequate water supply at this defensive line needs to be confirmed. Since branding is a hazard[23], several fire engine Companies need to be available for brand patrol downwind of this defensive line.

A major urban conflagration is a rare experience in modern times. The duration to control of a major urban conflagration can vary significantly. In San Francisco in 1906, this required about three days, although this would not be considered a modern event. The 1991 East Bay Hills fire raged for about 10 hours, and was declared contained after 48 hours (Routley, n.d.). Fires in the Kobe earthquake burned for about 24 hours.

Other Responses. The above discussion has covered only two key issues of many that the senior management of a fire department will have to respond to following a major earthquake. Other responses include hazmat release, structure collapse and associated search and rescue, etc. These other issues are all important, and further burden the senior management, but are beyond the scope of this Chapter.

5.4 Conflagration Tactics

While the above section discussed key issues for the overall command of a fire department following a major earthquake, there are other issues in a major earthquake which must be dealt with at the Company or Battalion level. These issues and considerations for response include:

The Earthquake. An earthquake is a surprise and a rare experience. Most people reach the conclusion that an earthquake has occurred via a process of elimination, and a fire department Lieutenant, Captain or Battalion Chief will be no different. In most cases, shaking in a major earthquake, while it may seem a lifetime, will have a duration of anywhere from a few seconds to perhaps as long as a minute. This is long enough for effective action on the part of people, after their initial surprise, and the most prudent action is to assure one's own safety. If the Company is in quarters (i.e., in the fire station), the key issue is '*is the fire station seismically reliable?*'. In most California larger urban fire departments have had engineering seismic reviews of their facilities, and have strengthened or retrofitted the most deficient. Fire officers

---

[23] Brands are pieces of burning material, such as pieces of burning roofing, which are carried aloft and downwind. When they fall to earth, they start fires. In a large conflagration, brands can number in the thousands and be carried a mile or more downwind. Standard procedure for a conflagration is to have several or more fire engine Companies perform Brand Patrol downwind, looking for and dousing brands as they fall to earth.

should know if their station has been reviewed, and whether it is seismically reliable (i.e., structurally reliable), or not. If the station is sound, then each firefighter should take actions to avoid personal injury from falling or flying objects. <u>The apparatus should not be approached during the shaking</u>, since the earthquake ground motions will be causing rocking of the apparatus, which can lead to major injury[24].

When the shaking has stopped, the fire apparatus should be removed out of doors to a pre-determined safe haven, such as the apron of the station, or adjacent to nearby open ground, such as a park or schoolyard. <u>Before leaving the station, load available hard suction hose on the vehicle</u> (this is discussed further below), as well as spare tools and equipment. As soon as possible, a radio check should be performed, as described above. A key element of this radio check is a visual surveillance of the observable neighborhood, for signs of fire, damage or human injury. Such observations, including the lack of any damage or injury, should be concisely reported in the radio check. Following the radio check and prior to any dispatch, the officer should provide comfort and assurance to citizens, and continue observations for signs of smoke (i.e., fire) or rising dust clouds (i.e., collapsed structures). If such signs are observed, the officer should of course report these to dispatch, and must use judgment as to whether to self-dispatch. As emphasized in the previous section, a danger for the department is the premature commitment of resources.

Initial Structural Fires. It takes time for a conflagration to develop, although high winds can greatly shorten this time. The initial stage following an earthquake is the first stage of fighting a conflagration – before it has developed, via suppression of individual structural fires before they can grow into a conflagration. The fundamental issue is the large number of individual fires all at the same time – too many for them all to be responded to, as they would under normal conditions. Because some fires are not responded to, they grow into multi-building fires, and then have the potential for conflagration. Thus, firefighting tactics following a major earthquake have as their highest priority, after life safety, rapid isolation of the fire. Property preservation is a low priority.

Extensive deployment of hose lines, and interior attack, are to be avoided if at all possible, and use of <u>the engine monitor is the preferred mode of attack</u>. Because time is of the essence, deployed hose lines will often be abandoned – there isn't enough time for hose pickup – the engine must move on to the next fire. After several such incidents, shortage of hose can become an issue.

> *Building triage is another tactic – in selected cases, a building can be left to burn, if (a) the potential for fire spread is minimal, or (b) the building is past saving, but spread to the neighboring building is likely. That is, suppressing fire in a burning building is not the goal anymore – isolation of the fire is the goal (assuming life safety is not an issue). In this regard, the following guidelines drawn from wildland firefighting, can prove useful (Brown, 1994):*

---

[24] In past earthquakes, parked fire apparatus have been observed to literally have their tires leave the ground. This is due to the ground motions 'tuning in' on the chassis suspension, greatly amplifying the rocking of the apparatus.

- *The foremost predictor of structure survivability is the composition of the roof. Experience argues that if a roof is starting to burn, the structure is probably not salvageable (Perry, 1990)*

- *The second most important predictor of structure survivability is the presence or absence of adequate defensible space (Perry, 1990). Structure triage is concerned with protecting structures from approaching fire. Numerous sources agree that the minimum radius of defensible space should be 30 feet.*

- *The third most significant structure-survivability predictor is a combination of slope and terrain. NFPA statistics, based on case studies, predict an unsuccessful outcome for structure defense when slopes surrounding the structure exceed 20 percent (NFPA, 1991). This is especially true considering the difficulties sloping terrain will cause firefighters in stretching and maneuvering hose lines.*

Other physical features of structures and land having great influence on the probability of success (or failure) in structure-protection operations include the following (Brown, 1994):

- *Access roads and driveways (dead-ends, length, width, slope, grade, surface, turnarounds).*

- *Exterior construction (noncombustible, fire resistive, or combustible).*

- *Projections, overhangs, and stilt construction (decks, eaves, etc.).*

- *Windows and other glazed openings (size, thickness, and protection).*

- *Vents and other openings into attics or foundations (presence or absence of screens).*

- *Fuel loading on land adjoining defensible space (type and amount of vegetation).*

- *Fuel stored within the defensible space (firewood, LPG, etc.).*

- *Aboveground power lines crossing over structures or defensible space.*

Water Supply. It is a given that a major earthquake will result in some damage to municipal water supplies, if not render them totally inoperable. Water distribution systems and their seismic performance are addressed in some detail in Chapter 6, but it is worth noting here that the Company Officer should be prepared for dry hydrants. As discussed in Chapter 2, this has been the case in many earthquakes, and fire service personnel have proven resourceful in accessing and using alternative water supplies. Drafting from swimming pools (as in the 1994 Northridge earthquake), nearby creeks, ponds, lakes or bays, industrial tanks, or sewers can sometimes provide adequate amounts of water, albeit with extra delay for set up, laying hoses, etc. Most fire departments today do not routinely carry hard suction hose lengths on engines, but have it in the station. Before departing the station, all available hard suction hose lengths should be loaded on the vehicle, to permit drafting.

Communications. Under the stress of the earthquake, and the numerous incidents occurring simultaneously, it almost certain that radio traffic will be congested. An additional complicating factor is possible damage to the department radio infrastructure – loss of antennae, repeaters and other equipment, so that parts of the fire department's jurisdiction may be effectively 'blacked out' for some time. Companies within line of sight of each other should still be able to communicate via radio, and via relay some departmental communications to blacked out may be possible. It is imperative that radio communications is kept to a minimum, and that radio chatter is avoided.

Defensive line. In the event of a large fire or conflagration developing, the choice of a defensive line is critical, and the Incident Commander (IC) may only get one chance for choosing the place to stand. In an urban environment, factors to consider in making this decision include:

- Position – the defensive line must be selected downwind of the fire, so as to be able to stop the fire's advance. Good radio communications should be assured.

- A broad boulevard, freeway, park, school ground or other natural firebreak offers the best opportunity. A width of at least 100 feet is necessary, and 300 feet preferable. In the 1991 East Bay Hills fire under high winds, the fire crossed Highway 24 which at that point was a natural firebreak more than 200 feet wide (the fire spread over the 8 lane freeway by branding).

- Required Apparatus. Each fire engine can protect a maximum radius of about 300 feet (this is the maximum fire stream that a Class A pumper can throw, under optimal conditions). Thus, if an advancing fire front is several city blocks wide, perhaps 1,500 feet wide, a defensive line at least 2,500 feet long will probably be required (additional length to protect the line's flanks). Since something less than optimal conditions will probably prevail, about 200 feet per engine should be allowed, which in this example would mean that at least 12 Class A engines would be needed. Additional engines may be needed to relay water to the front line engines – depending on circumstances, in many cases as many engines are needed to relay as there are on the line. Several more engines will be needed for brand patrol (discussed further below).

- Concentration – the defensive line must allow sufficient time and access to bring the entire department's available resources to bear in holding the fire. Dozens of fire engines may be involved – they need to be able to get by each other, and other vehicles, and position themselves prior to the arrival of the fire.

- Water supply – a reliable high volume water supply is necessary along the defensive line. In the above example, of 12 front line engines, about 15,000 gallons per minute (gpm) will be required – for example, this is the output of two large typical fireboats. If at all possible, redundant alternative water supply should be available. That is, the IC should not rely solely on one water supply (unless it is particularly abundant and reliable, such as a large lake, or bay). Particular prudence should be used in relying on the municipal water supply – the IC should be aware that as the fire approaches, and fully involved buildings collapse, their service connections will be wide open, each service connection thus constituting an open connection, and a drain on the municipal water supply. Although residential connections are individually small (typically $5/8^{th}$ inch), dozens of collapsed homes can significantly reduce pressure in the surround area. The IC should also be aware that in smaller residential areas in pumped pressure zones, the municipal water system will often be unable to provide flows of much more than 750 to 3,000 gpm, even if undamaged.

- Egress – in the event the stand is unsuccessful, evacuation of citizens, firefighters and their apparatus must be assured. At least two means of egress should be available, and the potential for the fire outflanking the defenders and preventing egress, at several points, should be considered.

- Topography and Wind – flat terrain is best. The defensive line should avoid ridge crests, or the foot of hills, due to unpredictable wind effects. Topography can also affect radio communications.

- Construction – a predominance of non-combustible buildings along the line is best, so that a broad commercial thoroughfare is preferred. Be aware of the predominant construction to the rear of the defensive line – if its primarily wood, especially with combustible roofing, branding may result in the fire jumping the defensive line, and threatening the defenders. Also be aware of especially vulnerable industrial facilities to the rear of the line, such as lumberyards, paint or chemical plants, etc, which are also vulnerable to branding.

Brand patrol. As emphasized several times, a major method for conflagration fire spread is via branding. Brands can result in multiple fires several city blocks (or more) downwind. The IC should be very aware of this, and allocate several fire engines for brand patrol to protect the line's rear.

Citizens. Manpower demands on the fire department's personnel will be tremendous. Fire officers should be prepared to accept assistance from citizens in stretching hose, carrying equipment, etc. The concept of each professional firefighter becoming an 'officer', leading a squad of citizen volunteers, should be considered.

5.5 Planning

It is accepted in the fire service that planning saves lives and is extremely beneficial. This is especially true for a major earthquake, which will probably be the greatest challenge most fire departments will ever face. The above discussion has highlighted a number of factors on which planning and exercises can focus. At a minimum, an effective fire department earthquake emergency plan element should include the following considerations:

The Earthquake. The earthquake emergency plan element should be based on the best available facts and data specific to the fire department and its jurisdiction. The plan or its annexes should not generalize, but rather be specific. This specificity starts with defining earthquake scenarios, which are the foundation for the rest of the earthquake emergency plan element. Likely earthquake scenarios, and maximum intensity events which might be considered, should be readily identifiable in most parts of the US. The US Geological Survey, state geological agencies, and regional university geological and seismological departments and observatories are all excellent resources for this task. The three national earthquake engineering research centers[25] are another regional resource. Figure 5-1 is an example of such specificity – in this case, a map of liquefaction zones taken from the City of Seattle's earthquake plan element.

Company initial response. As mentioned earlier, as soon as possible, each fire department Company should withdraw their apparatus to a safe haven. These should be pre-determined for each Company, and identified in the earthquake emergency plan element.

The Fires. Most experienced senior fire officers should know those areas of their jurisdiction that are 'conflagration-breeders'. These are the more congested districts in a city, especially densely built up areas of combustible construction, in which on a day-to-day basis an above average number of alarms, and multiple alarms, are being struck. Access in these areas will typically be more difficult than average. They will tend to be the older parts of the jurisdiction, with perhaps older infrastructure. Water pressures may be problematic, due to older infrastructure. Knowledge of these districts should be combined with knowledge of those parts of the jurisdiction where shaking is likely to be most intense (drawn from maps such as Figure 5-1), to identify the most likely districts where major conflagrations might occur following a major earthquake. These districts should be identified in senior officers' mental concept of operations, and appropriately indicated in the earthquake emergency plan element. As explained in Chapter 4, an approximate estimate of the number of fires to be dealt with following a major earthquake can be arrived at, using the simple rule of thumb provided in that Chapter, and repeated here in Table 5-1.

Table 5-1 Approximate No. of SFED per ignition, vs. MMI

MMI	VII	VIII	IX	X
1 Ign. Per SFED	12,000	7,000	3,000	1,000

Estimating the number of ignitions that a fire department will have to cope with is arrived at by two approximate rules: (a) that the total building floor area (in SFED) in a typical residential district with some commercial occupancies is about half the population in thousands, and (b) that the number of ignitions is per Table 5-1. Using these two rules, fire departments can quickly estimate the number of ignitions they may be confronted with. For example, for a town of population 50,000 subjected to seismic intensity MMI IX, the total SFED is then about 25,000, and the number of ignitions would be about 8. This rule of thumb can be applied to each, for example, battalion district of a city, and the total number of ignitions the fire department is likely to have to deal with arrived at by summation.

Water supply. A key aspect of the earthquake emergency plan element is the identification, enumeration, mapping and pre-arranging of alternative water supplies. Based on the conflagration prone districts identified as outlined above, each district should have identified

---

[25] The three centers are the Mid-America Earthquake Center (MAE, http://mae.ce.uiuc.edu), the Multidisciplinary Center for Earthquake Engineering Research (MCEER, http://mceer.buffalo.edu/default.asp) and the Pacific Earthquake Engineering Research Center (PEER, http://peer.berkeley.edu/).

alternative water supplies. As already discussed, these may be swimming pools, creeks, ponds, lakes, bays, industrial tanks, etc. Company and Battalion officers should visit and familiarize themselves with these sites, identifying how apparatus can access the water in such a manner to permit drafting. A book of maps and sketches for these sites should be kept in each Company's apparatus as well as be part of the earthquake emergency plan element, and each officer should review these with all firefighters under his command. Periodic exercises should demonstrate and reinforce how to access and draft from these locations. In selected cases, special hardware might be authorized to facilitate accessing these alternative supplies. Examples of special hardware would include Siamese connections to a school swimming pool or industrial tank, special suction connections or hydrants on a pier or shore to access lake or bay water, and fire lanes or ramps to similarly permit apparatus to easily access the waterside. Periodically, the department should exercise distant multi-engine relay pumping, moving high volumes of water a mile or more. These can be readily accomplished by for example drafting from one part of a bay or lakeshore, relaying along the shore road, and discharging back into the water body. Chapter 13 in this volume demonstrates this concept with 5 inch diameter hose, for example.

Firebreaks. Based on where fires are most likely to occur, and the locations of alternative water supplies, the jurisdiction should then be reviewed for feasible firebreaks and lines of defense. The attributes for a good line of defense (width, access and egress, water supply, neighboring construction) have been previously discussed. Company and Battalion officers should review their districts, and mutually agree on several crosscutting boulevards or other appropriate streets or open areas, as lines of defense. They should review what fires these lines would defend against, how water would be provided, where engines would be placed, and what problems traffic, fleeing citizens or other factors might pose. When satisfied as to the best lines of defense, these lines should be mapped out and appropriately identified in the earthquake emergency plan element. This planning and these lines of defense should be reviewed with all firefighters in the fire department.

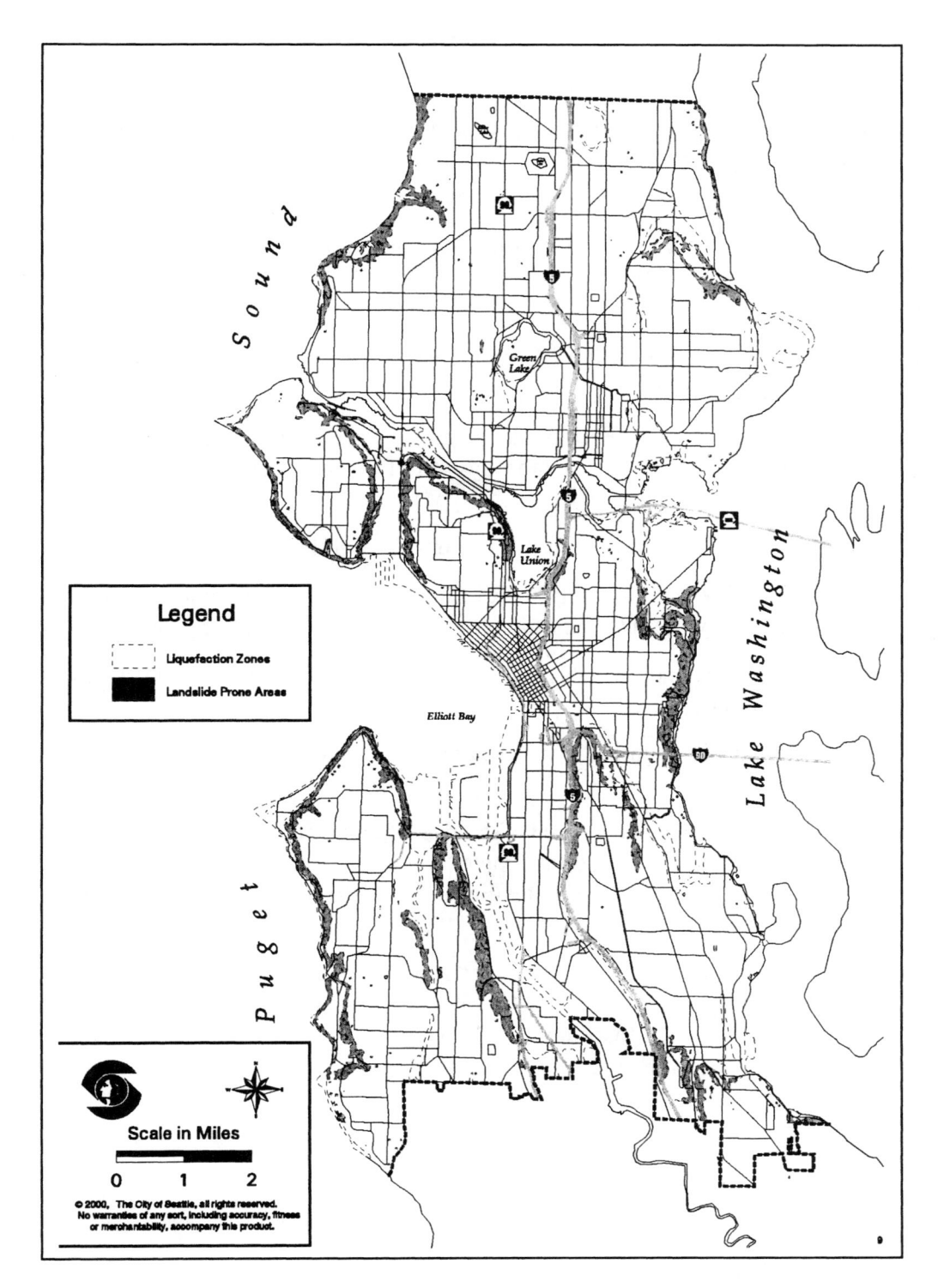

Figure 5-1 Map of liquefaction zones and landslide-prone areas, Disaster Readiness and Response Plan, City of Seattle. Available online at http://www.ci.seattle.wa.us/emergency_mgt/resources/plans.htm

5.6 References and Bibliography

Bryan, J.L., and Picard, R.C., eds. 1979. Managing Fire Services, published for the Institute for Training in Municipal Administration by the International City Management Association, Washington

Brown, K.1994. Structure Triage During Wildland/Urban Interface/Intermix Fires, Strategic Analysis Of Fire Department Operations, Lake Dillon Fire Authority Silverthorne,CO, submitted to the National Fire Academy

CFD. n.d. Standard Operating Procedures, City of Charlottesville, VA fire department, available online at http://www.cfdonline.org.

Chen, W. And Scawthorn, C. 2002. Earthquake Engineering Handbook, CRC Press, Boca Raton (contains chapters on past earthquakes, seismology, fire following earthquake, and emergency planning, for the more interested reader).

FDNY. 1997. Firefighting Procedures, fire department, City of New York.

ICMA. 1967. Municipal Fire Administration, 7th Ed., published for the Institute for Training in Municipal Administration by the International City Managers' Association, Chicago.

National Fire Protection Association. 1991. Standard for protection of life and property from wildfire (NFPA 299). Quincy, MA.

Perry,D.G. 1990. Wildland firefighting: Fire behavior, tactics &command (2nd Ed.). Bellflower, CA: Fire Publications, Inc.

Routley, J.G., n.d., The East Bay Hills Fire Oakland-Berkeley, California (October 19-22, 1991), Report 060 of the Major Fires Investigation Project conducted by TrlData Corporation under contract EMW-90-C-3338 to the United States Fire Administration, Federal Emergency Management Agency.

Scawthorn, C. and J. Donelan. 1984. Fire-Related Aspects of the Coalinga Earthquake. In Coalinga, California Earthquake of May 2, 1983: Reconnaissance Report. Report 84-03. Earthquake Engineering Research Institute, Berkeley.

6 Potable Water Systems

Most urban fire departments rely heavily on potable water systems for fire suppression. Potable water system failure has significantly hindered fire fighting in many recent earthquakes such as the 1989 Loma Prieta (San Francisco), 1994 Northridge, and 1995 Hyoko-Ken Nanbu (Kobe) earthquakes.

Many water utilities are implementing programs to prevent and mitigate earthquake damage to water systems. Eidinger has estimated that cost effective seismic mitigations for U.S. water utilities will ultimately cost about $10 billion (year 2002 dollars). Table 6-1 provides some examples for seismic mitigation programs now underway for various U.S water utilities, and Table 6-2 provides similar data for some larger Japanese water utilities.

Table 6-1 Seismic Upgrade Programs, Various US Water Utilities

Utility	Population Served	Capital Cost	Cost Per Person
EBMUD	1,200,000	$240,000,000	$200
City of San Diego	1,200,000	$46,000,000	$40
Los Angeles	3,500,000	$1,000,000,000	$285
Contra Costa Water District	430,000	$120,000,000	$280
Portland, Oregon	800,000		$10
Seattle, Washington	1,300,000	>$20,000,000	$20
St. Louis, Missouri	650,000	$20,000,000	$30
Memphis, Tennessee	800,000	$20,000,000	$25
County of San Diego	2,400,000	$700,000,000	$290
San Francisco Public Utilities Commission	2,400,000	$1,300,000,000	$540
20 SFPUC Suburban Water Agencies	1,700,000	$45,000,000	$26

Table 6-2 Seismic Upgrade Programs, Various Japanese Water Utilities

City	Population Served	Capital Cost	Cost Per Person
Hiroshima	1,140,000		>$300
Tokyo	11,000,000		~$300
Kobe	1,500,000	$1,360,000,000	$900
Yokohama	3,374,000		>$300
Osaka	2,600,000	$1,000,000,000	$380
Hanshin	2,000,000	$16,000,000	$8
Hachinohe	338,000	$50,000,000	$150

Recognizing that not all US water utilities have completed (or even initiated) seismic retrofit programs, and allowing that the level of seismic retrofit should be consistent with the relative seismic hazard and risk throughout the United States, it is estimated that a country-wide effort for seismic retrofit of water utilities will ultimately cost about $10 billion (year 2002 dollars). This forecast is based on the following projections in Table 6-3.

Table 6-3 US Water System Seismic Upgrades – Forecast

US Area	Population	Cost per Person	Total Cost
High Risk	40,000,000	$225	$9,000,000,000
Moderate Risk	15,000,000	$30	$450,000,000
Lower Risk	68,000,000	$5	$340,000,000
Very Low Risk	157,000,000	$0	$0
Total	280,000,000		$9,790,000,000

At the current time (early 2003), perhaps $1 billion of this $10 billion effort to retrofit US water systems has already been spent. Given the large capital expenditures for seismic upgrades that water utilities are likely to spend over the next 10 to 20 years, it would be prudent that these upgrades be implemented in a fashion to materially improve the capability of these systems to deliver water for fighting fires after earthquakes.

6.1 Potable Water System Facilities and Functions

Figure 6-1 shows a schematic diagram of a typical water distribution system.

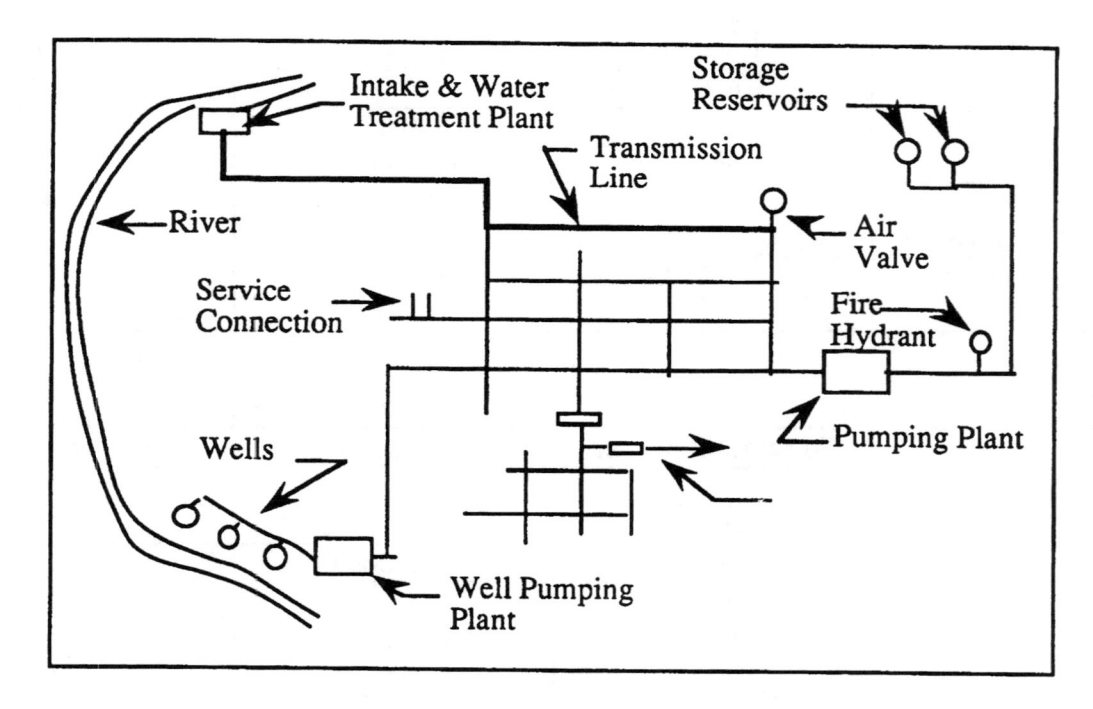

Figure 6-1. Schematic Diagram of a Typical Water System

Although some potable water systems provide water for only a small number of residents, those potable water systems most relevant to the fire following earthquake problem are larger systems that serve thousands or even millions of people. These potable water systems rely on a large number of components. These components consist of water treatment facilities and pumping stations (nodes) located at discrete locations and distributed facilities that may be spread out over a large geographical area, such as pipelines and communication lines. Potable water systems usually consist of the following:

- Water sources such as watersheds, river, lakes, reservoirs and/or underground aquifers,
- Transmission pipelines, canals and/or aqueducts to convey water from the source to the service area,
- Treatment facilities to remove impurities in the raw water,
- Storage facilities such as tanks and reservoirs,
- Distribution system piping to convey the water to end users,
- Pumping facilities to provide adequate water pressure,
- Valves, services, hydrants and other appurtenances to regulate the flow,
- Communication and control systems, and
- Administration, operation and maintenance facilities.

6.1.1 Water Sources

Rivers, lakes and underground aquifers are the primary sources for potable water systems. Intake structures that often only rely on gravity flow are typically used at reservoirs, lakes and rivers. Well water is typically pumped up through a vertical shaft from underground aquifers. In addition to acting as a water supply, water sources can also act as storage facilities to smooth out the variations in the precipitation that feeds these sources.

Water system sources are sometimes located hundreds of miles from the service area so that an earthquake that affects the service area may not affect the source, and visa versa. Within a potable water distribution system, there is typically enough water stored so that loss of a system's source may not significantly affect water availability for fire suppression for at least several hours. Although some systems rely on a single, primary source, most systems either have multiple sources or emergency sources such as inter-ties with neighboring systems. Many utilities purchase water from a wholesaler and do not directly control the source.

6.1.2 Potable Water Transmission

Although some potable water systems have well fields interspersed throughout the system, potable water is typically delivered from the primary sources via large diameter pipelines or open aqueducts. Transmission pipelines are usually 48-inches or larger and may be buried or supported aboveground. More-recently constructed transmission pipelines are welded steel pipelines that are less prone to earthquake damage than older transmission pipelines that used riveted steel, lockbar joints, masonry, concrete or concrete cylinder pipe. Aqueduct and canals are usually concrete-lined, trapezoidal sections.

In addition to the conveyance conduit, transmission facilities often include pumping stations to provide the energy needed overcome grade reversal or supply an adequate flow rate. Because transmission facilities are often spread over hundreds of miles, these facilities may pass through different seismic zones and be exposed to a variety of seismic hazards.

6.1.3 Potable Water Treatment

There are several different methods used to treat potable water. Screens are used to separate out debris. A coagulant such as aluminum sulfate is often introduced to bind suspended particles together so they settle out. Chemicals such as chlorine can be used to kill pathogens. Other chemicals are used, such as lime to control pH, fluoride to fight tooth decay, and carbon to remove particles, control hardness, and improve taste and odor may also be used. Filters can also be used to screen out particles. A newer technology, ozonation, can also be used to kill pathogens.

The most significant aspect of the water treatment process that may affect fire-department response is the release of hazardous chemical. Chlorine may be used at central treatment plants to kill pathogens in raw water and at facilities within the distribution system to maintain a chlorine residual in the water.

6.1.4 Water Storage

Water storage facilities are needed to smooth out seasonal variation in precipitation and demand, diurnal fluctuations in demand (working storage), for emergency purposes such as fire fighting (standby storage) and for operational needs such as providing adequate suction for pumps (operational storage). As mentioned previously, storage for seasonal variations in demand and precipitation is usually provided at the source by lakes, reservoirs and aquifers.

Reservoirs, standpipes and elevated tanks located throughout the distribution system are typically used to meet working, standby and operational storage needs. Reservoirs may include dams that retain water in concrete-lined basins, prestressed, post-tensioned and reinforced concrete tanks and squat (e.g.., height-to-diameter ratio less than one) welded- or riveted-steel tanks. Slender tanks, standpipes, and elevated tanks are usually constructed with welded-, bolted- or riveted-steel. It is still not uncommon for redwood tanks to be in use. In a few instances, standpipes, timber or concrete elevated tanks may still be in service.

Except in flat areas, most water distribution systems are divided into multiple pressure zones to ideally maintain water pressure between 40 and 80 psi in most areas. The storage capacity is usually designed to maintain a minimum pressure (e.g.., 20 psi) in the pressure zone being served by the tank after the working, standby and operational storage volumes have been depleted. A pressure zone's standby storage component is usually based on the water needed to fight a single, large fire located in the water pressure zone for a fixed amount of time (usually 2 hours in small residential areas, or four or more hours in large commercial or industrial areas). Because standby storage requirements are usually based on ignition of a single building or facility, a conflagration that involves multiple buildings or facilities can deplete the stored water. In small pressure zones (less than 2,000 to 3,000 people), fire flow demands can often greatly exceed normal potable water demands; due to the normal small turnover of water in the zone, local storage tanks are quite small, leading to rapid depletion under unusual circumstances like major fires or leakages due to broken pipes caused by earthquakes.

6.1.5 Water Distribution

A network of distribution pipes is used to deliver water to customer service connections. A wide range of materials such as cast iron, ductile iron, galvanized iron, steel, concrete, PVC and polyethylene are used for distribution piping. Newer pipelines, particularly those with restrained joints, built of more ductile materials such as butt welded steel, ductile iron or high density polyethylene, tend to be much more seismic resistant than older or more brittle pipelines. In addition to disrupting the flow of water to services, pipeline failures can drain connected pipelines and reservoirs.

In order to provide water to elevated areas, pump stations, typically reliant on commercial power and often without a backup power source, are used. Valves are located throughout a distribution system to control pressure or provide pressure relief, increase operational flexibility and isolate pipelines when they are repaired or break.

6.1.6 Emergency Operation Centers, Administration, Operation and Maintenance Facilities

The duties performed at emergency operation, administration, operation and maintenance facilities are critical to post-earthquake recovery. System operation and restoration will be directed from these facilities. These facilities often contain engineering drawings needed for restoration, communications facilities needed for coordination and remote operation of water system components, and repair parts and equipment.

6.2 Earthquake Performance of Potable Water Systems

Water systems have often performed poorly in past earthquakes and the lack of water pressure has severely hampered fire-fighting efforts. The following sections highlight some of the more prevalent damage modes to water systems, and a few pictorial examples are provided along with some damage statistics. Section 5.8 provides additional source references for damage to water systems in part earthquakes.

6.2.1 Source Facility Performance

Source facilities have generally demonstrated relatively good earthquake performance when compared to other water system facilities. Potential vulnerabilities include:

- Landslides or other earth movements that can raise the turbidity of surface waters above acceptable levels

- Loss of dam stability

- Contamination of watersheds or aquifers from natural or man-made sources

- Changes in aquifer geology that results in wells running dry

- Sanding or bending of well casings

- Structural and equipment failures of intake structures and facilities

- Loss of commercial power needed to operate facilities

- Seiche in water bodies that damages shoreline facilities (rare)

There may be a day or more of water stored within the distribution system to mitigate the effects if the water source is temporarily lost. However, because of the unique nature of many source facilities, repair may take weeks or months before the source can be returned to its pre-earthquake capacity.

6.2.2 Transmission Facility Performance

Fault crossings and areas subject to liquefaction, landslide and/or settlement are the areas where transmission facility failure is most likely. Buried transmission pipelines are most vulnerable from permanent ground displacements that may occur in these areas. In general, newer welded steel pipelines perform better than older, more brittle pipelines made of concrete or masonry.

In addition to permanent ground displacement hazards, aboveground transmission pipelines and/or their supports may fail from the energy imparted by ground shaking. In addition to fault crossings and areas with poor soils, particular areas of concern are at interfaces with other structures such as bridges that may move differentially with respect to the surrounding ground. Differential movement between different structure components such as bridge decks and abutments can also result in pipe failure.

Canals generally perform well although they may be subject to cracking, embankment failure or blockage by landslides. In addition to structure and equipment failures caused by ground shaking, pumping facilities are reliant on commercial power that is often unavailable after an earthquake.

Even if a water system is supplied by only one primary source, water supply reliability is much better if multiple transmission aqueducts, preferably that use different routes, are used to deliver water to the distribution system.

The following subsections provide a few examples of damaged transmission pipelines (those pipelines with diameter 30" and larger) from a few past earthquakes. Many more examples abound, the bulk of which are well documented in the various references at the end of this chapter.

6.2.2.1 1994 Northridge Earthquake

The Northridge earthquake damaged several transmission pipelines (36" diameter and larger) at a few dozen locations. A comprehensive review of the damage and repair times needed for these pipelines is provided by Lund in Eidinger and Avila (1999). The following highlights a few of the failures:

The Castaic Conduit from the water treatment plant was video inspected and had 35 leak repairs in the 54 inch, 39 inch and 33 inch modified prestressed concrete cylinder pipe.

At the south portal of the Balboa Tunnel, the 170 inch diameter State Water Project-West Branch branches into two 85 inch diameter by 13/16 inch thick welded steel inlet pipes. One of these two parallel pipelines ruptured.

The Los Angeles Aqueduct No. 1, a combination of riveted and welded steel pipe, concrete box and tunnel of various diameters, was damaged at four locations over its 233 mile long length.

Los Angeles Aqueduct No. 2 is a 177-mile long welded steel pipeline with welded bell and spigots joints. At Terminal Hill the 77 inch welded steel pipeline was shut down twice to repair two pulled mechanical couplings and an eight-inch long split in the wye branch stiffener.

FIRE FOLLOWING EARTHQUAKE

Figure 6-2. Wrinkled Los Angeles Aqueduct. This pipe remained in service

A collocation of nine lifelines occurred on Balboa Boulevard, north of Rinaldi Street, in Granada Hills, an apparent tension and compression zone. Located within the street were 3 gas, 3 water, 2 sewer and 1 oil underground lines; 34.5 kV and 4.8 kV power, telephone, and cable TV overhead lines; and ornamental street lighting. Ground movement caused the breakage of some of the underground pipelines, a fire occurred in the street which ultimately burned the overhead lines and five homes.

At this location there was a lateral spread on Balboa Boulevard south of Lorillard Street of approximately 9 to 12 inches. Figure 6-3 shows the compression wrinkling and tear to the 48-inch diameter welded steel water pipe that traversed this lateral spread.

Figure 6-3. Wrinkled 48" Balboa Boulevard Pipeline. This pipe leaked

6.2.2.2   1989 Loma Prieta Earthquake

On Tuesday, October 17, 1989 at 5:04 P.M. a magnitude 7.1 earthquake occurred in the Santa Cruz Mountains of California. The epicenter was about 10.5 miles northeast of Santa Cruz and 18.5 miles south of San Jose. Relatively modest damage occurred to large diameter water conveyance facilities, possibly because there were few such facilities in the areas subjected to the strongest shaking.

The following discusses the damage, method and time for restoration of some large water pipelines after the Loma Prieta earthquake.

A break occurred in a 60 inch modified prestressed-concrete cylinder raw water inlet line to the Sobrante Water Treatment Plant.

There was a compression failure of bell and spigot joints in a 20-inch cast iron pipe installed in 1916 under a 20-foot fill in a Oakland Park (Lake Merritt channel).

The imported supply to the Santa Clara Valley area is from the San Joaquin-Sacramento Rivers Delta via the South Bay Aqueduct of the State Water Project (SWP), the San Luis Reservoir via the Santa Clara Conduit of the San Felipe Project, a division of the U. S. Bureau of Reclamation's Central Valley Project (CVP), and from the Sierra Nevada via the San Francisco Hetch Hetchy Aqueduct. The earthquake caused a minor leak (400 gpm) in one of the two 66-inch diameter prestressed concrete parallel pipes of the Santa Clara Conduit where it crosses the Calaveras fault east of Gilroy. Several air valves in concrete vaults failed due to inverted pendulum action.

One air valve was broken on a 84-inch diameter prestressed concrete pipe north of Stanford University.

6.2.2.2.1   San Francisco Auxiliary Water Supply System

The San Francisco Fire Department operates a separate Auxiliary Water Supply System (AWSS) from the city's domestic system (see also Sections 2.4.2 and 12.2). The AWSS is for fire fighting purposes only. It was built following the 1906 earthquake and fire primarily in the northeast (business) portion of the city and has been gradually extended to other parts of the city.

The Loma Prieta earthquake caused many breaks in the domestic water supply system serving the San Francisco Marina District. That area is built primarily on artificial fill, and permanent ground movements up to a few inches were widespread. However, the AWSS supply main to that area remained intact.

The physical damage to the AWSS occurred in the area south of Market street were a 12 inch main and six fire hydrant assemblies were damaged. The damage caused leaks that resulted in the drainage of the 750,000-gallon tank serving the lower pressure zone. This resulted in only limited water pressure and little flow quantity for the AWSS in the Marina District.

After the earthquake a fire started in a corner four-story wood frame apartment building in the Marina District, in which the first and second stories had collapsed and the third and fourth stories were leaning toward the street. The fire started in the rear of the building and with the confusion of the earthquake was not reported until approximately one-half hour later. The fire engine first reporting connected to a AWSS hydrant in front of the building, but found no water pressure. The building was actually leaning over the fire hydrant. About one-half hour later the building exploded and the buildings across both streets began to burn or smolder, and the fire spread to the north and west within the block. Other engines were connected to a hydrant to the west and drafting from the Palace of Fine Arts lagoon. An engine attempted to draft from the Marina lagoon, but was unable to do so because of low tide.

The San Francisco Fire department has developed in recent years the Portable Water Supply System (PWSS) to supplement their AWSS. At the time of the earthquake it consisted of four hose tender trucks with 5,000 feet of five-inch diameter fire hose and appurtenances each. At 6:50 pm (local) the fire was continuing to burn and the Fire Department called for the fireboat Phoenix and all available PWSS hose tenders. The hoses were laid and connected to the fireboat. The fire spread was stopped at about 7:45 pm, by streams of salt water from the hose tenders. The two AWSS permanent salt water pumping plants were placed in service at 8:00 pm and continued to operate until 10:00 pm when the lower pressure zone was pressurized and the tank had been filled with salt water.

The report of the fire was delayed and never adequately communicated to the Fire Department. The AWSS pipe network was damaged due to a small number of breaks, however, the loss of system pressure in the Marina District may have been avoided if the operational status of the system could have been identified sooner, which would have expedited the placing into operation of the salt water pumps. Although the AWSS pipe network failed initially, the fireboat Phoenix and the PWSS backup system fulfilled its mission, and the fire was suppressed.

6.2.2.3 1995 Kobe Earthquake

The Kobe area of Japan occurred on January 17, 1995, at 5:46 am (local). This earthquake was by the worst earthquake to occur in Japan since the 1923 Kanto earthquake. The epicenter was located in Osaka Bay offshore from Kobe. However, the largest ground accelerations were felt in central Kobe and communities to the east. The magnitude was 6.9 (Caltech). Damages exceeded $100 billion and there were approximately 5,500 deaths. Significant damage occurred to buildings, housing, ports and lifelines. Most of the city is built on alluvial soils that wash down from the mountains just north of Kobe. Near Osaka Bay, there are considerable areas built on artificial fill. This area has high groundwater levels and was subjected to significant ground shaking, ground distortion in areas not particularly prone to liquefaction, and liquefaction in the made fills near the bay.

Most of the water supply to Kobe and three smaller cities comes from the Yodagawa (Yoda River), by the Hanshin Water Supply Authority, a wholesale water agency. This water supply serves a population of about 2.5 million. The water is pumped from the Yodagawa, treated and delivered into two gravity tunnels supplying Kobe and the other cities. Damage to the pump discharge lines required 12 days to repair. There was no functional damage of the two gravity concrete tunnels which suffered some cracking and was repaired by concrete patching. This was performed during the time of the restoration of the Kobe water distribution system (1,500 to 2,000 repairs) which took over two months. Some other cities' water distribution systems suffered comparable or higher damage rates to that in Kobe.

6.2.3 Water Treatment Facility Performance

Water treatment plants are vulnerable to a number of different failure mechanisms. Sloshing liquids in treatment tanks and basins create enormous loads on submerged elements and elements located near the surface. Baffles, piping, troughs and launders are often destroyed. Replacement of some elements such as generic baffles is relatively easy. Other elements such as specially designed clarifier equipment can take many months to replace.

Treatment plants are particularly vulnerable to hazardous chemical release if chemicals are not properly stored. At water treatment plants, inadvertent chemical release is usually more of an inhalation hazard than a fire hazard. Chlorine gas release has occurred during earthquakes at several treatment plants. Use of hypochlorite instead of chlorine gas can minimize the likelihood of chlorine release or use of a vacuum chlorine gas system can reduce the likelihood of chlorine release.

Common treatment-plant earthquake damage has been caused by ground shaking effects on buildings, equipment and nonstructural elements. Piping failure caused by either permanent ground displacements and/or differential movement between pipe attachment points such as at structure interfaces is another common failure mode. Some treatment equipment, such as pumps and motors rely on power and have been disrupted when commercial power was disrupted and backup power was unavailable

6.2.4 Water Distribution Facility Performance

Distribution pipeline breaks, which numbered in the thousands in the Northridge and Kobe earthquakes, have been the largest single contributor to loss of water pressure in potable water systems. In addition to denying service to areas directly served by broken pipelines, pipeline breaks also allow water to flow freely out of connected pipelines, and the storage tanks and reservoirs, pump stations and supply lines that feed the broken pipelines. Consequently, pipeline breaks have significantly affected water pressure/availability in areas that extend beyond the areas where the breaks occur.

Although seismic wave propagation can result in temporal differential displacements that result in pipeline failure, most buried pipeline damage occurs in areas of poor soils where liquefaction, landslide and/or settlement result in permanent ground displacement. Fault rupture zones also produce significant permanent ground displacements. If the ground shaking is violent enough, even stable soils can lurch and permanently displace. In general, ductile pipelines (e.g.., ductile iron, welded steel and high density polyethylene) with flexible, restrained joints (ductile iron) or with high quality butt welded (steel) or double lap welded (steel) or fusion welded (HDPE) perform better than brittle pipelines (e.g.., cast iron, concrete and pipelines with significant corrosion) with brittle and/or asbestos cement or PVC pipes with simple push-on unrestrained joints. However, even lower capable pipelines (cast iron with leaded joints, DI, AC or PVC with push on joints) can perform reasonably well if they are only exposed to moderate to moderately strong ground shaking without permanent ground deformations.

A restrained joint designed specifically for seismic applications has preformed extremely well in Japan. There were no reported failures of pipelines that used this type of joint during the Kobe Earthquake. This joint permits axial movement and rotation to help accommodate ground displacement. EBAA Iron "flextend" type joints can provide this type of flexibility, but the extra cost of such installations can make them prohibitive for all but the most extreme case.

All pipe interfaces with tanks, pumps and other equipment, and support structures are subject to failure if differential movements occur and adequate flexibility is not provided. Areas dependent on pumped water have lost water pressure due to commercial power outages when backup power has not been available. Earthquakes have also caused water hammer within distribution piping that was strong enough to blow out fire hydrants. Air and vacuum valves supported on slender necks are very vulnerable to earthquake motions. Although pump station buildings and valve vaults have usually performed well in earthquakes, pump, valve and other equipment supports have failed due to differential movement between equipment.

The following sections summarize the performance of smaller diameter pipeline water distribution systems in several recent earthquakes.

6.2.4.1 Loma Prieta 1989

Figure 6-4 shows a map with damage to buried pipelines in the San Francisco Bay Area due to the 1989 Loma Prieta earthquake. This map shows damage at 916 locations for water (including a few wastewater) utilities.

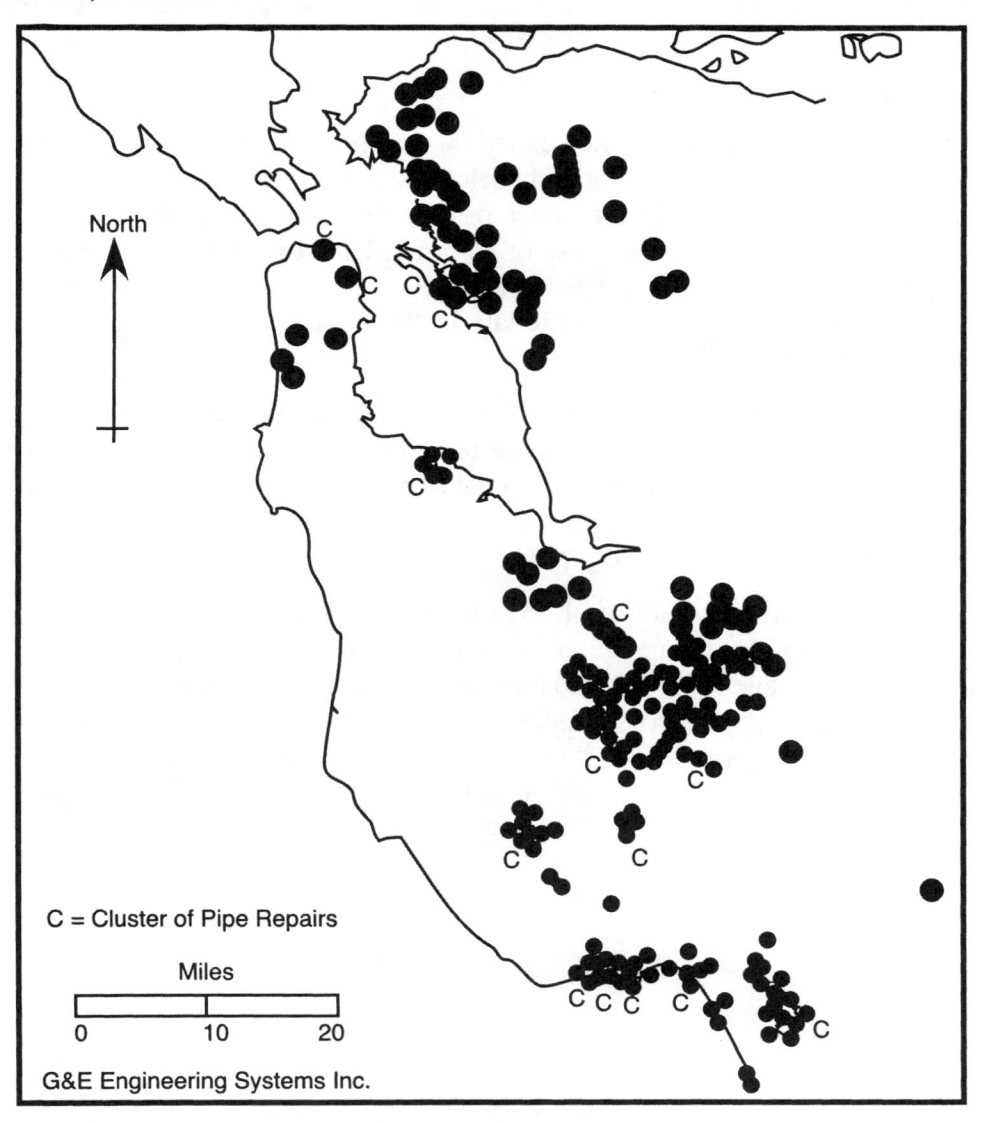

Figure 6-4. Buried Pipeline Damage in the San Francisco Bay Area, 1989 Loma Prieta Earthquake

6.2.4.2 Northridge 1994

A GIS-based analysis of the pipeline damage to the Los Angeles Department of Water and Power (LADWP) water system was performed by O'Rourke and Jeon, (1999). Key points of the analysis include:

The analysis was for cast iron, ductile iron, asbestos cement and steel pipe, up to 24-inch diameter. The pipeline inventory included: 7,848 km of cast iron pipe; 433 km of ductile iron pipe; and 961 km of asbestos cement pipe.

A total of 1,405 pipe repairs were reported for the LADWP distribution system. For distribution pipe where attributes could be reconstructed from available data, there were: 673 repairs for cast iron pipe; 24 repairs for ductile iron pipe; 26 repairs for asbestos cement pipe, and 216 repairs for steel pipe.

6.2.4.3 Kobe 1995

The 1995 Hanshin-Awaji earthquake (often called the Hyogo-Ken Nanbu (Kobe) earthquake) was a M 6.7 crustal event that struck directly beneath much of the urbanized city of Kobe, Japan. At the time of the earthquake, the pipeline inventory for the City of Kobe's water system included 3,180 km of Ductile Iron pipe (push on joint), 237 km of "Special Ductile Iron" pipe (with special flexible restrained joints), 103 km of high pressure steel welded pipe, 309 km of cast iron pipe with mechanical joints, and 126 km of PVC pipe with push-on gasketed joint (Eidinger and Avila, 1999).

Importantly, the "Special Ductile Iron" pipe in Kobe had no reported failures. This pipe has joints that could permit axial movement and rotation to help accommodate ground displacement. Unfortunately, there has not been sufficient demand to make this joint available in the United States, although there have been a few applications of conceptually similar joinery using EBAA-Iron "flex-tend" joints (and other similar products).

The City of Kobe's water system suffered 1,757 pipe repairs to mains. The repairs could be classified into one of three types: damage to the main pipe barrel (splitting open); damage to the pipe joint (separated); damage to air valves and hydrants; the damage rate was divided about 20% - 60% - 20% for these three types of repairs, respectively. Figure 6-5 shows the time needed to restore service to various districts in Kobe (called sections or centers in the figure). It took 90 days to restore the last customer. Figure 6-6 shows damaged pipeline being repaired.

City of Kobe Water System Service Restoration

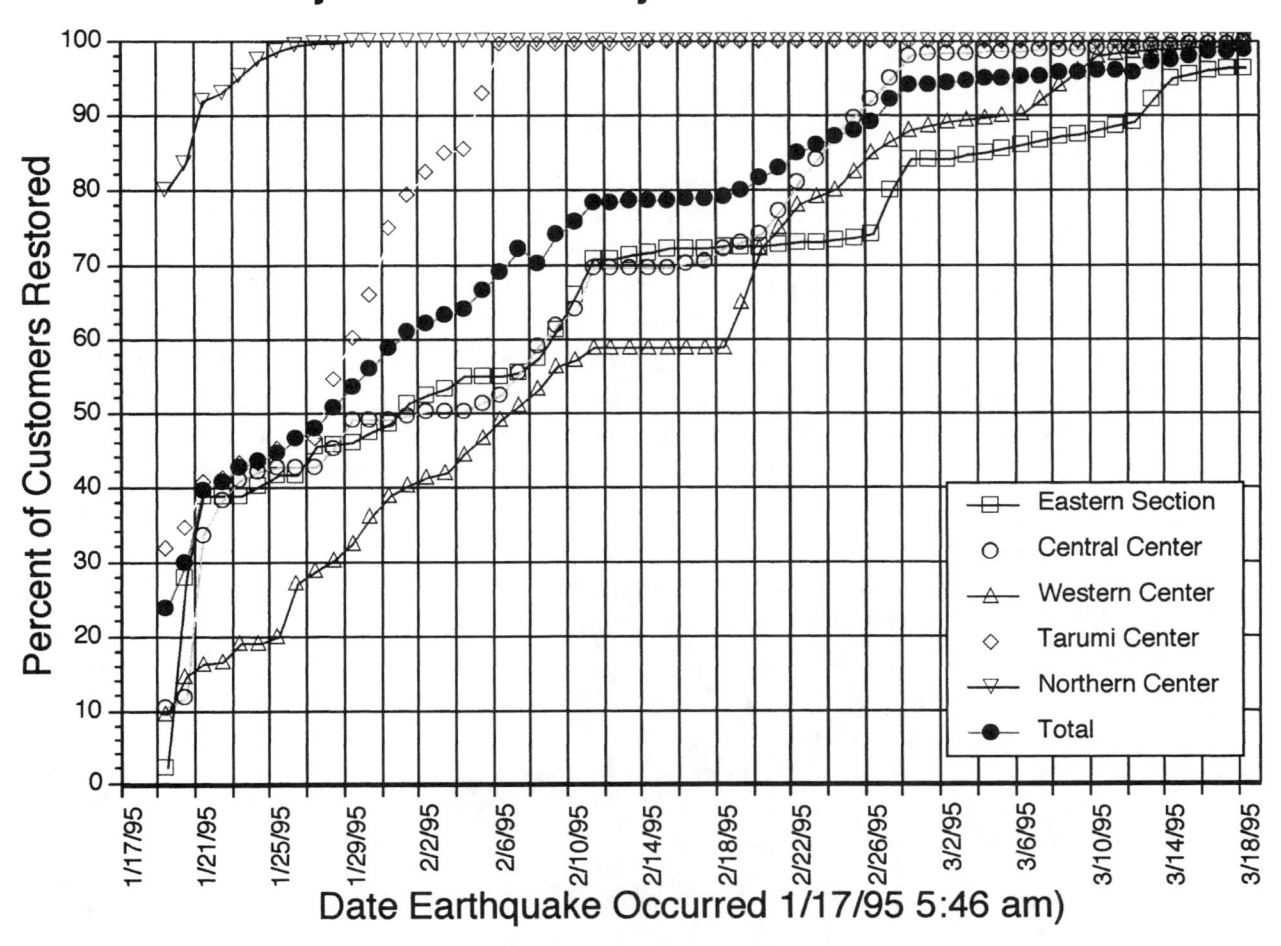

Figure 6-5. City of Kobe Customer Water Service Restoration Time

Figure 6-6. Damaged Pipeline, Kobe 1995

6.2.4.4 Izmit, Turkey 1999

The August 17, 1999 Kocaeli earthquake (moment magnitude M_w 7.4) occurred on the North Anatolian fault in northwestern Turkey. The Kocaeli earthquake has also been called the Izmit earthquake in various publications.

A population of about 1.8 million people lived within the zone of strong shaking caused by this earthquake. While there was just slight to modest damage to the three water treatment plants in the area (all had been designed, more or less, to the equivalent of California seismic Zone 4), there was widespread and major damage to transmission and distribution pipelines. Damage to a 2.2 m diameter butt-welded steel transmission pipeline where it crossed the Anatolian fault is documented in Eidinger (2002). At other locations, there were multiple failure of an older transmission pipeline subjected to permanent ground deformations (faulting and/or lateral spreads). But the most significant damage was to distribution pipelines, largely due to liquefaction settlements and strong ground shaking. Figure 6-7 shows damage to a 12" diameter asbestos cement pipe, the most common type used in the region.

Figure 6-7. Damaged AC Pipeline, Adapazari, Turkey 1999

6.2.4.5 Chi Chi 1999

The M_w 7.7 Chi-Chi (Ji-Ji) earthquake of September 21, 1999 in Taiwan led to 2,405 deaths and 10,718 injuries. Potable water was lost to 360,000 households immediately after the earthquake, largely due to damage to buried pipelines. Figure 6-8 shows damage to a 2-meter diameter pipeline that was bent due to fault offset (the thrust up side was on the right side of this pipe).

There was about 32,000 km of water distribution pipelines in the country; perhaps a quarter or more was exposed to strong ground shaking. The largest pipes (diameter ≥1.5 meters) were typically concrete cylinder pipe or steel, with ductile iron pipe being the predominant material for moderate diameter pipe and a mix of polyethylene and ductile iron pipe for distribution pipe (≤8 inch diameter).

Based on (as yet) incomplete analysis of the pipeline damage, it appears that about 48% of all buried water pipe damage is due to ground shaking. The remainder is due to liquefaction (2%), ground collapse (11%), ground cracking and opening (10%), horizontal ground movements (9%), vertical ground movement (16%), other (4%) (Shih et al, 2000).

Figure 6-8. Damaged Steel Transmission Pipe, Chi-Chi Earthquake 1999

6.2.4.6 *Gujarat (Kutch) India, 2001*

The M_w 7.7 Gujarat (Kutch) earthquake of January 26, 2001 in India led to about 17,000 deaths and about 140,000 injuries. Potable water was lost to over 1,000,000 people immediately after the earthquake, largely due to damage to wells, pump station buildings and buried pipelines.

There was about 3,500 km of water distribution and transmission pipelines in the Kutch District; perhaps 2,500 km was exposed to strong ground shaking. It is estimated that about 700 km of these pipelines will have to be replaced due to earthquake damage.

6.2.4.7 *General Form of Pipeline Fragility Curves*

Eidinger (2001a) organized the available damage to buried pipelines into a set of fragility curves. The fragility curves for buried pipe are expressed as a repair rate per unit length of pipe, as a function of ground shaking (peak ground velocity, PGV) or ground failure (permanent ground deformation, PGD).

Most of the empirical evidence for earthquakes prior to 1989 is for the performance of small diameter (under 12 inches) cast iron pipe. This is because cast iron pipe was the most prevalent material in use in water systems for earthquakes that occurred some time ago (like San Francisco, 1906). More recent earthquakes (Loma Prieta 1989 and Northridge 1994) have yielded new empirical evidence for more modern pipe materials, like asbestos cement, ductile iron and welded steel pipe. Still, as of 2003, we do not have a complete empirical database for all pipe materials under all levels of shaking.

A pipe repair can either be due to a complete fracture of the pipe, a leak in the pipe, or damage to an appurtenance of the pipe. In any case, these repairs require the water agency to perform a repair in the field.

The pipe repairs predicted using the fragility curves are for repairs in buried pipe owned by the water agency. This includes the pipe mains in the street, pipe laterals that branch off the main to fire hydrants, and service connections up to the meter owned by the water agency.

Buried pipe from the water agency's meter up to the customer's structure may also break. This pipe is usually very small diameter (under 1 inch diameter), and is generally the responsibility of the customer for repair. If this pipe breaks, then water will leak out of the water main until someone shuts off the valve at the service connection. Damage to these service lines had a major role in hampering fire fighting activities in the 1906 earthquake and the 1991 Oakland Hills fire.

Buried pipes can be damaged due to seismic wave passage effects (ground shaking) and earthquake induced ground failure (permanent ground deformation).

Wave passage effects refers to the transient vibratory soil deformations caused by seismic waves generated during an earthquake. Wave passage effects cover a wide geographic area and affects pipe in all different types of soil. Strains are induced in buried pipe because of its restraint within the soil mass. In theory for vertically propagating shear waves, peak ground strain is directly proportional to peak ground particle velocity (PGV), and therefore PGV is a natural demand description.

Ground failure effects are the permanent soil movements caused by such phenomena as liquefaction, lurching, landslides, and localized tectonic uplifts. These tend to be fairly localized in geographic area and potential zones can be somewhat identified *a priori* by the specific geotechnical conditions. Ground failure can be very damaging to buried pipe because potentially large localized deformations can develop as soil masses deform and move relative to each other. Such deformations can cause fracture or pullout of pipe segments embedded within the soil. Permanent ground displacement (PGD) is used here as the demand description. It is recognized that the PGD descriptor ignores the variation in the amount of ground displacement and the direction of ground displacement relative to the pipeline; if the analyst is interested in this kind of detail, then site-specific analytical methods should be used instead of area-wide vulnerability functions.

Repairs can be to a variety of system components including in-line elements (e.g., pipe, valves, connection hardware) and appurtenances (e.g., service laterals, hydrants, air release valves). Table 6-4 shows that the ratio of service lateral repairs to pipe repairs can vary widely, and the numbers of service repairs can even exceed the numbers of pipe repairs (in one of four cases reported; but note that in Japan, the length of service laterals can be quite long, whereas typical U.S. water utilities own only a few feet of service lateral up to the meter connection). Unrestricted flow from service laterals due to collapsed or burned structures can play an important role in depressurizing the pipe network, as has happened in the 1906 earthquake and the 1991 Oakland Hills fire.

Table 6-4 Reported Statistics for Main Pipe and Service Lateral Repairs

Earthquake	Number of Service Lateral Repairs	Number of Main Pipe Repairs	Ratio Service: Pipe
1995 Hyogoken-nanbu (Kobe) (Shirozu, et al, 1996)	11,800	1,760	6.7:1
1994 Northridge[1] (Toprak, 1998)	208	1,013[2]	1:4.8
1989 Loma Prieta (Eidinger, et al, 1995)	22	113	1:5.1
1971 San Fernando (NOAA, 1973)	557	856	1:1.5
Notes 1. Numbers of field repair records. 2. Includes repairs to hydrants.			

Based upon a regression on the empirical dataset, coupled with considerations for the empirical record and engineering mechanics, a linear regression model is best suited for evaluation of pipeline damage due to wave propagation. Figure 6-9 shows the empirical data set and various regression models. See Eidinger (2001a) for a complete discussion and comparison of this model with other types of models.

Refined analyses can be performed to assess the influence of pipe material, pipe diameter and earthquake magnitude. Ductile iron and steel pipe have been found to be less vulnerable than cast iron, by less than a factor of two; and asbestos cement was the best performer. These trends are not in keeping with conventional thinking which ranks brittle materials such as cast iron or asbestos cement more vulnerable than ductile materials such as steel or ductile iron by more than a factor of three (e.g., HAZUS, 1999). There is also some evidence that large diameter pipes have lower damage rates than small diameter pipes. Longer duration of strong motion shaking during an earthquake could cause more pipe damage due to cumulative cyclic damage. With these and other considerations outlined in Eidinger (2001a), the fragility model is adjusted to account for common pipeline materials in Table 6-5.

In a similar manner, vulnerability functions have been developed for damage to pipelines due to permanent ground deformations (wave propagation effects are masked within the more destructive effects of PGDs).

Statistical analysis of the empirical data was carried out in a similar way as that described above for the wave propagation data. The dataset and regression curve are shown in Figure 6-10. The repair rates are about two orders of magnitude greater than those for wave propagation thus indicating the extreme hazard that PGD poses for buried pipe. Portions of water systems that experience ground failure are likely to be mostly inoperable immediately after the earthquake.

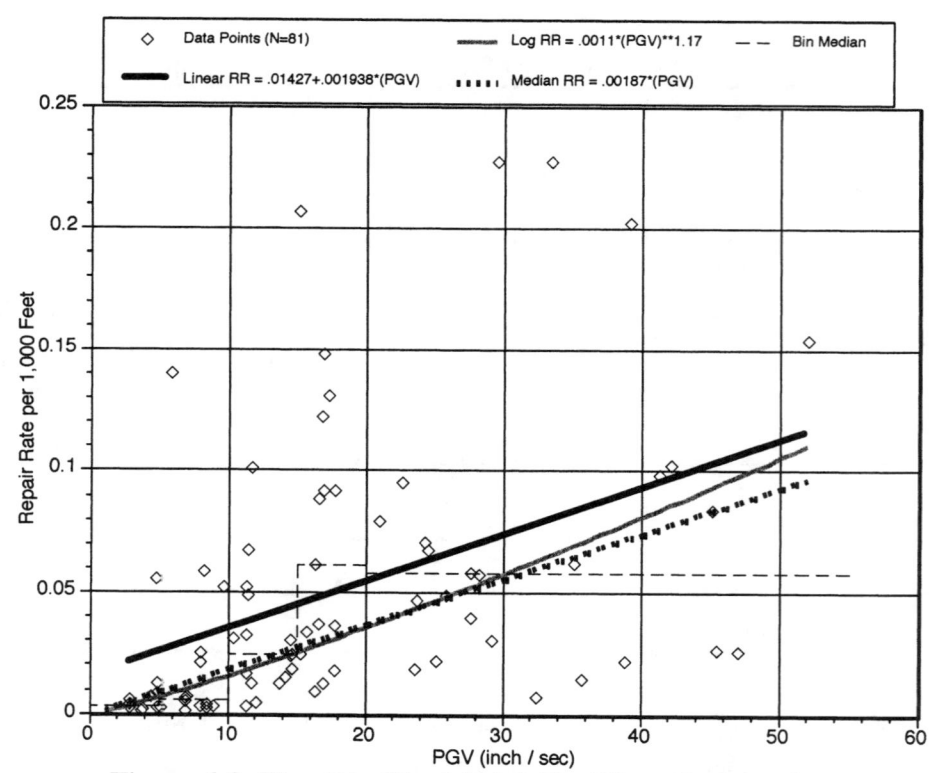

Figure 6-9. Pipe Fragility Model, For Wave Propagation

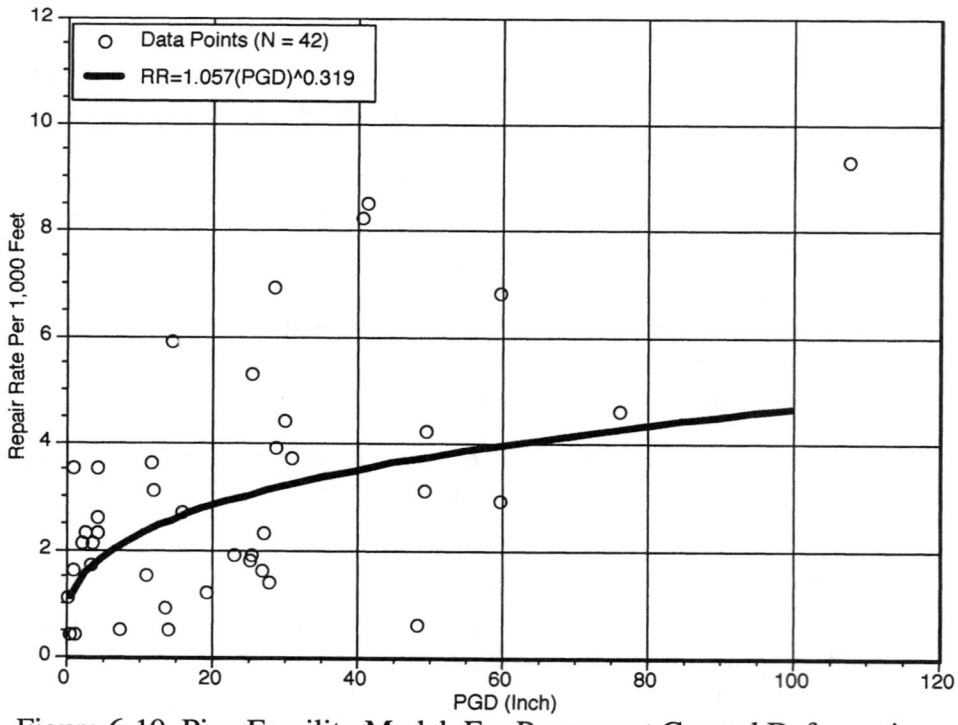

Figure 6-10. Pipe Fragility Model, For Permanent Ground Deformations

Table 6-5 provides the recommended "backbone" pipe vulnerability functions (also sometimes called damage algorithms or fragility curves) for PGV and PGD mechanisms. These functions can be used when there is no knowledge of the pipe materials, joinery, diameter, corrosion status, etc. of the pipe inventory; and when the evaluation is for a reasonably large inventory of pipelines comprising a water distribution system.

Table 6-5 Buried Pipe Vulnerability Functions

Hazard	Vulnerability Function	Lognormal Standard Deviation, β	Comment
Wave Propagation	RR=0.00187 * PGV	1.15	Based on 81 data points of which largest percentage (38%) was for CI pipe.
Permanent Ground Deformation	RR=1.06 * PGD$^{0.319}$	0.74	Based on 42 data points of which largest percentage (48%) was for AC pipe.
Notes 1. RR = repairs per 1,000 of main pipe. 2. PGV = peak ground velocity, inches/second. PGD = permanent ground deformation, inches 3. Ground failure mechanisms used in PGD formulation: Liquefaction (88%); local tectonic uplift (12%)			

The user can use the fragility curves in the table to predict damage to buried pipes due to ground shaking, liquefaction and landslide. The fragility curves in the table are "backbone" fragility curves, representing the average performance of all kinds of pipes in earthquakes. There are many reasons as to why various types of pipe will behave differently in earthquakes. Table 6-6 and Table 6-7 present recommendations as to how to apply the fragility curves in Table 6-5 to particular pipe types. By diameter, small means 4 inch to 12 inch diameter, and large means 16 inch diameter and larger. Table 6-6 and Table 6-7 are for pipelines installed without seismic design specific to the local geologic conditions. To apply these tables, the pipe vulnerability functions in Table 6-5 are adjusted as follows:

$$RR = K_1(0.00187)PGV \text{ (for wave propagation)}$$

$$RR = K_2(1.06)PGD^{0.319} \text{ (for permanent ground deformation)}$$

Table 6-6 Ground Shaking - Constants for Fragility Curve

Pipe Material	Joint Type	Soils	Diam.	K_1
Cast iron	Cement	All	Small	1.0
Cast iron	Cement	Corrosive	Small	1.4
Cast iron	Cement	Non corr.	Small	0.7
Cast iron	Rubber gasket	All	Small	0.8
Welded steel	Lap - Arc welded	All	Small	0.6
Welded steel	Lap - Arc welded	Corrosive	Small	0.9
Welded steel	Lap - Arc welded	Non corr.	Small	0.3
Welded steel	Lap - Arc welded	All	Large	0.15
Welded steel	Rubber gasket	All	Small	0.7
Welded steel	Screwed	All	Small	1.3
Welded steel	Riveted	All	Small	1.3
Asbestos cement	Rubber gasket	All	Small	0.5
Asbestos cement	Cement	All	Small	1.0
Concrete w/Stl Cyl.	Lap - Arc Welded	All	Large	0.7
Concrete w/Stl Cyl.	Cement	All	Large	1.0[26]
Concrete w/Stl Cyl.	Rubber Gasket	All	Large	0.8
PVC	Rubber gasket	All	Small	0.5
Ductile iron	Rubber gasket	All	Small	0.5

---

[26] Unpublished work by J. Eidinger, D. Ballantyne and T. O'Rourke using new empirical data from the 1989 Loma Prieta earthquake suggest K1 = 0.2 to 0.6 for large (60" and larger) diameter prestressed concrete cylinder pipe, depending on age, corrosion and other factors; or K1 is higher when using only 1994 Northridge data.

Table 6-7 Permanent Ground Deformations - Constants for Fragility Curve

Pipe Material	Joint Type	K_2
Cast iron	Cement	1.0
Cast iron	Rubber gasket	0.8
Cast iron	Mechanical restrained	0.7
Welded steel	Arc welded, lap welds (large diameter, non corrosive)	0.15
Welded steel	Rubber gasket	0.7
Asbestos cement	Rubber gasket	0.8
Asbestos cement	Cement	1.0
Concrete w/Stl Cyl.	Welded	0.6
Concrete w/Stl Cyl.	Cement	1.0
Concrete w/Stl Cyl.	Rubber Gasket	0.7
PVC	Rubber gasket	0.8
Ductile iron	Rubber gasket	0.5

In principle, the fragility functions in these three tables could be used for buried natural gas pipelines. However, experience has shown that damage to natural gas pipelines of the same style of construction as water pipelines, subjected to the same level of earthquake hazard, has been higher than what would be predicted using these tables. For example, cast iron gas pipes were reported (Hamada, et al, 1986) to have a trend of higher repair rates than the weaker asbestos cement water pipes in the Nihonkai-Chubu quake because gas leaks were detected much more accurately (implying that many water pipe leaks go undetected). This trend has also been observed in other earthquakes.

6.2.5 Water Storage Facility Performance

Water storage tanks have been occasionally shown to be vulnerable to earthquakes. Ground-supported tanks can slide and/or uplift if they are not adequately anchored. This movement can sever attached piping and result in tank drainage. Steel ground-supported tanks are susceptible to "elephant" foot buckling of the shell or separation of the shell and base if the tank uplifts and/or the shell is not adequately designed to resist the compressive forces from the contained water, roof and shell. These compressive forces can also cause failure of concrete tank walls.

Elevated steel tanks that were not adequately designed to resist ground-shaking forces have toppled. Because the ground shaking forces usually result in eccentric forces that cause the tank structures to twist, elevated tanks often fall nearly within their footprint instead of falling over laterally. When the tanks hit the ground, shrapnel can be projected hundreds of feet.

Soil liquefaction, landslide, settlement and fault movements have damage reservoirs formed by earthen embankments. These embankments are often defined as high-hazard dams and subject to stringent governmental regulations that require earthquake-resistant construction. Covered reservoirs have had their roofs and roof supports damaged from sloshing of reservoir water.

Because one function of water storage reservoirs is to supply water for fire suppression, water storage reservoir failure can have a significant impact on fire suppression capability. This impact will be amplified if the water supply has also been compromised.

6.2.5.1 *Empirical Performance of Tanks in Past Earthquakes*

In order to examine the empirical performance of tanks, Eidinger (2001a) examined the performance of 532 tanks that have undergone ground shaking of at least 0.10g in past earthquakes. Table 6-8 summarizes the empirical database.

Table 6-8 Earthquake Characteristics for Tank Database

Event	No. of Tanks	PGA Range (g)
1933 Long Beach	49	
1952 Kern County	24	
1964 Alaska	39	0.20 to 0.30
1971 San Fernando	27	0.20 to 1.20
1972 Managua	1	0.50
1975 Ferndale	1	0.30
1978 Miyagi-ken-ogi	1	0.28
1979 Imperial Valley	24	0.24 to 0.49
1980 Ferndale	1	0.25
1980 Greenville	1	0.25
1983 Coalinga	48	0.20 to 0.62
1984 Morgan Hill	12	0.25 to 0.50
1985 Chile	5	0.25
1986 Adak	3	0.20
1987 New Zealand	11	0.30 to 0.50
1987 Whittier	3	0.17
1989 Loma Prieta	141	0.11 to 0.54
1989 Loma Prieta (Low g)	1,670	0.03 to 0.10
1991 Costa Rica	38	0.35
1992 Landers	33	0.10 to 0.56
1994 Northridge	70	0.30 to 1.00
Total (excl. low g)	532	0.10 to 1.20

Table 6-9 provides the breakdown of the number of tanks with various damage states. The damage states are: DS=1 no damage; DS = 2 slight damage; DS = 3 moderate damage, including roof damage due to sloshing; DS = 4 extensive damage; DS=5 complete (collapse) damage. The tank remains able to hold water for DS=2 and DS=3, but rapidly loses all water contents for DS=4 and DS=5. The value in the PGA column in Table 6-9 is calculated as the average PGA for all tanks in a PGA range; the ranges were set in steps of 0.10g.

Fragility curves were calculated for a variety of fill levels in the tank database. Table 6-10 provides the results. In the table, "A" represents the median PGA value (in g) value to reach or exceed a particular damage state, and Beta is the lognormal standard deviation. N is the number of tanks in the particular analysis.

Table 6-9 Complete Tank Database

PGA (g)	All Tanks	DS = 1	DS = 2	DS = 3	DS = 4	DS = 5
0.10	4	4	0	0	0	0
0.16	263	196	42	13	8	4
0.26	62	31	17	10	4	0
0.36	53	22	19	8	3	1
0.47	47	32	11	3	1	0
0.56	53	26	15	7	3	2
0.67	25	9	5	5	3	3
0.87	14	10	0	1	3	0
1.18	10	1	3	0	0	6
Total	532	331	112	47	25	16

Table 6-10 Fragility Curves, Tanks, As a Function of Fill Level

DS	A, g	Beta	A, g	Beta	A, g	Beta	A, g	Beta	A, g	Beta
DS≥2	0.38	0.80	0.56	0.80	0.18	0.80	0.22	0.80	0.13	0.07
DS≥3	0.86	0.80	>2.00	0.40	0.73	0.80	0.70	0.80	0.67	0.80
DS≥4	1.18	0.61			1.14	0.80	1.09	0.80	1.01	0.80
DS=5	1.16	0.07			1.16	0.40	1.16	0.41	1.15	0.10
	All Tanks N=531		Fill < 50% N=95		Fill ≥ 50% N=251		Fill ≥ 60% N=209		Fill ≥ 90% N=120	

The following trends can be seen in Table 6-10:

- Tanks will low fill levels (below 50%) have much higher median acceleration levels to reach a particular damage state than tanks which are at least 50% filled.

- Tanks with low fill levels have not been seen to experience damage states 4 or 5 (elephant foot buckling with leak or other damage leading to rapid loss of all contents; collapse). Thus, no values are given.

- Tanks with fill levels 90% or higher have moderately lower fragility levels than tanks with fill levels 50% or higher. Most water system distribution tanks are kept at fill levels between 70% and 100%, depending upon the time of day. If no other attributes of a given water storage tank are known, then the fragilities for the 90% fill level or higher should be used.

- The Beta values are mostly = 0.80. This reflects the large uncertainty involved in the tank database. It is recognized that Beta values would normally be in the 0.30 to 0.45 range for a tank-specific calculation. If the user does no tank-specific calculations, then it is recommended that beta be set to 0.80.

The empirical database was analyzed to assess the relative performance of anchored versus unanchored tanks. All tanks in the empirical database were designed prior to the AWWA D100-1996 code. Since fill level has been shown to be very important in predicting tank performance, only tanks with fill levels ≥ 50% were considered in this analysis. Table 6-11 provides the resulting fragility curves.

Table 6-11 Fragility Curves, Welded Steel Tanks, As a Function of Fill Level and Anchorage

DS	A, g	Beta	A, g	Beta	A, g	Beta
DS≥2	0.18	0.80	0.71	0.80	0.15	0.12
DS≥3	0.73	0.80	2.36	0.80	0.62	0.80
DS≥4	1.14	0.80	3.72	0.80	1.06	0.80
DS=5	1.16	0.80	4.26	0.80	1.13	0.10
	Fill ≥ 50% All N=251		Fill ≥ 50% Anchored N=46		Fill ≥ 50% Unanchored N=205 [1]	

Note [1]. The low beta values (0.12, 0.10) reflect the sample set. However, beta = 0.80 is recommended for use for all damage states for regional loss estimates for unanchored steel tanks with fill ≥ 50% unless otherwise justified.

As seen in Table 6-11, the empirical evidence for the benefits of anchored tanks is clear. The median PGA value to reach various damage states is about 3 to 4 times higher for anchored tanks as for unanchored tanks. It should be noted, however, that the anchored tank database (N=46) is much smaller than the unanchored tank database (N=251), and fill levels may not have been known for all tanks in the anchored tank database. Some of the anchored tanks in the database are relatively smaller (under 100,000 gallon capacity) than most other tanks in the database. In the 2003 Mw 6.5 San Simeon earthquake, 6 water tanks experienced PGA=0.30g to 0.50g; 1 was anchored, 5 were unanchored. The anchored tank had no damage (DS=1). Two unanchored tanks had moderate (DS=3) and three unanchored tanks had extensive (DS=4) damage. With these considerations, the empirical evidence strongly suggests that anchored tanks outperform unanchored tanks.

6.2.6 Administration, Operations and Maintenance Facility Performance

Administration, operations and maintenance facilities are sometimes overlooked. Unavailability of these facilities can significantly hinder post-earthquake recovery efforts. Significant structural and nonstructural damage often occurs to these facilities. An important part of the operations system is the system control and data acquisition system. These systems have been disrupted when commercial power is disrupted, and even with battery backup stop operating after a few hours.

6.3 Direct Impacts of Potable Water System Earthquake Performance

The most significant impact of water system damage is that water will not be available for fire suppression and potable water uses. Those areas most likely to be without water are areas with poor soils and in fault rupture zones. Areas that rely on pumps that do not have a backup power source are also likely to lose water pressure. As pipe breakage drains tanks and adjoining areas, water pressure may be lost in other areas with relatively little damage.

Using the models presented in Table 6-5, to Table 6-7, a hydraulic analysis of the damaged pipe distribution system can be performed. This analysis must be considered at two stages: immediately after the earthquake, after pipes are damaged but before the water utility has had a chance to valve-out leaking pipelines; and several hours (perhaps a day) after the earthquake, once the water utility has valved out the leaking pipes.

In-lieu of performing these sometimes complex hydraulic analyses, a number of researches have developed simplified models to forecast the serviceability of pipeline networks under various amounts of pipe damage. Figure 6-11 shows the results. In applying Figure 6-11, it should be recognized that the horizontal axis represents pipe breaks (leaks have much less impact), and that the extent of the grid (number of pipe loops) also plays an important role. For example, a single break in a long transmission pipeline would put the entire pipe out of service (Index = 0%), whereas a single break in a equally long length of pipe in a small well-looped network would possibly result in an Index near 100%.

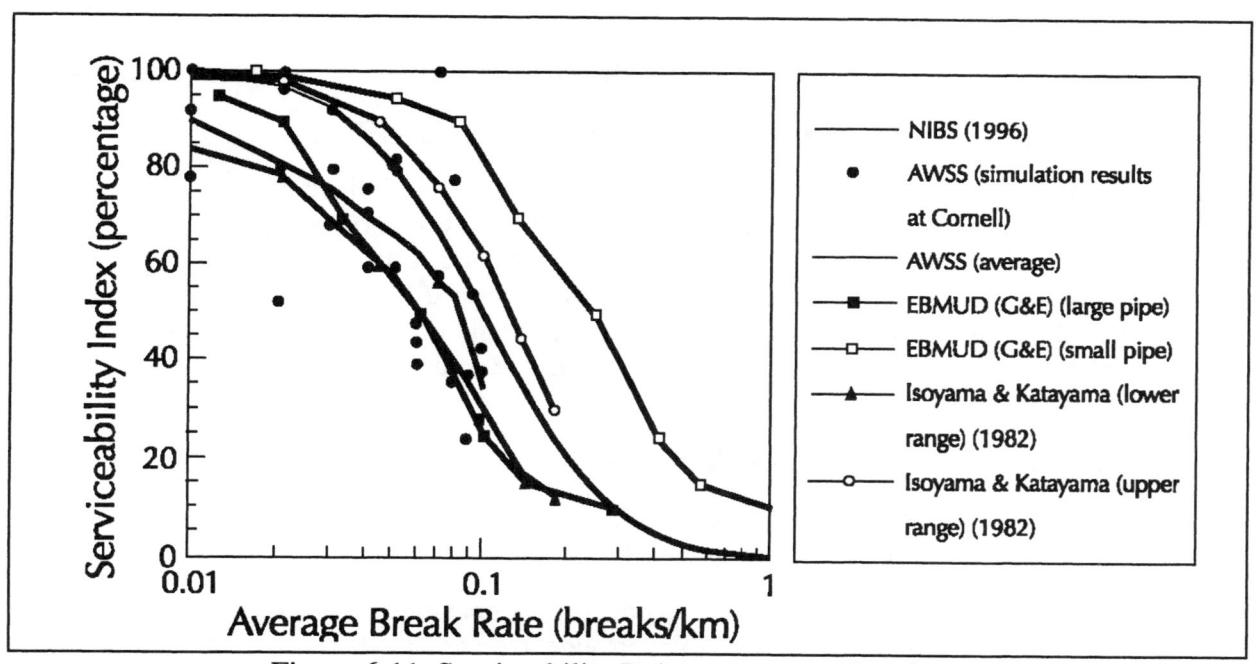

Figure 6-11. Serviceability Index vs. average break rate

The model shown in Figure 6-11 can be used to establish the capability for fire suppression, assuming that suitable fire department resources are at the fire scene. For example, an Serviceability Index over 90% would relate to a "good" water system, and an index below 20% would relate to a "poor" water system.

Hazardous materials used at treatment plants may also be released. Chlorine gas release has been the most common hazardous material incident. Although water system facilities do contain materials and equipment that may be fire ignition sources, the potential for ignition is probably no greater than the potential at a light industrial and warehouse facilities.

Sudden water release from broken pressurized water mains or storage tank failure, may flood, erode and/or damage adjacent facilities and roads. Water release volumes can be high enough to cause flooding and/or personal injury. Collapsed buildings, tanks and shrapnel, and equipment and other building content failure can also result in personal injury.

6.4 Secondary Impacts of Potable Water System Earthquake Performance

Potable water is needed for several purposes: public health, sanitation, fire suppression, business and industrial needs. In modern, industrialized countries, drinking water and sanitation facilities are usually trucked into earthquake disaster areas so there are minimal public health impacts. However, it is difficult, if not impossible, for business and industrial facilities to operate without water. The loss of make up water for emergency-generator cooling systems can impact first responders and other critical facilities, such telephone company central offices. The loss of make up water for air conditioning systems can disrupt critical computer facilities in banking and insurance industries that can have an immediate national impact. Consequently, business and industry is often forced to shutdown or operate at a reduced capacity until water service can be restored. Secondary economic losses can be significant and greatly exceed direct repair costs.

6.5 Potable Water System Earthquake Mitigation Strategies

Earthquake damage mitigation for potable water systems begins with the concept that the mitigation plan must address the systemic nature of the water system of concern. Upgrading different system components without regard to the operational characteristics of the system may improve individual component performance but will not necessarily significantly improve the system's overall performance and post-earthquake water availability. Because of the resources needed to implementing a comprehensive plan that improves a water system's post-earthquake functionality will require many years.

This comprehensive plan should include the following elements:

A seismic vulnerability assessment to identify vulnerable system components and determine how water availability would be affected throughout the system

Development of realistic system performance goals for the design-level earthquake(s). Performance goals could be deterministic (like: provide average day demand to 90% of all customers within 3 days after a maximum credible earthquake). Performance goals may need to reflect the cost needed to meet the goals, in that it may be prohibitively costly to meet certain goals. Benefit cost analyses can shed insight as to how much seismic upgrade is cost effective (see Chapter 14 of this report). Each water utility can establish its own performance goals with the understanding of what is cost effective.

Development of a plan to meet the performance goals.

The seismic vulnerability assessment could be based on a selection of scenario earthquakes, or based on probabilistic earthquakes, such as those minimum suggested by the Uniform Building Code. In either case, it will be most important to estimate the potential for permanent ground displacements (from fault offset, liquefaction, landslide, etc.) throughout the water system. Structural analysis techniques can be used to evaluate structures such as water treatment plants, tanks, pump stations, pipeline crossing, buildings, etc. Fragility equations based on empirical data may be more appropriate for estimating buried pipeline damage, and can often be used to estimate performance of structures. Once the vulnerabilities of the components have been determined, consultation with those with extensive knowledge on system operation and/or hydraulic models can be used to estimate system performance.

Post-earthquake system performance goals also need to be defined. A generic set of performance goals is presented in Ballantyne (1994). A compilation of actual performance goals adopted by utilities such as EBMUD and CCWD are provided in Eidinger (1999). The specific performance goals for a particular water system need to be realistic and reflect the seismic threat, the system vulnerability, and available resources to mitigate the vulnerability. The development of these goals should be coordinated with the fire department so the fire department is aware of likely water availability scenarios.

Once the system has been assessed and performance goals have been established, a strategy for upgrading the system so that the post-earthquake performance goals are likely to be reached needs to be developed. The most cost-effective plan will likely include a combination of approaches:

- Developing seismic design standards for new facilities
- Upgrading existing facilities to improve earthquake performance as special projects or in the course of normal refurbishment
- Implementing operation and control strategies to mitigate earthquake damage effects
- Developing and testing emergency preparedness and response plans

Building new facilities so they are seismic resistant is often less expensive than upgrading facilities once they have been constructed. The water utility owner needs to be especially vigilant to make sure that new water facilities have been designed and constructed to meet the design level earthquake ground motions. There are not sufficient seismic code requirements for many water system components such as pipelines and water facility equipment and components. It may be appropriate for the water utility to develop its own seismic design and construction criteria for those elements not covered in codes and standards.

Upgrading existing facilities to meet modern seismic design criteria can be effective but is often expensive. If the facility is near the end of its useful life or if there is another facility or strategy that can be used, it may not be cost effective to upgrade the facility. Replacement of buried pipelines only for seismic reasons is usually not cost effective. Using operation and control strategies, and/or emergency preparedness and response is usually more cost-effective than wholesale pipeline replacement.

Operation and control strategies can be developed to mitigate earthquake damage effects. Valves can be used to direct water flow to needed areas and/or isolate areas with heavy damage and keep these areas from affecting neighboring areas. However, valve system installation and maintenance can be a significant expense. Transportation system disruptions and personnel shortages can make it difficult for crew to operate manual valves in a timely manner.

Alternatively, it is a serious decision to install automatic valves (e.g.., seismically or flow activated) that may isolate an area where post-earthquake fires may be burning. Remote-controlled valves overcome some of the shortcomings of manual and automatic valves but are more expensive and may overwhelm already busy system operators. Consequently, developing appropriate operation and control strategies requires coordination with water utility operations personnel, local planners and the fire department.

Emergency preparedness and response is another technique that can be used to mitigate earthquake damage effects. In addition to developing a sound emergency preparedness and response plan, some water utilities have begun to buy flexible water hose that can be used to temporarily span broken pipeline sections.

It is difficult and expensive to disinfect distribution piping and other facilities used to convey potable water. Consequently, using potable water system piping to convey untreated water from a nonpotable water source would probably only be considered in a dire emergency. An argument can be made that putting disinfected raw water into the pipeline system (i.e.., lake water with heavy chlorination) and thereby bypassing complete treatment (flocculation, sedimentation and filtration) is a reasonable approach under a severe emergency, and would be definitely preferable than letting a City go without any sort of piped water.

Some cities have developed saltwater systems dedicated to fire suppression (see Sections 12 and 13).

Several water utilities have adopted the following mitigation strategies:

- Buried pipes. Install new pipelines to better seismic standards, where the new pipes are known to traverse soils prone to PGDs. Sometimes, add isolation valves adjacent to soils prone to PGDs. Sometimes, re-route the pipes or provide new pipes to go around the soils prone to PGDs. Sometimes, provide suitable manifolds (hydrants or special-purpose outlets) and use flexible hose to create an above-ground temporary pipeline to be used until such time that the buried pipeline is repaired. Sometimes, replace air and vacuum valves that are subject to inertial overloads and/or have been subjected to corrosion.

- At-grade unanchored steel tanks. Provide suitable flexible joints for all site entry (and sometimes bottom entry) pipes. Sometimes, anchor the tanks.

- At-grade unanchored prestressed concrete tanks. Sometimes, re-wrap the tanks and add a suitable anchorage system.

- Elevated tanks. Sometimes, improve the tank support system to resist seismic loads. Sometimes, base-isolate the tank.

- Pump Stations. Suitably anchor all electrical equipment. Restrain emergency generator batteries. Snub vibration isolated equipment (emergency generators, air compressors). Sometimes, upgrade the building (especially older unreinforced masonry buildings. Sometimes, provide flexible pipe joints between the building and the outside-buried pipes (especially if there are local poor soils or if the building is pile supported, with the potential for differential settlements). Sometimes, provide emergency standby power and/or quick connect couplings to allow rapid mobilization of portable emergency generators.

- Treatment Plants. Similar to pump stations, plus: sometimes, improve baffles in settling basins. Provide emergency standby power.

6.6 Summary

The most significant water system issue with respect to fire following earthquake is that potable water is usually the primary agent used by fire departments for fire suppression. In most large earthquakes, there are widespread areas that are left without water service. Although storage tank, pump station and water treatment plant failures have impacted water availability after earthquakes, numerous pipeline failures that overwhelm the system have had the most significant negative impact on water availability. These pipeline failures make it impossible to distribute water within the failure area and can also drain storage tanks and adjacent areas. In recent major earthquakes, potable water system failure has greatly hindered fire-fighting efforts.

Water facilities in each pressure zone are usually sized to provide enough water to fight a single fire at the largest facility within the pressure zone. Multiple fires in a single pressure zone and/or a large conflagration may overwhelm the water system.

Because natural gas pipelines and water pipelines are vulnerable to the same earthquake hazards and failure modes, both types of pipeline failures often occur in the same geographical area. The same strong ground motions that damage water systems can also result in ignitions from other sources. Therefore, there is a strong likelihood there will be numerous fires in the same area where water pressure is lost.

Mitigating earthquake effects on water system functionality requires a comprehensive assessment and program that considers the systemic nature of the water system. To be cost effective, a mitigation program will likely include a combination of approaches such as upgrade of existing facilities, development of operation and control strategies to mitigate damage effects, implementation of seismic design standards for new construction, and emergency preparedness and response planning. Because only limited resources are usually available for water system seismic mitigation, implementation of a successful mitigation program requires many years to complete.

6.7 Credits

Figures 6-1, 6-4, 6-5, 6-7 Eidinger. Figure 6-2, 6-3. Lund. Figure 6-9, 6-10, Eidinger 2001a. Figure 6-11, O'Rourke 2002. Tables 6-1, 6-2, Eidinger (2003).

6.8 References

ATC, A Model Methodology for Assessment of Seismic Vulnerability and Impact of Disruption of Water Supply Systems, 1992, Report ATC-25-1, funded by the Federal Emergency Management Agency. Prepared by EQE, C. Scawthorn and M. Khater, Redwood City.

Ballantyne, D.B. and Crouse, C.B., 1997, Reliability and Restoration of Water Supply Systems for Fire Suppression and Drinking Following Earthquakes, NIST-GCR-97-730, National Inst. For Standards and Technology, Gaithersburg.

Eidinger, J., Economics of Seismic Retrofit of Water Transmission and Distribution Systems, in Proceedings,2003, Sixth US Conference on Lifeline Earthquake Engineering, ASCE, TCLEE, Long Beach, August.

Eidinger, J., Ballantyne, D.,Lund, LeVal, 1998. Water and Wastewater System Performance in the January 17, 1995 Hyogoken-Nanbu (Kobe) Japan Earthquake, in Monograph 14, A. Schiff, Editor, Technical Council on Lifeline Earthquake Engineering, ASCE, September.

Eidinger, J. 1998. Lifelines, Water Distribution System, in The Loma Prieta, California, Earthquake of October 17, 1989, Performance of the Built Environment - Lifelines, US Geological Survey Professional Paper 1552-A, pp A63-A80, A. Schiff Ed. December.

Eidinger, John and Avila, Ernesto (editors), 1999. Guidelines for the Seismic Upgrade of Water Transmission Facilities, ASCE, TCLEE Monograph No. 15, January.

Eidinger, J., Elliott, T., Lund, L, 2000. Water and Wastewater System Performance in the August 17, 1999 Izmit (Kocaeli) Turkey Earthquake, in Monograph 17, A. Tang, Editor, Technical Council on Lifeline Earthquake Engineering, ASCE, March.

Eidinger, J. 2001a, Seismic Fragility Formulations for Water Systems, G&E Report 47.01.01 Revision 1, July 12, 2001 (http://home.earthlink.net/~eidinger), also Revision 0, American Lifelines Alliance, April.

Eidinger, J., 2001b, Water and Wastewater System Performance in the January 26, 2001 Gujarat (Kutch) India Earthquake, in Monograph 19, J. Eidinger, Editor, Technical Council on Lifeline Earthquake Engineering, ASCE, June.

Eidinger, J., and O'Rourke, M., 2002, Performance of a Pipeline at a Fault Crossing, in Proceedings, Seventh U.S. National Conference on Earthquake Engineering, EERI, Boston, MA.

Eidinger, J., 2003, Potable Water System Performance in the June 23, 2001 Atico (Moquegua) Peru M 8.4 Earthquake, in Monograph 24, C. Edwards, Editor, Technical Council on Lifeline Earthquake Engineering, ASCE, 2003.

Heubach, William F. (editor), 2003, Seismic Screening Checklists for Water and Wastewater Facilities, American Society of Civil Engineers, TCLEE Monograph No. 22.

O'Rourke, M. 2002. Buried Pipelines, chapter in Earthquake Engineering Handbook, Chen, W.F. and Scawthorn, C. (editors), CRC Press, Boca Raton.

O'Rourke, T. D., and Tawfik, M.S. 1983, Effects of Lateral Spreading on Buried Pipelines During the 1971 San Fernando Earthquake, Earthquake Behavior and Safety of Oil and Gas Storage Facilities, Buried Pipelines and Equipment, PVC,-Vol.77, New York, ASME.

O'Rourke, T.D., and Jeon, S-S., 1999, Factors Affecting the Earthquake Damage of Water Distribution Systems, in Optimizing Post-Earthquake Lifeline System Reliability, TCLEE Monograph No. 16, ASCE.

Schussler, H., 1906, The Water Supply System of San Francisco, California, Spring Valley Water Company, San Francisco, California, July.

Wang, L., 1990, A New Look Into the Performance of Water Pipeline Systems From 1987 Whittier Narrows, California Earthquake, Department of Civil Engineering, Old Dominion University, No. ODU LEE-05, January.

7 Gas Systems and Fire Following Earthquake

7.1 Introduction

The use of natural gas, like any flammable fuel, carries some risk of fire or explosion. The history of natural gas use throughout the world has shown it to be a safe fuel for consumer and industrial applications when buildings, natural gas systems, and appliances are constructed, installed, and maintained properly. However, when potentially threatening conditions arise—such as an earthquake capable of damaging the gas system—gas utilities, gas customers, and government agencies should consider steps to maintain a high level of safety.

This section of the report describes the hazards and operational characteristics of a typical gas utility system, along with a summary of recent earthquake experience in urban areas of California. Also addressed are potential types of earthquake damage to gas systems and their potential impact on building owners and surrounding communities. Finally, several alternatives for improving the safety of gas systems are described, along with associated benefits and drawbacks.

7.2 Understanding the Natural Gas Distribution System

7.2.1 Natural Gas Basics

Natural gas is a fossil fuel extracted from deep underground wells. It is a physical mixture of various gases, typically containing 85 to 95% methane, 7 to 12% ethane and small amounts of propane, butane, nitrogen, and carbon dioxide. The proportions vary from field to field and sometimes from well to well.

Natural gas is odorless and colorless when it comes from the wellhead. As a safety measure, an odorant is added so gas leaks can be detected. Commonly known as *mercaptans*, the odorant is a blend of organic chemicals containing sulfur. The odor of the mercaptans can be detected long before there is sufficient gas to cause a fire, explosion or asphyxiation.

Unlike propane, natural gas is lighter than air. Natural gas typically has a specific gravity of 0.6, meaning that it weighs about 0.6 times as much as air.

Not all mixtures of gas and air will burn. Some mixtures have too little gas, while others have so much gas there is not enough air (oxygen) left to burn. The two cutoff points between combustible mixtures and non-combustible mixtures are called the Explosive Limits.

- The Lower Explosive Limit (LEL) for natural gas is approximately 5%. At concentrations below the LEL, there is insufficient gas to cause a fire or explosion.

- The Upper Explosive Limit (UEL) for natural gas is approximately 15%. At concentrations above the UEL, there is insufficient air to cause a fire or explosion.

The ideal mixture for combustion of natural gas is approximately 10% and the ignition point is 1209° F.

7.2.2 The Natural Gas Delivery System

Gas utilities install and operate a network of mostly underground pipelines to deliver natural gas from the gas well to residential, commercial, industrial and agricultural customers, as shown in Figure 7-2. The pipelines operate at various pressures throughout the system. Gasses are compressed higher when entering transmission pipelines and regulated (pressure control) lower when entering distribution pipelines and supplying customers. Depending on the operating pressure, size of the pipe, year of installation and other factors, pipe material can be steel, plastic, cast iron or copper.

Natural gas is delivered to a gas distribution service area or local distribution company via a number of metering and/or pressure-regulating stations along the transmission pipeline. Gas is supplied to customers through a grid of distribution pipes, valves, and connections typically located underground with telecommunications, electricity, water, sewer, storm drains and other utilities.

Small-diameter gas service lines connect the gas distribution pipe to one or more customers at a gas meter typically installed near the customer's facilities. The gas meter assembly has a manual gas service shutoff valve, a pressure regulator to reduce pressure from the gas main pipe to standard delivery pressure, a gas meter to measure the volume of gas, and a service tee that allows a utility to bypass the meter without entering the structure if an alternate gas supply is available (e.g. compressed gas cylinder). Customer meters may not have a pressure regulator if they are fed from a low-pressure distribution system. The customer's natural gas houseline piping is attached to the service tee, which is typically considered the utility point of delivery and defines the physical boundary between utility and customer facilities.

Gas utilities routinely conduct surveys for gas leaks and categorize leaks as Grade 1, Grade 2, or Grade 3. A Grade 1 leak represents an existing or probable hazard and requires immediate action. A Grade 2 leak is not hazardous to life or property at the time of detection but requires scheduled repair. A Grade 3 leak is non-hazardous at the time of detection and is expected to remain so. For a large gas distribution system, several hundred Grade 2 or Grade 3 leaks may exist at any one time.

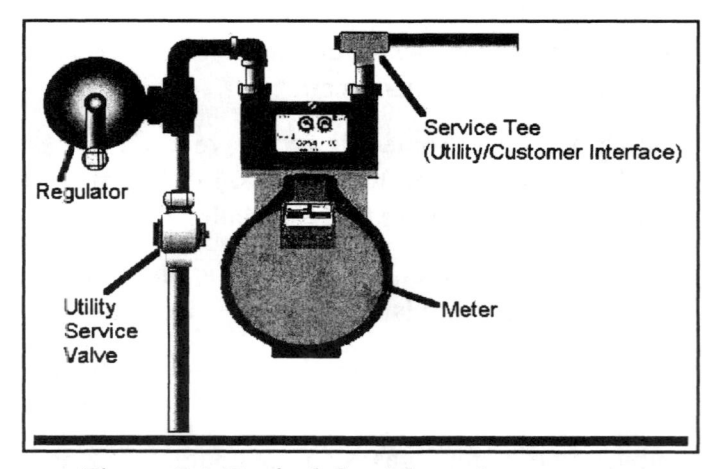

Figure 7-1 Typical Gas Meter Arrangement

FIRE FOLLOWING EARTHQUAKE

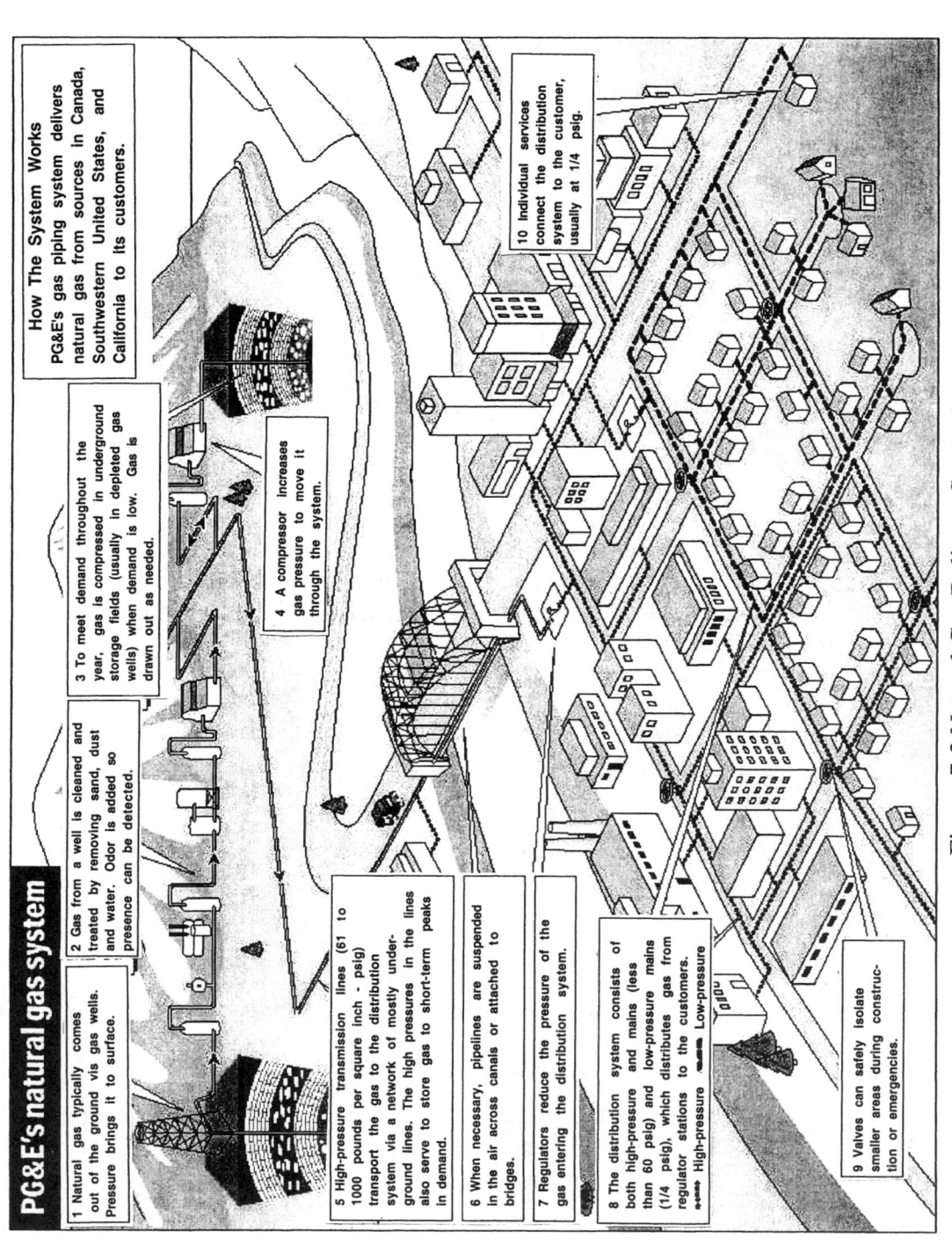

Figure 7-2 Natural Gas Delivery System
(provided by Pacific Gas & Electric Company)

7.3 Earthquake Performance of Natural Gas Systems

Given a good basis of understanding of the relative number of natural gas fires and their associated causes, alternatives can be assessed to improve natural gas safety. However, it is important to distinguish fire ignitions from general fire damage. Any fire has an initial ignition that can come from any source. The size of the fire and the damage it causes is highly variable and depends on a multitude of factors. These factors can lead to a single ignition causing a fire that destroys an entire city block, or a fire that is quickly extinguished without fire department assistance.

Historic earthquakes may not be representative of current types of buildings, appliances, natural gas systems, water delivery systems, transportation systems, population densities, emergency services, social impacts and other factors associated with future earthquakes. Similarly, drawing meaningful conclusions about natural gas safety from earthquakes in other parts of the world is often tenuous; significant differences exist in the pipeline materials, operating pressures, types of gas appliances, and building construction used outside of the United States. Nevertheless, lessons can be inferred from historic and recent foreign earthquakes. These lessons enhance our understanding of the role natural gas can be expected to play in fire ignitions in the future.

Some caution is necessary when extrapolating information from past earthquakes. Nearly every major earthquake in California has demonstrated some seismologic characteristic that was previously unknown or considered insignificant. Similarly, future earthquakes may produce quantities and types of infrastructure damage not previously observed. In particular, the number of fire ignitions experienced in past earthquakes may not be a reliable indicator of future ignitions because of the complex relationship between such variables as ground shaking severity, time of day, and damage sustained by the infrastructure. The following conditions, when combined, pose the greatest risk for severe post-earthquake fire damage:

- Buildings are unoccupied or individuals are not present to mitigate damage to gas systems or control small fires.

- High building density or dense, fire-prone vegetation.

- High wind, low humidity weather conditions, and low moisture content of combustible materials.

- Damage to water systems that severely limits fire-fighting capabilities.

- Reduced responsiveness of fire-fighting resulting from impaired communications, numerous requests for assistance, direct damage to fire stations, restricted access because of traffic congestion and damaged roadways, and delays in mutual aid from neighboring fire districts.

It is highly unlikely that all of these conditions will be present when earthquakes occur. The three most recent California earthquakes to strike in or near an urban region serve as examples of what might be expected in future earthquakes in the United States. Ground motions sufficient to damage buildings are most likely to impact utility and customer gas systems and create the potential for gas-related fire ignitions. Although people are advised in an emergency to shut off their gas service only when they observe or suspect gas appliance or structural damage, or can hear or smell leaking gas, many customers shut off their gas as a precaution, which increases service restoration calls. Gas restoration efforts following major earthquakes require massive mobilization of properly trained service personnel.

For example, in the 1994 Northridge earthquake, the total number of customers left without service immediately after the main shock and subsequent aftershocks exceeded 150,000, with approximately 133,000 of the service interruptions initiated by customers as a precautionary measure. Approximately 15,000 of the interrupted services were found to have leaks of unspecified severity when service was restored.

More than 3,400 employees, 420 provided by other California gas utilities as part of mutual assistance agreements, were mobilized to restore gas service. Service was restored to approximately 120,000 customers within 12 days. Approximately 9,000 customers remained without service one month after the earthquake because of building damage or an inability to access the customer's building or facility.

Natural gas also may be a contributor to the post-earthquake fire risk. The number of fire ignitions caused by earthquakes will be an order of magnitude less than the number of buildings damaged to the point of total loss or near collapse. Table 7-1 provides a summary of post-earthquake fire ignitions related to natural gas and other causes. The total number of fire ignitions in future earthquakes may be larger or smaller than in past earthquakes. However, gas-related fire ignitions can be expected to be 20% to 50% of all post-earthquake fire ignitions. While an earthquake may produce numerous leaks in the customer's gas system, the potential for fire ignition associated with natural gas leaks will be low compared to the number of leaks.

7.4 Causes of Post- Earthquake Ignition of Natural Gas

Damage to natural gas systems can cause gas leaks within customer facilities. The amount of leakage depends on the severity of damage and the operating pressure of the gas system. In many cases for residential appliances, damage may include partial or complete fracture of threaded pipe connections, flexible tubing, pipefittings, or damage to combustion-product vent piping. The displacement of unanchored gas equipment or gas equipment without a strong foundation or footing can be large enough to sever or damage the gas supply line to the equipment or damage the equipment itself. The absence of a flexible connection between the gas supply line and unsecured equipment increases the likelihood of damage from equipment movements.

There are two primary risks to public safety from damage to a natural gas system sufficient to cause release of natural gas. If the leakage is sufficient to create a flammable air-gas mixture and an ignition source is present, there is a risk of fire, or, in rare cases, explosion. The life safety risks from a gas-related fire are greatest if a fire is initiated in a damaged or collapsed building that has not been evacuated. In many cases, if electric power service is interrupted by the earthquake, an ignition source may not be present to ignite leaking gas. For this reason, careful coordination is needed with restoration of electric service by the electric utility to reduce the possibility of triggering additional ignitions by restoring electric power. Another potential life safety risk can result if gas service is restored improperly in the presence of gas leaks that are not first detected and repaired. Improper service restoration may also fail to correct inadequate venting conditions that might lead to the accumulation of carbon monoxide in a structure.

The risk of a gas-related fire in residential structures following earthquakes is generally very low because of the numerous conditions necessary for gas ignition (see, for example, Williamson and Groner, 2000). The ignition of leaking gas requires an ignitable mixture of gas and oxygen between the approximate range of lower (5%) and upper (15%) explosive limits and an ignition source. This can occur in the presence of a pilot light or when a light switch is turned on or off. For natural gas that is lighter than air and tends to disperse, the rate of gas leakage capable of igniting is related to the air exchange rate in the area of the leak. The likelihood of ignition is higher in conditions where poor air mixing allows formation of pockets of higher concentrations of gas.

Based on a review of the causes of fire ignitions in recent earthquakes, the following points summarize fire ignition scenarios involving gas or electric service. These scenarios incorporate the necessary presence of a fuel source and an ignition source.

- The earthquake interrupts electrical service to a structure and an electric-powered device is displaced or damaged and comes into contact with a quantity of fuel. When electric power is restored to the building, the device causes the flammable fuel to ignite. Example: A high-intensity light falling onto a polyurethane mattress.

- A hot water heater or other appliance is overturned or moved, rupturing the gas houseline or appliance connector, and the released gas is ignited by a flame or spark.

- A gas pipe in a building is broken due to building damage and the released gas ignites.

- A gas pipe in a building is broken and an electric spark from damaged electrical wiring is present, igniting the released gas.

- Bottles and/or open cans of flammable liquids are thrown to the floor by the earthquake, and an open gas flame or an electric spark ignites the vapors from the spilled liquid.

- Cooking oils and other kitchen fuels are spilled during the earthquake, and either electrical or gas-based cooking equipment ignites them.

- An open flame from a candle or Bunsen burner contacts a quantity of fuel.

- Arcing power lines or transformer damage ignites grass or brush.

- A fire is ignited by turning on light switches or the automatic switching of electrical equipment in the presence of a gaseous fuel.

Life-safety consequences from post-earthquake fires depend on the ability of individuals to evacuate buildings following earthquakes. Building layouts differ as to whether occupants must use shared paths of emergency egress or by a direct, unshared route. In multi-unit occupancies (R-1 occupancies), common paths of egress and limited means of escape make it more likely that persons can be trapped after earthquakes. The greater the number of occupants in a building, the greater is the likelihood an individual will be trapped in an emergency. Damage to exterior doors of apartment and condominium units often prevent occupants from exiting safely. In buildings of more than two or three stories, the escape paths usually include enclosed stairways whose doors can be jammed by the racking deflections of the doorframes caused by the earthquake. Frequently, the elevators in these buildings are also unusable. Some older buildings may have exterior fire escapes, but they may not be well attached after earthquakes. Single-family residential units (R-3 occupancies), on the other hand, cannot, by code, be more than three stories high, and their windows are usually constructed in such a way that they can serve as secondary exits. More and easier pathways exist for escape in R-3 occupancies than in R-1 occupancies. In addition, if the R-3 structure is properly tied to its foundation, it is less likely to lose its means of escape than the larger and more complex R-1 structures.

7.5 Options to Reduce Gas Fires Following Earthquakes

Many individuals and community leaders perceive the primary risk of post-earthquake fires is related to damage to the natural gas system. As indicated in the previous summary, this perception does not agree with actual experience in recent earthquakes. However, the role of gas in post-earthquake fire is important and does deserve attention. The most devastating damage from an earthquake is conflagrations, or uncontrolled, rapidly spreading fires, particularly in an urban center with high building density. Damage to the natural gas system is only one potential source of post-earthquake fires and is often not the primary contributor.

Several options are available to improve the earthquake performance of natural gas systems and increase public safety. Individual customers are often more concerned about protecting their property and the safety of those on their property. Gas utilities are generally most concerned about assuring reliable service to their customers. Government considerations may include protecting the community at large and maintaining a level of commerce necessary to meet the needs of the community while balancing investment in earthquake risk reduction with the other community needs.

7.5.1 Customer Actions to Improve Natural Gas Safety in Earthquakes

Many beneficial alternatives exist for gas customers to improve the safety of natural gas systems in earthquakes. These include improving appliance integrity and structural integrity and using gas flow limiting devices. Each alternative has advantages and disadvantages related to implementation costs, level of safety improvement, and collateral benefits for non-earthquake emergencies (see Table 7-2). Because every situation is different, deciding which alternative will improve safety is best done on a case-by-case basis. Approximate costs for implementing the measures in Table 7-2are provided in Table 7-3. The use of any one measure may or may not achieve the desired level of safety. In some cases, professional assistance may be required to determine what seismic safety measures offer the best protection.

7.5.2 Utility Actions to Improve Natural Gas Safety in Earthquakes

In addition to customer actions, natural gas utilities can take steps to minimize the consequences of damage to the gas supply system. Damage to the natural gas transmission and supply system generally has little direct impact on safety. One reason is that pipelines are often located beneath city streets, where the potential consequences to the public are low, even if leaking gas were to ignite. Benefits to gas customers and the community from pipeline improvement projects include greater overall reliability of service and reduced interruption in gas service to business and manufacturing sectors of the community following earthquakes.

Most natural gas utilities that operate in regions where major earthquakes are likely recognize the vulnerabilities of the natural gas transmission and distribution system. The most significant risk for earthquake damage exists in older distribution systems that were not constructed of welded steel or medium- and high-density polyethylene pipeline materials commonly used today. Pipelines in older distribution systems may have been constructed of bare steel pipe, cast iron, or copper, and their complete replacement is enormously costly. The installation of a large distribution main in a dense urban area using typical construction methods can cost $1 million to $3 million per mile.

Since February 3, 1999, the US Department of Transportation has required gas distribution utilities to notify customers that excess flow valves are available whenever a new service line is installed or an existing service is exposed. This requirement is intended to improve safety when a gas service line is severed during construction or trenching. Although not specifically designed to reduce earthquake-related risks, excess flow valves installed on service lines can reduce the potential for gas to be released at locations where severe earthquake ground disturbance is possible. Excess flow valves installed on service lines will not actuate from a break in the gas line downstream of a gas pressure regulator. For this reason, customers with services employing a pressure regulator (the vast majority of cases) do not reduce their risk of property damage by installing excess flow valves on service lines.

Other utility actions that are sometimes raised include the use of "smart" meters and remotely actuated valves to isolate portions of the natural gas supply system shortly after a major earthquake. The term "smart meter" is commonly used to refer to any customer meter that performs functions other than simply recording the volume of gas passing through the meter. The most sophisticated smart meters in use can be found in Japan where the practice of routinely installing smart meters originated. In addition to meter gas, smart meters can provide other services including

- Automatic transmission of gas usage to the utility to eliminate manual meter reading

- Monitoring unusual gas flow and sending an alarm to the utility or customer

- Automatic shut off of gas if the meter experiences a preset ground motion parameter

- Automatic shut off of gas if a signal is received of a hazardous condition such as leaking gas or high carbon monoxide levels

The primary obstacles for the widespread use of smart meters in the United States are initial equipment cost (smart meters can be 10 to 15 times more expensive than a typical gas meter), the installation cost, and the required supporting communications infrastructure (if remote communication with the utility is desired). Relying on smart meters as a means to reduce natural gas ignitions following earthquakes is further complicated by the lack of a reliable ground motion parameter to trigger the meter shut-off mechanism. The need for replacing existing meters with smart meters is also questionable considering that most of the gas control functions of a smart meter can be provided by various components of hybrid gas shutoff valve systems currently available for use with the existing gas meter.

Automatic isolation of the natural gas supply system is sometimes proposed as a mitigation strategy, primarily by those unfamiliar with gas system operations. Isolation of the gas supply system shortly after an earthquake does not reduce the risk of post-earthquake ignition of leaking gas. Since gas is supplied under pressure, substantial volume of gas remains in an isolated system. Some reduction in risk can only be achieved by rapid venting of the gas system although even venting the gas to atmosphere requires considerable time and is not likely to reduce the potential for gas leakage from damaged customer facilities shortly after the earthquake. Venting the gas system also dramatically increases the magnitude of the service restoration effort. For this reason, some indication of what portions of the gas system need to be isolated and vented would be needed before such a drastic measure would be taken. (This seems to imply that isolation and venting is appropriate in some cases, but you also make a case that is does not help because of the time required for it to actually reduce pressure at leak. There is an issue of benefit from aftershocks that is not raised, but seems to be of very limited benefit.) This additional delay further reduces the effectiveness of isolation as a means to reduce ignitions shortly after the earthquake. Finally, in certain jurisdictions, there may be environmental impediments to purposely releasing large amounts of methane to the atmosphere. For these reasons, isolation and venting of portions of the utility gas system is rarely beneficial except in very unique situations (e.g., locations where only a small portion of the system is to be isolated and vented immediately following an earthquake regardless of whether or not any damage has occurred).

7.6 Community Preparedness and Response Planning

Improving earthquake safety plans involves a complex process of identifying risks, evaluating safety alternatives under various scenarios, and developing and implementing effective strategies and plans. Several actions that a community can take are summarized in Table 7-4. Good information is necessary to define the issues, understand their magnitude, identify contributing factors, and consider alternatives. Decisions on what actions are appropriate for a particular community involve weighing each alternative's potential benefits with its expected consequences. Highest priority should be given to improving the safety of the most vulnerable areas, structures, and gas system installations. In general, communities should initially consider less expensive and more cost-effective strategies.

Challenges include:

- Balancing the needs of individuals (e.g., imposed financial burden, improved property protection) against the needs of the community (e.g., reducing probability risk of earthquake conflagration, other demands on community resources).

- Assessing potential earthquake fire risks from natural gas with other earthquake and non-earthquake risks.

- Balancing the costs of specific actions with the likely benefits, while assuring that costs do not impede effective actions and are not unreasonably high by both social and financial measures.

- Balancing strategies that address hazard prevention versus hazard response.

- Balancing potential benefits with the adverse consequences of any alternative.

It is important to identify areas of high risk for fire spread, regardless of the cause of the fire. These areas can be selected based on the construction type and age of the structures, the amount of flammable material in either the building stock or the natural environment, the location of firefighting resources, access restrictions, water supplies, and prevalent local wind conditions. In most cases, these areas will be known to the local fire departments. Prioritizing high-risk fire areas is often beneficial in formulating implementation schedules and budgetary requirements.

Many of the actions shown in Table 7-4 have an additional advantage in that they improve the fire safety of the community for any fire occurrence, not just earthquake-related fires.

Decisions on what actions are appropriate for a particular community involve weighing each alternative's potential benefits with its expected consequences. Highest priority should be given to improving the safety of the most vulnerable areas, structures, and gas system installations. Determining the value of reducing post-earthquake fire ignitions can be a very complex task. The potential damage from post-earthquake fires should consider the potential for fire spread and available firefighting resources. These considerations require some means to incorporate regional variation in earthquake ground motions, wind conditions, building construction, building density, earthquake damage that might impair firefighting response (e.g., damaged fire houses, loss of power for communications, congestion of telephone and radio communications, damage to water systems providing water for fire fighting, damage to roads and bridges that could delay access), and mutual aid resources. Because of the complexities involved in assessing post-earthquake fire damage in large urban areas, computer models support relatively rapid assessment of earthquake damage for a variety of initial conditions. HAZUS, a computer program developed by the Federal Emergency Management Agency, is one model that performs seismic loss estimates on a regional scale. Developing the basic information necessary to implement computer-based models of post-earthquake fire damage can be a significant effort and is generally only economically feasible for studying large urban areas.

Some caution is warranted when using the results from any computer model used to assess post-earthquake fire damage and the value of reducing ignitions related to natural gas. The underlying assumptions incorporated in the model must be fully understood to properly interpret the output. In particular, substantial uncertainty is associated with every aspect of the modeling process and the cumulative effect of this uncertainty needs to be captured in the resulting cost and benefit estimates to allow an informed decision to be made.

7.7 Key Issues

- Individuals should continue to be encouraged to shut off their gas only when they suspect damage.

- Property damage is the primary risk to residential buildings from post-earthquake fire since residential structures commonly have numerous escape routes that allow rapid egress following earthquakes.

- Unnecessary gas shut off by users (intentionally or through the use of earthquake actuated automatic gas shutoff devices) can create lengthy (several weeks to several months) and costly re-light effort. In cold climates this can be a life-safety hazard for frail individuals and secondary losses due to frozen water pipes in dwellings.

- Gas leaks can be expected to be associated with 20% to 50% of fire ignitions after an earthquake.

- Only a small number of gas leaks will likely lead to ignition after earthquakes.

- An energized electrical system following an earthquake plays a significant role in igniting leaking gas and other combustible materials.

- The use of seismically activated gas company isolation valves and system venting will have little impact on reducing gas related fires immediately after an earthquake. Area-wide use of such mitigation methods may require several months to restore customer gas services.

- Communities wishing to reduce earthquake fire hazards should weigh the benefits and consequences of any potential action. Highest priority should be given to improving the safety of the most vulnerable areas, structures, and gas system installations. In general, communities should initially consider less expensive and more cost-effective strategies and strategies that are beneficial for all fire hazards, not just those related to earthquakes.

7.8 References

Williamson, R.B., and N. Groner, 2000. Ignition of Fires Following Earthquake associated with Natural gas and electrical Distribution Systems, Pacific Earthquake Engineering Research Center.

Table 7-1 Earthquake Ignitions in Selected Earthquakes

Earthquake	Magnitude	Earthquake Fire Ignitions	Gas-related Fire Ignitions
1964 Alaska	9.2	4-7	0
1965 Puget Sound	6.7	1	?
1971 San Fernando	6.6	109	15
1983 Coalinga	6.2	1-4	1
1984 Morgan Hill	6.2	3-6	1
1986 Palm Springs	6.2	3	0
1987 Whittier	5.9	6	3
1989 Loma Prieta	7.2	67	16
1994 Northridge	6.7	97	54
1995 Kobe	6.9	205	36
2003 Denali	7.9	1	1

FIRE FOLLOWING EARTHQUAKE

Table 7-2 Valves and Alarm Devices That Assist in Limiting Natural Gas to Customer Facilities

Aspect	Manual Shutoff Valves	Earthquake Actuated Valves	Excess Flow Valves (at Meter)	Excess Flow Valves (at Appliance)	Methane Detectors	Hybrid Systems
Basis of Operation	Gas service shutoff valves are installed by utility at all gas meter locations and allow gas to be shut off manually.	Automatically shuts off gas when motion is sensed and the valve's trigger level of motion is exceeded.	Automatically shuts off gas if damage results in leakage down-stream of device, above the valve's designed shutoff flow-rate.	Automatically shuts off gas if damage results in leakage downstream of device, above the valve's designed shutoff flow-rate.	Sensor detects the presence of natural gas and initiates an audible alarm.	A system of modular devices that could include a main control unit, sensor inputs, and control and alarm outputs.
Installation and Maintenance Requirements	None since the valve exists as part of utility piping system.	Requires installation by a qualified person.	Requires installation by a qualified person. Needs to be sized for a specific appliance load and re-evaluated if the load changes.	Can be installed by building owner. Needs to be sized for a specific appliance load and re-evaluated if the load changes.	Can be installed by building owner.	Typically requires installation by a qualified person depending upon modules (required for installations associated with gas shutoff mechanisms).
Benefits	All gas services already have valves installed. Guidance on the use of manual valves is currently provided to customers in many public information documents.	Shuts off gas when the level of ground shaking might be sufficient to damage the gas piping system. Valves must be certified by the state to meet ASCE 25-97.	Shuts off gas only in cases when a hazardous condition exists, i.e., leak down-stream of device. Valves must be certified by the state to meet CSA 3-92.	Shuts off gas only in cases when a hazardous condition exists, i.e., leak down-stream of device. Valves must be certified by the state to meet CSA 3-92.	Alerts customer when potentially dangerous gas concentrations are present, allowing time for action.	Systems are modular and can be customized for specific applications. Each module has specific functions (e.g., vibration sensing, flow sensing, methane detection).
Potential Drawbacks	Can only be used if someone is present, knows where the valve is and has access to it, and has a wrench suitable to close the valve.	Gas can be shut off even if a hazardous condition does not exist. Aftershocks could cause the device to repeatedly activate after service has been restored. Device may activate from a vibration unrelated to an earthquake.	Will not shut off gas if leakage is below the valve's designed shutoff flow-rate, even if a hazardous condition exists. May not actuate as installed if the downstream load changes and the device is not modified.	Does not provide protection for damage upstream of the device. Will not shut off gas if leakage is below the valve's designed shutoff flow-rate, even if a hazardous condition exists. May not actuate as installed if the downstream load changes and device is not modified.	Customer is required to be on premises to hear and act on alarm to mitigate a hazardous condition. Alarm may occur for vapors other than natural gas.	Each module has specific functions (e.g., vibration sensing, flow sensing, methane detection).
Other Issues	Operation of a manual valve may be difficult if the valve is stuck, or impossible for customers who are handicapped elderly, or injured.	Widespread installation will produce extensive gas outages and delay service restoration. Not sensitive to changes in gas flow-rates or pressure conditions.	Available with and without bypass flow (allows automatic reset). Not sensitive to motion.	Available with and without bypass flow (allows automatic reset). Needed at each appliance to be effective. Not sensitive to motion.	California performance standards and certification requirements do not exist.	One or both California performance standards and certification requirements apply or do not exist for individual modules.

Table 7-3 Approximate Costs for Actions to Limit Natural Gas Flow After Earthquakes

Device[2]	Hardware Cost	Installation Cost[1]
Restrain Individual Gas Appliance	$15-$50	$0 - $100
Manual Shutoff Valve and Wrench	$5-$20[3]	$0
Earthquake Shutoff Valve	$100 - $300	$100 - over $300[4,5]
Excess Flow Valve at Meter	$20 - $100	$100 - over $300[4,6]
Excess Flow Valve at Appliance	$5 - $15	$0 - $100
Methane Detector	$25 - $75	$0
Hybrid System	$150 - over $500[7]	$100 - over $500[8]

NOTES:
1. All costs are approximate and do not include permit and inspection fees that may range from $25 to more than $100, depending on jurisdiction. Installations that can be performed by the building owner are assumed to have no cost.
2. Significant differences exist in the operation of the various devices listed. See Table 6.2 for more information.
3. Cost of a suitable wrench.
4. Installation costs do not include cost of a gas system survey, which can cost more than $200.
5. Higher installation costs may occur if substantial modifications of plumbing and valve support are necessary.
6. Higher installation costs may occur if substantial plumbing modifications are necessary.
7. Costs for hybrid systems depend on the number and type of components installed.
8. Higher installation costs can be incurred for hybrid systems that require installation of wiring to connect multiple sensing units.

FIRE FOLLOWING EARTHQUAKE

Table 7-4 Summary of Community Actions to Improve Natural Gas Safety

STRATEGIES	Reduce Gas Release	Improve Fire Fighting Capability	Improve Earthquake Response	Regulation
Provide information to the public through government offices, mail inserts, and the Internet.	n		n	
Present information on earthquake risk and risk reduction measures and provide recommendations to the community (e.g., homeowner associations, schools).	n		n	
Increase public engagement in earthquake response simulations.	n		n	
Organize neighborhood groups to assist in simple earthquake response measures (e.g., checking on neighbors, pooling emergency supplies).	n	n	n	
Provide improved firefighting response by improving water system reliability or addition of fire stations.		n	n	
Define high-risk fire areas within the jurisdiction and hold workshops to publicize the potential risk.		n	n	
Modify zoning regulations to reduce the potential for uncontrolled spread of fire • Limit building density • Require minimum street widths (to provide access and fire break) • Require brush growth management		n		n
Assess potential impact of reducing fire ignitions for future earthquake scenarios • Scenario modeling (e.g., HAZUS) • Qualitative assessment based on past experience and knowledge of firefighting capacity		n	n	

FIRE FOLLOWING EARTHQUAKE

STRATEGIES	Reduce Gas Release	Improve Fire Fighting Capability	Improve Earthquake Response	Regulation
Develop, implement and communicate an earthquake preparedness plan for the community.	n		n	
Provide training on proper procedures for manual gas shutoff, restraining appliances and installing devices to limit gas flow or provide warning of unsafe conditions; provide public with a list of trained individuals.	n		n	
Adopt ordinances to encourage or require installation of devices to limit gas flow into buildings following earthquakes • At time of sale or transfer of property • All new buildings • During major alterations or additions • All new and/or existing buildings	n		n	n
Modify building regulations to decrease likelihood of earthquake fire ignition and spread • Gas houseline installations that can accommodate earthquake-related building displacements without leaking • Structural retrofits • Automatic sprinklers • Fire-resistant construction	n		n	n
Require or encourage disclosure of potential gas system vulnerabilities at time of sale and develop appropriate disclosure forms.	n	n		n
Create public and private funding sources to support voluntary incentive or subsidy programs.	n		n	n
Create new funding sources or redirect existing funds for mitigation measures, training, and education.	n		n	n

8 Electric Power and Fire Following Earthquake

8.1 General Description of Power Systems

For the purposes of this monograph, power systems are divided into four parts: power generation, power transmission, power distribution, and ancillary facilities and functions. Although the equipment and its configuration are generally similar throughout the country, there can be differences, the explanation of which are beyond the scope of this document. Rather than providing a full description of power systems, only those parts that are needed to understand fire-following-earthquake issues are included. While not part of a utility's power system, emergency power will also be discussed because of its importance to first responders.

8.1.1 Power Generation

Power is generated as it is used, that is, power is generally not stored. As a result, power systems have sophisticated control systems to adjust generator output in balance power consumption. Depending on the region of the country and the progress of power system deregulation, large power generation stations are owned and operated by the utility to which they provide power. In addition, there may be large independent power producers, particularly in regions that have deregulated, that contract with utilities to provide them power or that sell power to power brokers. Another source of power is from power plants associated with industrial facilities. These plants tend to have smaller capacity units and tend to be newer. Because they are generally dispersed throughout the service area of a utility and their output is usually injected into the power grid at distribution system voltages, they may be able to provide power locally even when an earthquake has disrupted the high-voltage grid.

8.1.2 Power Transmission System

The power transmission system has a grid configuration of power circuits (each circuit consists of three conductors) in which substations are located at grid intersections. The transmission system typically operates at several voltages, such as 500 kV, 230 kV, or 166 kV. Power is transferred between transmission voltages at substations by means of power transformers. The redundant character of the grid configuration enhances system reliability by providing power to substations even if one of the circuits between substations is taken out of service for some reason. The power grid is supplied with power from generating stations. Large power generation stations inject power into the transmission system. The transmission grid also provides power to the distribution system (see below). Due to the historical evolution of the power system in some regions, each of the main transmission voltages may have a grid configuration. Frequently the 765 kV and 500 kV systems form a backbone configuration that tends to be linear in form rather than forming a fishnet-type grid.

Substations, either transmission or distribution are very similar in form and function. The operation of both will be discussed here even though distribution substations operate at lower voltages.

Substations have several functions:

- Connects circuits operating at the same voltage
- Connects circuits operating at different voltages using power transformers
- Protects substation equipment and circuits connected to the substation by opening circuit breakers when circuit conditions are outside of preset levels. This is similar to a circuit breaker in a house that opens to prevent a circuit from overloading
- Measures voltages and currents with monitoring equipment in the substation to track system status and indirectly control protective circuit breakers
- Communicates data describing the status of equipment and circuits at the substation to the utility's control center. Some equipment and functions at the substations can be controlled remotely from the control center

8.1.3 Power Distribution System

The power distribution system, which typically operates at voltages less than 60 kV, can be divided into two parts, a grid and feeders. The grid system is similar to the transmission system, that is, it consists of a grid of power circuits that intersect at substation. In addition to the functions described above, the distribution substation also provides power to distribution feeders.

The second part of the distribution system is the feeders that emanate from distribution substations like branches of a tree to carry power to utility customers: homes, businesses and industry. Typically, the feeders do not form a grid so that if a line opens, all of the users further away from the substation will loose power. The downtown areas of some cities have a grid distribution system of feeders. Three types of protective devices are used on feeder circuits. Each feeder will have a circuit breaker at the substation to protect for an over-current condition. The feeder may have circuit reclosers along the feeder that function in the same way as the circuit breaker in the substation. The circuit breakers and reclosers are typically programmed to open when the current exceeds a preset level, to pause for several seconds and then to reclose. It may repeat this sequence two to four times. If the overload condition persists, the unit will open and lock open and require manual intervention to reclose. The third type of protective device is a fuse cutout, which serves the same function as a fuse in a house circuit. When it blows due to an overload, it must be manually replaced.

8.1.4 Ancillary Facilities and Functions

A few of the ancillary functions and facilities not discussed in the previous sections are the control center, the network control system, service centers, Trouble Call Center, the communications network, and the emergency-operating center.

8.1.4.1 Control Center

Each utility has a primary control center that monitors the status of the system and controls several functions. The control center typically has a system status board that shows a schematic diagram and status of the entire system including all network circuits, substations (including circuit breaker status), and generating stations. Distribution feeder circuits are not shown. The control center also manages power dispatch; that is, sends commands to generating stations to adjusted their output to match the demand. Within the control center there may be a group that contracts for power on the spot market to purchase power that cannot be met by long-term contracts with generating facilities. The private branch exchange (PBX) which controls most of the telephones of the utility's telephone network is usually located at the control center. Facilities for the Emergency Operating Center (EOC) are often located at the control center. The control center is usually provided with and emergency generator for back up power and uninterruptible power supplies (UPS) for the system computers and the PBX. A UPS supplies power without interruption when the normal supply is disrupted to critical systems, such as computers that cannot tolerate even short power disruptions.

8.1.4.2 Service Centers

Distributed throughout the service area are service centers. They dispatch repair crews, maintain spare parts (primarily for the distribution system), provide a home base for the vehicles that it dispatches, maintain plans for the system in its service area, and have a base-station radio to communicate with its crews in its service area. Typically the need for a service call is provided by a message from the trouble call center (see below) or the control center. Typically, different crews repair the distribution lines and damage to substations. Spare parts for substations may be stored at the substation or at a central storage facility.

8.1.4.3 Trouble Call Center

The trouble call center is a function that most utilities have centralized at two locations. Customers reporting a power outage, downed line, or other dangerous condition will generally be routed to one of these facilities. Trouble Call Center personnel will then communicate, often by fax or email, to the appropriate service center so that they can respond.

8.1.4.4 Communication Networks

Power companies typically have extensive communication resources. There are usually utility owned microwave links between the control center, major substation, service centers, and generating stations. Recently, many utilities have added a system of optical fiber links between many of these facilities. Most large utilities have their own telephone network using a private branch exchange (PBX) (a telephone switch that may be the same as that used by a telephone company) to manage the telecommunication network. It usually has a radio system for communication for its mobile repair crews that includes base-station radios at service centers and repeaters throughout the service area to amplify the signal so that crews can be contacted anywhere in the service area. There is also equipment that allows signals to be transmitted over the power lines (carrier system), although this is usually used for protective systems rather than for voice communications.

8.1.4.5 Emergency Operations Center

A utility's emergency response plan provides for an emergency operations center. This is a room or series of rooms, often located at the control center that has special communication resources and a physical layout to facilitate the response to a disaster, such as a major earthquake. These systems are often modeled after the Incident Command System developed and used by fire service. Within California, a modification of this system is mandated for all state and local emergency operations centers.

8.1.5 Emergency and Backup Power Systems

Many facilities have installed emergency power systems because of the need for continued function of the facility or the economic costs associated with the loss of power. Emergency operation centers, utility control centers, and other critical facilities typically have emergency power.

There are two types of emergency power, uninterruptible power supplies (UPS) and backup power generators. When equipment, such as a computer, or a function, such as telephone service, cannot tolerate even short power disruption, they are provided with an UPS. That is, when commercial power is disrupted, the UPS will provide the needed power so that the equipment continues to function without interruption. Most UPS systems are based on batteries, although there are other technologies.

For many situations, a delay of 10 to 15 seconds is acceptable so that a back up generator can be used and the delay is caused by the time for the unit to start and get up to speed. Engine-generators can typically supply power for longer periods, as this only requires a larger fuel storage system. Most engine-generator systems use diesel fuel although gasoline and bottled gas are also used. In general, municipal gas supplies are not recommended, as an earthquake can disrupt them; however, historically these systems have been very robust. In addition to a fuel supply, engine-generators also need a control panel and a cooling system. Cooling can be achieved with a self-contained radiator and fan, remove closed system radiator and fan, or for larger system a cooling tower may be needed. Evaporative cooling towers also require makeup water.

8.2 Earthquake Performance of Power Systems

8.2.1 Overview

There are several perspectives that can be taken in evaluating power system performance after a damaging earthquake. Since the occurrence of a damaging earthquake that affects any given area is very rare (typically much less than once every 25 years), the loss of power for a day is generally tolerable when measured against storm related disruptions that can occur once a year or more. Historically, power disruptions from earthquakes in the United States have been less than 24 hours, excluding disruption in remote rural areas. Disruptions from severe wind or ice storms have been in excess of a week.

From the perspective of equipment damage, the performance has generally been poor in that a substantial amount of equipment operating at 230 kV and above is frequently damaged. The overall good system performance, as measured by power disruptions, can be attributed to the highly redundant design of power networks and the dedication and resourcefulness of maintenance personnel.

From the perspective of fire following earthquake, power systems are problematic, as they are a major ignition source. As discussed in Section 8.2.2, there are often area-wide power disruptions after a significant earthquake. This can have two impacts, one on sources of ignition, which is discussed in Section 8.3.1, and another on the significant secondary impacts discussed in Section 8.4.

All of the earthquakes that have affected large metropolitan in the United States have had moderate magnitudes. There is a danger in assuming that the response from these earthquakes will be similar to future, larger events. Thus, the one or two day power outages observed in the Loma Prieta or Northridge earthquakes may be significantly longer for earthquakes that release 50 to 200 times more energy observed in historic earthquakes.

8.2.2 Power Transformer Sudden Pressure Relay

Power transformers are the most costly equipment item in a substation. It is arguably the most important because its function of transferring power between circuits at different voltages, for example, between the transmission and distribution system, is necessary to provide service. Other damaged switchyard equipment can be bypassed so that substation operation can continue. Because of the high cost of transformers, they have several protective systems. One is the sudden pressure relay that is designed to sense the build up of pressure in the transformer case due to an internal short circuit. However, earthquake induced vibrations can cause oil in the transformer to slosh and trip the relay. This causes the transformer to be taken out of service, which can disrupt power over a large area. When this occurs, utility personnel will usually inspect the substation for damage and possibly run tests on the transformer. This can take several hours. The disruption to the network causes customers to lose power and in the case of step-up transformers at generating plant, the generator can be isolated from the network. In either case the output of the generating station will have to be curtailed because of the drop in power demand. When a thermal generating plant is taken off line for several hours, it may take a day or longer to get it back on line.

8.2.3 Distribution System Damage

In a recent damaging California earthquake, the damage to the distribution system was compared to that from a severe storm. This damage was primarily to the low-voltage feeders that radiate out from distribution substations. Most earthquake disruption to the feeders is due to the touching of energized lines that causes a fuse cutout to blow, or a line to burn and break the conductor causing the line to drop to the ground. Had the line not been energized these things would not happen.

In some cases, the protection devices will cause the dropped line to be de-energized, but it can remain energized if it comes in contact with dry ground. It can also arc and start a fire on whatever flammable material it comes in contact with. This last effect is the most significant from a fire following earthquake perspective. Downed lines that remain energized are a safety hazard. Because of the large number of blown fuse cutouts and downed lines, restoration of service is very labor intensive. In urban areas, power is usually restored within a day; however, in rural areas and in areas with rough terrain, where access may be impeded, restoration may take several days. In Section 8.2.2 it was noted that area-wide blackout can result from the action of sudden-pressure relays. Experience indicates that there can be major distribution system damage described above before the power is lost. If a downed line did not trip protective devices before there was a general loss of power, it is an ignition source when power is restored.

8.2.4 Service Center Communications

After a damaging earthquake, there is telephone system congestion (See Section 8), so that information about dangerous situations is not reported to the Trouble Call Center.

Congestion on the utilities telephone system may prevent the Trouble Call Center from contacting the service centers so that they can respond. Also, there have been situations where communication between the trouble call center and critical service centers has been disrupted due to equipment damage. Finally, local and regional governmental emergency operation centers may need to communicate with the utility, but congestion may prevent this.

8.2.5 Utility PBX Congestion

Just like the commercial telephone systems they emulate, the sudden increase in telephone traffic can cause the power-company PBX to become overloaded so that communication between facilities can be disrupted. While power companies have several alternative means of communication, in the aftermath of a damaging earthquake there have been problems in communication.

8.2.6 Emergency Power

Because of the high reliability of power systems, emergency power systems are seldom subjected to the test of a real power disruption. Unfortunately, when called upon during a loss of power they often are unreliable. As a result control centers, emergency operating centers, service centers, and radio repeater station have experience power outages immediate or shortly after earthquakes where power has been disrupted.

The causes of uninterruptible power supply failure are usually the failure of a battery rack, inadequate restraint of batteries to their rack, and poor maintenance so that the batteries are not fully charged or have aged and deteriorated. The causes of failure of engine-generators are too numerous to mention here, but the main causes are the following:

- Failure or discharge of the battery used to start the engine-generator
- Contaminated fuel that clogs fuel filter or injectors
- Inadequate generator-support system or poor installation that fails in the earthquake
- Failure of fuel line from the main fuel tank
- Failure of cooling system lines to the cooling tower
- Failure of the municipal water system so that there was no supply of makeup cooling water

There are two other issues related to the fuel supply for emergency generators. Historically, the fuel tanks for engine-generators were buried and tended to have a capacity to run the generator for a week or more. Environmental Protection Agency regulations now require these tanks to be double walled. Many of the replacement tanks are now placed above ground and have much smaller capacity. Coupled with this is the vulnerability that it may not be possible to get re-supplied with fuel due to disruptions in road, shortage of fuel tanker trucks, and inadequate supply. For example, of the three major diesel fuel suppliers on the peninsula near San Francisco, only one has emergency power to get their fuel out of their storage tanks.

8.2.7 Reliable Source of Fuel for Emergency Vehicles

After the 1989 Loma Prieta earthquake commercial flights were diverted from San Francisco and Oakland International Airports to San Jose. While San Jose had jet fuel in storage tanks for the planes, there was no emergency power to get the fuel from the storage tanks to the airplanes, so they could not depart. The situation was the same for emergency response vehicles.

8.3 Direct Impacts of Power System Performance on Fire Following Earthquake

Probably the most significant direct impact of power systems on fire following earthquake is that electric power is a major fire ignition source. In addition to dropped distribution lines, power circuits in damaged houses are another major ignition source. Even without major structure damage, fallen light fixtures and items falling on electric stoves that have been left on can also be ignition sources. There have been cases where as many as two-thirds of all ignitions after an earthquake has been attributable to the power system.

One of the prized tenets of most utilities is the reliability of service, that is, continuity of service and rapid restorations should there be a disruption. To meet this objective, power is frequently restored to areas without coordination with other first responders. As a result, if a fire should start, the fire service may not be in a position to respond quickly.

Many buildings and businesses have centralized security and fire detection systems and status boards near the building entrance for the use of the fire service. When power is lost and restored, some of these systems generate a false fire alarm or reset an alarm that was triggered prior to the loss of power.

8.4 Secondary Impacts of Power System Disruption on Fire Following Earthquake

Many in the first-responders community are not adequately prepared for the loss of power; even when emergency power is provided, it is often unreliable. Many facilities, such as hospitals, have emergency power generators but they typically only supply critical needs when the earthquake may require the hospital to meet extraordinary demands.

The loss of power can disrupt the fuel supply for emergency response vehicles because fuel storage tanks, be they at the organization's garages or commercial gas stations, are below ground and the gas pumps are seldom provided with emergency power.

The loss of power can be disruptive to communication systems (wire-based, wireless and radio systems, See Chapter 9).

Water Systems may lack emergency power for pumping to maintain water pressure or power for system control and data acquisition (SCADA) systems for monitoring and remote control of the water system (see Chapter 6).

The loss of power may severely disrupt transportation systems due to the disruption of traffic signals. Downed power lines may block roadways. In many urban areas, light rail and subway systems would not be operative. (See Chapter 10).

The loss of power may place added demands on the fire service to rescue people trapped in elevators and in subways systems.

8.5 Key Issues Related to Fire Following Earthquake

The key impacts of power systems to fire following earthquake are their contribution to fire ignition sources and the secondary effects of the loss of power on other lifelines and the emergency response community.

8.6 Mitigation Strategies

8.6.1 Seismic Shutdown of Selected Feeders

The action of transformer sudden pressure relays that disrupt power during an earthquake has been shown to reduce damage to the distribution system as well as reduce sources of ignition. Several years ago the suggestion to utilities that power be disrupted intentionally by using seismic switches was met with strong resistance. First, any fast-acting switch would cause some unnecessary shutdowns. That is, the system would be shut down even when the subsequent ground shaking did not cause significant damage. Second, it would be undesirable to disrupt the power to some users, such as hospitals and some manufacturing facilities where in-process products would be damaged or ruined. Even if disrupting critical customers could be avoided, the utility would have to be protected against liability.

Advances in the science of evaluating strong motion records now allows better prediction of the severity of ground shaking based on the evaluation of the first part of a ground motion. Advances in electronics now allow low cost measuring and analysis devices to be designed and placed in distribution substation to control specific feeders. It would still be necessary for public utility commissions or for legislatures to promulgate rules that would protect the utilities when they deliberately shut off power to reduce damage, disruption, and the potential for catastrophic fire following earthquake. Utilities would also have to accept balancing reduced reliability (more power disruptions) for reduced damage, reduction of longer disruptions, and reduction of risk of fire following earthquake.

8.6.2 Coordinated Restoration of Electrical Service

Ideally, fire service personnel should be informed and allowed to preposition personnel and equipment to areas where power is being restored so that if a fire should start, it can be quickly extinguished. The distribution system should be inspected for downed lines before power is restored in regions where houses are dispersed in woodlands when it is dry and windy. There is a need to coordinate communications between electric utilities, gas companies, and fire services on an on-going basis.

8.6.3 Enhance Communications to Service Centers

Government emergency operations centers (EOC) may become aware of situations in which a utility's response is needed. There is a need for these organizations to have a reliable means of communication with the utility just after the earthquake occurs. At the present time government EOCs are staffed by members of the Amateur Civil Emergency Service (RACES) or Amateur Radio Emergency Service (ARES) personnel on a voluntary basis. These services have provided the only means of communication when other means have failed. Adding this resource to electric power service centers should be considered.

As the Trouble Call Center function is consolidated, it is important that there be redundant systems so that should one be rendered inoperable due to the earthquake, an alternate center should automatically get calls. Also, as systems become automated and computerized, redundant links between service centers and Trouble Call Centers should be established.

8.6.4 Fuel for Emergency Vehicles

A source of fuel for fire engines should have backup power or a manual pump so that equipment can be refueled.

8.6.5 Seismic Qualification of Substation Equipment

The IEEE Standard 693, Recommended Practice for Seismic Design of Substations should be used in the purchase order specifications as equipment meeting this standard should perform better in earthquakes so that the extend and duration of power system disruptions due to equipment damage should be reduced.

8.6.6 Seismic Upgrade for Key Power System Structures

Many utilities have important structures, such as those that house the control center or substation control houses, that were build with antiquated seismic elements in building codes. Key structures that affect power system performance should be seismically evaluated and if deficient given high priority for seismic upgrading.

8.6.7 Department of Energy Emergency Fuel Supply

The Department of Energy (DOE) has established a program to supply energy to National Defense/Emergency Preparedness organizations. Participants in the Energy Service Priority (ESP) program can get priority service for the restoration of electrical service or for the supply of fuel. Organizations should generally be registered prior to the emergency.

8.7 References

"IEEE Recommended Practice for Seismic Design of Substations," Institute of Electrical and Electronic Engineers, 345 East 47th Street, New York NY, 10017, IEEE Std 693-1997.

"Guide to Improved Earthquake Performance of Electrical Power Systems," ASCE, Manual No. 96.

"Hyogoken-Nanbu (Kobe) Earthquake of January 17, 1995 – Lifeline Performance", ASCE, A. J. Schiff Ed., Monograph 14, 1998.

"Hokkaido Earthquake Reconnaissance Report", Earthquake Spectra, Supplement A to Vol.11, April 1993.

9 Communication Systems and Fire Following Earthquake

Functioning communication systems have often been identified as the single most important element to an effective post-earthquake emergency response. This section describes communication systems, their earthquake performance, vulnerabilities and methods to mitigate problems. Preventing disruption of communications will improve the post-earthquake emergency response and reduce the risks and damage of fire following earthquake.

9.1 General Description of Communication Systems

Communication systems are divided into the following major categories: the public switched network, the wireless network, private networks, mobile radio, emergency services, and other communication systems (e. g. pagers, satellite, and internet). Each communication system is a network consisting of nodes and links that interconnect subscribers to deliver the functions of the system. Different types of equipment are installed in the nodes to perform the necessary functions while the links are copper cables, optical fiber cables or radio waves (such as microwave). Today the different communication systems are required to interconnect and interoperate with each other.

9.1.1 Public Switch Network

The public switched network is the traditional telephone system that is often referred to as the wireline system after the introduction of the wireless network. Today the wireline system primarily uses optical fiber links (trunks) to form a network that interconnects the nodes. The redundant character of the communication network provides high system reliability. For simplicity, in this monograph the nodes will be referred to as central offices, although there are different classes of offices. Central offices contain switches needed to make the connections to complete a call between the caller and the called party. Central offices also contain the facilities (hardware and software) for the Internet and transmission.

Some of the links (trunks) between central offices use microwave transmitted via land-based towers and satellites. The trunks that connect central offices are composed of copper wires, optical fibers, or microwave links. Most trunks are buried, but some are supported on poles. Highway bridges are often used to carry the trunks over rivers. Long trunks have repeaters; that is, equipment is inserted at intervals to amplify the signal to maintain signal quality. The lines from the central office to homes are still primarily copper, but due to demand for data speed and volume these lines are being replaced with optical fiber.

Advances in technology in telecommunications systems have resulted in specialized computers replacing the electro-mechanical equipment in central offices. The introduction of computers has made possible many features that were not possible with the old system (e.g. consumer features, such as caller ID, call forwarding, and call waiting, etc.). The configuration of the network has also changed; it is a network of multiple layers of functionality. The control layer provides the system with a fast response if the called number is busy, so that the time required to optimize the route is not wasted. Other features that are important from an emergency response perspective include the availability of Essential Line service (also known as high priority services) and the use of the Government Emergency Telecommunications Service (GETS) that are described below.

The capacity of the switching equipment in a central office is designed to at least meet the normal peak call volume of the area that the central office serves. Thus, it cannot accommodate all subscribers at the same time, as this would be extremely costly and would not be practical. If demand increases above the switch's capacity, as occurs after an earthquake, the system becomes congested and users must wait to get access to the system. If a call is attempted when the system is congested, there may be a dial tone delay, the subscriber will be given a fast busy signal (circuits busy), or the subscriber will receive a recorded message to call later.

To improve access to critical users such as fire service and doctors, some lines can be given a special status, called Essential Line Service, which gives them priority access to the central office. If a called party is distant from the caller, the call will have to be routed through different central office, and the call will have to compete for access to the trunks at each central office along the route. Thus, an Essential Line Service only improves access to the first central office. Although modern switches can implement the Essential Line Service feature, not all telephone companies provide this service. It is usually a free service for qualifying users.

To provide priority service beyond the first central office to critical users, the National Communication System has created the Government Emergency Communications Service (GETS). The GETS system allows qualified government and emergency responders to subscribe to the GETS system that will provide priority connection to the called party once access is gained to a central office. About 85% of all access lines in the United States can provide this service.

For most users there is no redundancy between the user and the central office. If the line connecting them to the central office is severed or the central office is out of service, the user will not be able to use the telephone. However, because of system redundancy, the loss of a central office will have little impact on the system from a network perspective. The services provided by a central office to a user depend on the hardware and software deployed in the switches.

Switching equipment used at central offices is inherently redundant to ensure reliability; that is, one duplicated system is always ready to take over. When the function of the primary system is transferred to the standby system an alarm will be set off at the control center to notify maintenance personnel.

Telephone switches must have a continuous supply of power. To assure this, central offices are provided with backup batteries with the capacity to operate the equipment for 8 to 10 hours when commercial power fails. In addition, the batteries are configured in parallel so that they are redundant. Power rooms at the central offices are also designed with high reliability. In general, in key offices, a backup engine-generator will be available to provide power for several days or longer. Switching equipment requires a controlled environment (temperature, humidity and dust) for their proper operation and improved reliability. Most large HVAC systems require both electric power and water supply to operate. The disruption of the HVAC system for several hours (depending on outside temperature) will impact the functions of the switching equipment and any electronic equipment.

Wireline and wireless telephone companies maintain trouble call centers where service problems can be reported. These centers are typically provided with emergency power and there may be two sets so that an alternate is available should one be closed for some reason. The center will dispatch a repair crew in response to trouble reports, such as a loss of power, a damaged pole or downed line.

Within the telecommunication industry regulatory bodies have established design criteria, standards, and operating procedures that provide for earthquake protection as well as other hazards, such as fire and floor loading. In particular, communication switchgear is required to be seismically qualified by shake-table test standards that have been in effect for several decades. These test requirements are more conservative (severe) than any other seismic standard to date. As a result, there has been no structural damage to this equipment in earthquakes if it is installed in accordance with manufacturers' recommendations. Therefore the probability of switchgear equipment failure either due to earthquake or fire is small.

9.1.2 Wireless Network

The wireless network is almost the same as the wireline network from a network perspective. The wireless system takes a different form as it approaches the user. Signals from central offices go to cell sites. Each cell site consists of antenna, antenna support, and an enclosure for radio transmitters, receivers, switchgear, transmission equipment, backup battery and possibly an engine-generator. The cell site communicates to users' handsets via radio links, either analog or digital. Cell-site antennas dot the landscape and in urban areas are usually located on top of tall buildings so that each antenna has a large area of coverage. The wireline and wireless networks are independent, but are interconnected at many central offices and mobile telephone exchanges (the wireless equivalent of a central office) so that a cell-phone user can call a person on a wireline telephone, and vise versa. Isolated cell sites usually have a small engine-generator that can provide long-term backup power. For cell sites in urban buildings, the equipment may be indoor or outdoor type. The indoor type generally requires a controlled environment while the outdoor type usually comes complete with its own environmental control. These sites seldom have engine-generators and their batteries can typically support the site for only about three hours when commercial power is lost.

Cell sites and the central office that serve an area can get congested when the demand for service is greater than capacity. The equivalent of the Government Emergency Telecommunications Service system for cell phones (Wireless Priority Service) is being implemented. Like GETS, the National Communications System will control subscription to the Priority Wireless Service.

Inadequate installation of cell sites, particularly in urban areas where they are located in existing structures, can contribute to failure in an earthquake. Anchor bolts on roofs often cannot be used because they would penetrate the waterproof roof coating and they may not be allowed in rented spaces. Although support frames with multiple support points can be engineered, this was frequently not done during the days of rapid expansion as in the late 1990's.

The Short Message Services (SMS) available on the wireless network allows a short text message to be sent to the called party. This system reduces the time needed to transmit the call as sending text is much faster than voice and the called party is on the line a shorter time. When more cellular users use this feature after an earthquake, more system capacity will be freed up for emergency services. Another advantage of the Short Message Service is that if the called party phone is busy, the message is stored and can be retrieved later, although the storage capacity of messages is limited.

9.1.3 Private Networks

There are many private telephone networks including universities, hospitals, large companies, and many utilities (e.g. power companies). These systems are referred to as Private Branch Exchanges (PBX). These networks are typically wireline systems and often use switches that are identical to those found in wireline central offices. They usually have links to the wireline network. They may lack some of the redundancies found in a telephone company facility and their installation may not be to the same high standard. The capacity of the backup battery of private networks is often only a few hours and there may be no engine-generator backup power. In some cases there may not even be a backup battery.

When a PBX fails the phones that are connected to the PBX system will not have any dial tone (that is, calls cannot be made). However, many PBX systems have a few circuits directly connected to the central office, the phones that are on these lines will work if the central office is functional. In most cases, the fax machine lines are the direct lines to central office.

9.1.4 Mobile Radio Systems

Radio systems are used by first-responders (i.e., fire, police and rescue), utilities and other organizations that must dispatch personnel. Each of these systems usually consists of a base station (with antenna), repeaters, and mobile units. The base station is a fixed radio transmitter that is usually provided with an antenna strategically located so that it has broad coverage. These systems work better with "line-of-sight" to communicate between a transmitter and receiver. For large service areas or hilly regions, additional transmitters, called repeaters, are required to assure that good communications can be maintained throughout the entire service area. The base-station antenna and repeaters are often located on hilltops. After an earthquake, power to the repeater sites is often disrupted and physical access may be limited so that an engine-generator is needed to provide long-term backup power.

Some organizations are replacing their radios systems with cells phones because of the low capital cost, no need for maintenance, good coverage of the service area, and generally high reliability. The problems with the loss of power and congestion with cell-phone systems after an earthquake are often not appreciated by the utility personnel and first responders who use them.

Mobile radios may be vehicle mounted or handheld. Handheld units require batteries that are typically rechargeable modular units and are not compatible with standard dry cells. Thus, the continued operation of these units requires that charged replacement battery packs be available.

Mobile radio systems can also become congested. Not only can usage of existing units in any give system increase, but also there can be a gathering of additional units to aggravate the situation, as in the case of a large fire where additional units are dispatched.

Amateur radio associations have organized groups that have provided vital communications services after disasters using their own equipment. At the present time government emergency operations centers are staffed by members of the Amateur Civil Emergency Service (RACES) or Amateur Radio Emergency Service (ARES) on a voluntary basis. Through prearranged agreements volunteers report to designated sites with their own radio equipment shortly after any disaster in which it appears that their services may be needed.

9.1.5 Fire-Service Communications

It is impossible to characterize the communications systems for a typical fire service because of the large differences in the size of fire departments and because the decisions about equipment are made locally. In small towns and rural areas the fire service consists primarily of volunteer personnel where as large cities may have scores of vehicles and hundreds of professional firefighters.

Professional firefighters in a fire truck wear earphones and have a microphone to communicate with other members of the unit. The commander is able to communicate with the dispatcher using a push-to-talk key. Some radio systems can access the wireline network through a dispatcher. As noted above, congestion can occur on fire-system radios. In the high-stakes, high-tension situation of fighting a fire, units attempting to get assess to a congested system can talk over each other therefore communication deteriorates. Implementing a system of supervisory control of the network can help this situation. That is, one unit on the network acts as a gatekeeper to control access to the network. A technical solution is to use a so called "trunking" system that uses a computer to transparently switch to less-used radio channels so that all members designated to be in a talking group can communicate with each other. The talking group can be created so that all units working a fire can communicate.

A dispatcher affiliated with a Public Service Answering Point (See below) will usually dispatch fire personnel. However, in the aftermath of an earthquake it is not uncommon for units to be dispatched by an individual walking to a nearby fire station to request assistance. This can be done in lieu of a call to 911 because the telephone system does not respond due to system congestion. Some fire departments install call boxes outside their fire stations. If the crews have already been dispatched, the call box, which contains a telephone that is connected directly to the fire dispatch center, can be used to report the fire. The use of fire-pull boxes distributed throughout cities has largely been replaced with the 911 system. While these systems were reliable if properly maintained, maintenance was costly.

Many structures have sophisticated fire-alarm systems and status-board systems located near the building entrance for fire service use. These systems may have smoke and heat detectors, fire-suppression water-flow detectors, and automated call systems (via the Public Switch Network) to provide a warning to appropriate authorities. The call may go to a private security service or less frequently directly to a fire department.

9.1.6 Other Means of Communication

There are other means of communication that may be important for fire following earthquake. Some of these means of communication may only be one way.

9.1.6.1 Public Service Answering Points

When a 911 call is placed on a wireline system, it is directed to a Public Service Answering Point (PSAP). Densely populated counties and many cities have their own PSAPs. In many locations the sheriff's office handles this function. Dedicated PSAPs typically have two or more equipment positions staffed by operators who identify the caller's need and either transfers the call to the police, fire, or medical service for dispatch or dispatch these services directly. Enhanced positions automatically display the address from which the call is being placed.

A 911 call placed on a wireless system may be routed to the highway patrol or to a PSAP, depending on how the local jurisdiction is organized.

9.1.6.2 Satellite Systems

Many organizations, such as state emergency organizations, maintain satellite communication systems. Of particular note is the system established by the Office of Emergency Services (OES) in California which has its own communication system that includes wireline, wireless, and satellite (cellular, microwave, and short-wave) networks. This integrated communication system also connects to the wireline network to provide alternate communication resources and enhanced reliability. Mobile units are also available to deal with extreme cases. An extensive emergency operations center with monitoring stations for utilities, civil services and government agencies is set up and ready to be activated in the case of an emergency.

9.1.6.3 The Internet

The Internet was designed to operate without central control and it has been very robust to communication system disturbances. Individuals who access the system through modems attached to a computer may experience access problems due to telephone system congestion. Individuals with Digital Subscriber Line (DSL) have a dedicated line and should be able to get access to the Internet. Many emergency response organizations maintain web sites that post up-to-date information on the disaster and can provide other useful information. However, as the number of DSL subscribers increases, congestion (a different form of circuit busy) will occur.

9.1.6.4 Emergency Alert System

The Emergency Alert System consists of broadcast radio stations throughout the country that are selected to provide good coverage and generally have emergency power.

National, regional, or local messages can be broadcast using this system. This system replaced the Emergency Broadcast System.

9.1.6.5 Earthquake Information Systems

There are three systems in the United States that are designed to provide information on the size and location of an earthquake shortly after it has occurred. The systems in Northern and Southern California develop this information and make it available within two minutes after the earthquake and also provide an intensity map for the earthquake based on observed ground motions. Notices are distributed by pager and over the Internet. For the United States outside of California information is posted within two hours after an earthquake. Emergency responders can use this information for early response planning. The World Wide Web address for data for Southern California is http://www.trinet.org/shake/, for Northern California is http://quake.wr.usgs.gov/research/strongmotion/effects/shake/, and for the entire United States is http://earthquake.usgs.gov/recenteqsUS/.

9.1.6.6 Radio and TV Broadcast

Radio and television stations are a good source of information following an earthquake. In order to capture audience, broadcast media often provide continuous news at damage and fire sites. This coverage can be useful to first responders, and television may be of particular value to fire fighters, as aerial views from helicopters may provide an overview and perspective not otherwise available.

9.2 Earthquake Performance and Vulnerability

9.2.1 Overview

The overall performance of communication systems within the United States and Canada has been very good as measured by facility or equipment damage. However, even a moderate earthquake can cause significant communication system disruption due to congestion. In the 1989 Loma Prieta earthquake the call volume increased by a factor of 4 to 10 times of norm and remained high for at least 3 to 4 days. The epicenter of the earthquake was located about 60 miles south of San Francisco; the loss of normal power and subsequently loss of the emergency power resulted in the disruption of all communication to the local-government emergency operations center except that provided by amateur radio operators.

In large earthquakes electric power has typically been disrupted over a wide area (See Section 7). This, coupled with the poor reliability of emergency power systems has increased vulnerability of communication system to disruption.

9.2.2 Earthquake Performance of Wireline Systems

In the 1971 San Fernando earthquake electro-mechanical switchgear in a central office fell over. Since then, there has been no earthquake damage to modern switchgear in the United States or Canada, although there have been many instances in which system congestion has severely impaired system performance. There has been damage to building support services (e.g., HVAC and water systems and cases in which the external supply of water needed for the HVAC system was disrupted.). There has been damage to central office batteries and engine-generators have experienced a broad range of problems. In most cases normal power was restored or alternate emergency power was supplied so that central office operations were not disrupted. In one case, some switch functions had to be shut off to conserve power. In a few cases fires started as a result of the earthquake, damaged overhead communication lines; one such fire occurred in the 1995 Northridge earthquake that destroyed several optical fiber trunk lines and disrupted service at a remote central office. There were two reports of fires within central offices after earthquakes; however, these fires were put out and did not affect service.

There were major equipment building collapses and equipment damage in the 1985 Mexico, in the 1988 Armenia, Soviet Union, in the 1999 Kocaeli, Turkey and in the 1999 Chi Chi, Taiwan earthquakes. In 1994 Kobe, Japan earthquake there was extensive damage to communication facilities caused by fires that destroyed several large areas of the city.

Most central offices in older communities were constructed many years ago using antiquated seismic requirements in building codes. In many locations particularly outside of California, construction may predate seismic code requirements. Although central offices are structurally strong due to windowless design and large floor loading capacity, they may still be vulnerable to earthquake damage or collapse. The introduction of computer based switches, which require much less space than the older systems means that as service demands increase new equipment can be accommodated in the existing structure. Also, should a new structure be constructed, the old structure would have to be retained during the transition for continuity of service. The relocation of a central office would also require great expense in relocating all of the underground trunk lines from the existing structure. For these reasons there are strong incentive to continue to use the existing seismically vulnerable building.

There is relatively little seismic performance data for optical fiber cables. A single optical fiber has a capacity much larger than a large cable of copper conductors. In some locations bundles of as many as 144 optical fibers have been installed along a single route in a single conduit. This concentration of transmission capacity and the lack of route diversity increase the vulnerability of the system.

9.2.3 Earthquake Performance of Wireless Systems

System congestion is also a problem for the wireless telephone system. Congestion on the wireless system occurs at the cell site and at the central office. In some previous earthquakes the wireless system worked marginally better between wireless handsets. With an influx of handsets, congestion on the wireless system will be more sever. Also, at critical locations, such as near the regional emergency response center, the wireless system became congested.

The operating life of cell sites on batteries after the loss of commercial power is typically three hours. Area-wide black out can be expected in severe earthquakes (See Section 7), for blackouts longer than three hours the cell sites will fail unless temporary power is made available to the cell sites

9.2.4 Earthquake Performance of Mobile Radio Systems

Mobile radios have become congested when large numbers of vehicles and hand-held radios congregate at a large disaster scene.

There have been several earthquakes in which remote base-station antennas and repeaters went off line because equipment was damaged or the site lost power. These sites are often located on local peaks or ridges, where earthquake ground motions tend to be amplified (i.e., more severe). As the sites are often remote and surrounded by steep slopes, ground slumping and landslides may disrupt power lines leading to the site. The landslides also make roads to the sites impassible.

In a severe earthquake, the emergency response phase can last for over two days. As with cell phones, radio-handset batteries quickly run down and cannot be recharged due to the disruption of commercial power.

There have been base-station radios that were unsecured on tables and fell to the floor in an earthquake. While not all were rendered inoperable, this shows a need for proper installation.

9.2.5 Public Service Answering Points

Several public Service Answering Point facilities have lost commercial power and their engine-generator either failed to start or stopped running shortly after it started. While the telephones continued to ring because the central office supplied their power, the control positions did not work so that the operator could not identify which phone was ringing and could not dispatch crews.

9.3 Direct Impacts of Communication System Performance on Fire Following Earthquake

9.3.1 Inability to Access Emergency Services Through 911 System

The 911 system is the primary means available to the public to report dangerous situations (e.g., fire, a downed power line, a gas leaks), and to request medical assistance). If the telephone appears to be inoperative due to system congestion, the services that can be provided by calling 911 will not be available. In most communities this is the main means that the fire service is made aware of a fire, therefore they cannot respond rapidly without this system of communication.

9.3.2 Dispatch Function Disrupted by Damage to Radio Communication Systems

Damage to base stations and repeaters makes it difficult to dispatch first responder crews as well as utility repair crews.

9.4 Secondary Impacts of Communication System Disruption on Fire Following Earthquake

The disruption of communication will slow many aspects of the emergency response. Fire departments may not find out about a fire as quickly therefore suppression is more difficult. It will take longer to clear downed power lines that cross roads so that fire vehicles may not be able to get fires, and gas-company service personnel may not be able to get to valves to isolate broken gas mains.

9.5 Summary of Key Issues Related to Fire Following Earthquake

Key issues include:

- The function of the three main communication methods (i.e. wireline, wireless, and mobile radio) may be impaired due to congestion

- Portable radio handsets have rechargeable batteries that cannot be recharged when commercial power is lost

- Some critical systems and facilities, (e.g., telephone company central offices and Public Service Answering Points centers) have unreliable emergency power

- The installation of some equipment (e.g., switchgear in cell sites, HVAC equipment at central offices, and base-station radios), can be inadequate and results in equipment damage in earthquakes

- Structures housing communication equipment may predate good seismic construction practices and are vulnerable to severe damage or collapse

9.6 Mitigation Strategies

Mitigation strategies are listed in the order of potential impact on system performance, ease of implementation (technical, administrative and political), and cost of implementation. Because these criteria depend on local needs, the ranking should be considered tentative.

9.6.1 Communication System Congestion

9.6.1.1 Improve Broadcast Radio Notices

After damaging earthquakes there are often public service radio announcements asking citizens to limit calls to emergency use. However, this does little to deter people calling to find out if family and friends are safe. An unscientific survey indicates that once a person is contacted the length of the call is typically over 15 minutes as "war stories" are exchanged. Public service announcements should emphasize the effects of congestion and the need to keep calls short.

Many telephone sets and handsets have an automatic redial feature. During times when there is telephone system congestion, this feature should be disabled, as it can add to system congestion and degrade system performance further. Public service announcements could suggest that this be done.

9.6.1.2 Limit the Duration of Calls

Although this would require Public Utility Commission regulations or legislation, after a governor declares an emergency, telephone companies could limit the duration of calls. This could be preceded by radio public service announcements and a telephone message that calls will be limited. The duration of calls could be adjusted based on telephone company measures of congestion. These restrictions would not apply to Essential Service or Government Emergency Telecommunications Service calls.

9.6.1.3   Request an Essential Service Lines at Fire-Dispatch Center

The fire dispatch center may have to use the wireline telephone system to contact other organizations or facilities. To improve access to the central office when the phone system is congested, an Essential Service line should be requested from the telephone company. This service is usually provided free.

9.6.1.4   Obtain Government Emergency Telecommunications Service for Key Personnel

Key qualified fire-department personnel should be provided with a Government Emergency Communications Service access card to improve the ability to use the wireline telephone system when it is congested. This service is provided free, although there is a $0.15 per minute use fee. Qualification requirements and request forms can be obtained from the National Communication System web page under programs. Their web site is http://www.ncs.gov.

9.6.1.5   Improve and Get an Alternate Access to Emergency Services

Better public information should be made available about requesting emergency services when telephone lines are congested. When a handset is lifted to make a call, the caller should check and wait for a dial tone. Dialing before there is a dial tone will assure that the call will not be completed. The delay in the dial tone may be a few seconds to about a minute.

If a 911 call does not get through, it may be that the 911 office is overwhelmed or that it is not functioning for some reason. Emergency services, such as fire, police and medical assistance, can also be reached by dialing the appropriate 7-digit number that should be listed in the front of telephone directories.

9.6.1.6   Use Public Phones to Call Emergency

Most public phones are on priority service; that is, calls from public phones have priority in getting access to the central office. Calling emergency services (particularly 911) from public phones is free.

9.6.2   Batteries for Handsets

Spare batteries for handheld radios should be available and their chargers should be connected to circuits with backup power. An adapter for car cigarette-lighter charger should be part of the spare kit for emergencies.

9.6.3   Add Call Box Outside of Firehouses

After an earthquake the phone system may appear not to operate and people near a firehouse may walk to the firehouse for assistance. If the units had been dispatched so that the firehouse was empty the call box could be used to connect directly to fire dispatch center to request assistance. A sign that is displayed prominently near the call box should indicate that if the firehouse is not occupied, it can be used to contact fire personnel.

9.6.4 Seismic Installation of Base-Station Radios

The installation of base-station ratio transmitters should be evaluated for seismic ruggedness and upgrade if needed.

9.6.5 Congestion on Radio Systems

A policy and procedure should be developed to manage congestion on radio communication system following an earthquake (or other disaster) to allow the fire service to make the best use of its resources. One approach is to use a supervisory control system in which one person acts as a gatekeeper to make access to the system more orderly and prevent many people from trying to shout over each other. There are also technical solutions; such as trunking systems contain a computer that automatically switch to an unused channel so that all people in the talk group can talk to each other.

9.6.6 Backup Power for Cell Site

The minimum duration of backup power carry-through capacity should be set by the public utility commission or by public policy through legislation. It should consider the importance of this means of communication in emergencies and the difficulty in providing backup power during an emergency, such as an earthquake. Cell sites in areas where access is not easy should include a small generator in addition to battery backup power. The level of backup-power capacity should be accessible from the control center to facilitate power management during critical power disruptions.

9.6.7 Assure that Important Structures will have Adequate Seismic Performance

The trouble-call centers, control centers, and older central office buildings designed without or to old seismic requirements should be seismically evaluated and retrofitted if needed.

9.6.8 Review and Seismically Upgrade Central Office Building HVAC

The seismic installation of building HVAC system components should be evaluated and retrofitted if needed.

9.6.9 Cell Site Installation

The installation of cell-site equipment located in commercial building should follow manufacturers, recommendations or provided with adequate support. That is, the equipment should be braced if it cannot be anchored in the recommended manor.

9.6.10 Adding External Emergency Utility Hookups at Central Offices

Emergency external water and power hookups should be added to important central offices so that mobile units can quickly connected to the facility, should normal and emergency services fail.

9.6.11 Requesting Service Priority for Critical Facilities

There is a Federal Communication Commissions (FCC) program administered by the National Communication System (NCS) in which National Security and Emergency Preparedness (NS/EP) organizations can pre-register to participate and get

Telecommunication Service Priority (TSP). Under this program priority restoration and provisioning can be obtained.

In a joint program of the Department of Energy (DOE) and the Office of the Manager National Communications System (OMNCS), priority restoration of electric service can be obtained under the Telecommunications Electric Service Priority (TESP) program.

A document that describes these programs and gives information for applying to them can be down loaded from the World Wide Web. A search using search engines on most web browsers (such as Google) for "TSP_TESP.pdf" will locate the document.

Under a Department of Energy (DOE), the Energy Service Priority (ESP) program was established for NS/EP organizations in which they could priority service to restore power or fuel supplies. Generally, requests to participate in this program must be made prior to the need for services. Communications companies can get restoration of electric power under the Telecommunications Electric Service Priority program described above. However, the ESP program may be used to get fuel supplied for backup engine-generators.

10 Roadway Systems and Fire Following Earthquake

10.1 Overview

A roadway system includes freeways, state and interstate highways, arterials, city streets, bridges, overpasses, intersections, elevated interchanges, tunnels, retaining structures, etc. Roadway systems are spatially distributed, large-scale networks, with large number of components, some of which are very vulnerable to earthquakes. The failure of one or more components might cause traffic incidents and significantly impact the traffic flow capacity under emergency conditions.

The planned and coordinated process to detect, respond to, and remove traffic incidents and restore traffic capacity as safely and quickly as possible is known as traffic incident management (TIM) (AASHTO, 2002). TIM involves the coordinated interactions of multiple public agencies and private sector partners. The implementation of TIM in an urban and rural setting can be quite different, however, under emergency conditions such as immediately after a major earthquake, city, county and state transportation agencies collaborate in both settings.

The California Department of Transportation (Caltrans) has in place an Incident and Disaster Management Plan with the California Highway Patrol (CHP). In an emergency situation, e.g., immediately after an earthquake, Caltrans' emergency response priorities include traffic control (through traffic management center, TMC), damage assessment and route recovery (through structures and maintenance division). In addition to traffic control, damage assessment and route recovery, Caltrans would also provide traffic control devices such as barricades, cones, highway signs, and arrow boards, traffic management and highway maintenance equipment. The city would take on the responsibility of clearing debris while Caltrans evaluate would determine if roadway closures are justified for public safety. In areas with traffic problems the CHP is responsible for short-term traffic control during the time before Caltrans employees arrive. Ideally, the TMC would inform the local fire departments about the road and bridge closures, however, after the Northridge earthquake the information exchange between the TMC and fire departments was limited. One of the initiatives in the San Francisco Bay Area in California is the Silicon Valley Smart Corridor. The corridor will have a wide incident management program that will be accessible by the police departments (Wang, 2003).

10.2 Direct Impacts of this Roadway Lifeline on FFE Issues

Successful confinement of fires following earthquakes depends on the ability of fire fighting teams to respond in a timely manner, to mobilize adequate forces and to allocate the necessary resources. Lack of adequate access to fire areas due to roadway and bridge failures, traffic congestion or roadway closures caused by debris will hamper the fire suppression efforts.

FIRE FOLLOWING EARTHQUAKE

10.3 Secondary Impacts on other Lifelines

10.3.1 Road and Bridge Damage also Damages Water Lines

Damage to roads and bridges can cause water mains to fail, resulting in the loss of water to extinguish fires. However, the trend is to put flexible joints on pipes between the abutments and the superstructure and at other locations where differential movement could cause the pipeline to break.

10.3.2 Road and Bridge Damage also Damages Communication Lines

Bridges are also used to carry communication cables and failure of the bridge or even subsidence behind abutments have damaged communication cables and severely disrupted communications.

10.3.3 Road and Bridge Damage Delays or Prevents the Delivery of Services and Supplies

Mobilization of emergency response crews and supplies, and supplies necessary for communications and operations, such as diesel fuel to emergency generators, can be hampered by roadway system closures due to bridge or roadway damage.

10.3.4 Utility Service Crews Unable to Restore Services

Water, power, gas, and communication company service crews can be delayed or prevented from getting to damage sites to restore service.

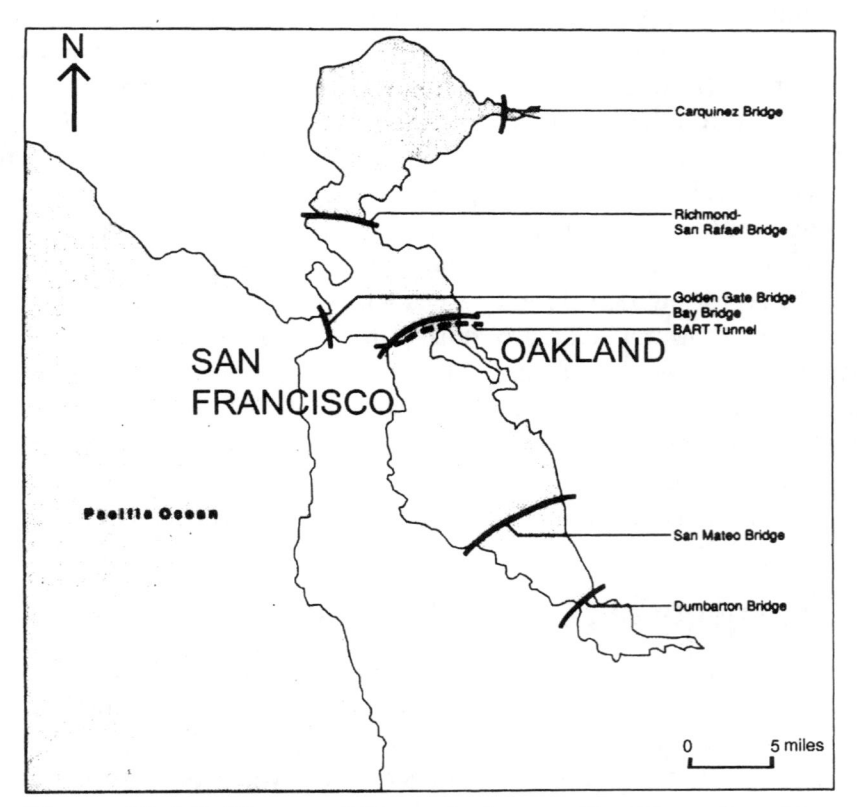

Figure 10-1 Bridges and Tunnels Across San Francisco Bay

10.4 Earthquake Vulnerabilities of Roadway Systems

Either bridge or road damage can close a traffic corridor for an extended period. The collapse of one span on the San Francisco to Oakland Bay Bridge took a month to repair. In addition, liquefaction of the approach to the bridge also required a month to repair. Each type of damage was sufficient to prevent traffic from moving between Oakland and San Francisco,
Figure 10-1.

Commuters were fortunate that the BART light-rail system was available (closed for a day after the earthquake for inspection and minor repairs) as well as ferry traffic to move them across the Bay. Truck drivers were less fortunate since they were forced either to cross the Bay to the north on the Richmond-San Rafael Bridge or to cross to the south on the Dumbarton Bridge; each route adding about two hours to their trip. The Bay Bridge collapse also severed an optical cable that impacted communication in the Bay Area.

10.4.1 Past Earthquake Performance with Direct Impacts on FFE

Damage to roads and bridges can result from earthquake hazards besides ground shaking. These include surface faulting, landslides, slumping, liquefaction, subsidence, lateral spreading, or from a tsunami that can be generated thousands of miles away. In areas of high seismicity, bridges and roads often have to be placed across faults and in unstable ground, and are subject to considerable seismic risk.

Bridges are often closed as a result of damage due to ground shaking and deformation. Settlement of the embankment behind bridge abutments is one of the most common causes of road closure,
Figure 10-2. Although it is usually not severe and can be repaired relatively quickly, the settlement of embankments can hinder accessibility to damage areas and to fires following earthquakes.

Figure 10-2 The Copper River Bridge #2 was closed after the 1964 Great Alaska earthquake due to approach settlement and shifted truss spans (NAS, 1973).

After the 1998 Loma Prieta earthquake, access to the Marina District in San Francisco was affected from closure of many streets due to liquefaction as well as building damage and water main breaks.

Landslide damage to roads can be particularly devastating on mountain passes and urbanized or semi-urbanized areas with steep topography. After the 1989 Loma Prieta Earthquake, landslides in the Santa Cruz Mountains blocked many streets, roads, and highways. Twelve miles of Route 17 (as well as Bean Creek Road and Old San Jose Road) were closed due to landslides for 32 days following the 1989 Loma Prieta earthquake, Figure 10-3, isolating many small communities (Yashinsky, 1998).

A similar situation was observed within the Santa Monica mountains (on roads such as Mulholland Drive) after the 1994 Northridge earthquake (Perkins et al. 1997). Communities in the mountains of Taiwan were isolated for several months following the 1999 Chi-Chi earthquake, due large landslides that covered the roads.

Landslides can be a problem for fire crews trying to get into damaged areas. Following the 1992 Cape Mendocino California earthquake, there was an ignition in Petrolia's general store and post office. Even worse, the Petrolia fire crew couldn't get their fire

Figure 10-3 Landslide blocking Route 17 after the Loma Prieta Earthquake.

trucks out of the firehouse due to racking of the doors. They radioed for mutual aid from the nearest California Division of Forestry fire station 34 minutes away in Ferndale. However, landslides on the steep, mountain road delayed them by more than an hour, by which time the building was a smoldering ruin (Turner, 1999).

In addition to the closures due to ground shaking and deformation; the threat of building collapse or structural damage to highway or railway structures, HAZMAT releases, or water and gas pipeline leaks after an earthquake can also lead to closure of roadways. The threat of freeway and building collapses are a source of particular concern within dense urban areas. Road closures due to damage to adjacent structures along freeways and streets prevent crews from getting to the damaged area and to fires following earthquakes. This is particularly true in countries like Japan that use a narrow right-of-way. After the 1993 Hokkaido Earthquake, a fire started on the Island of Okushiro. *'A brigade of 10 men immediately responded and attempted to reach the fire by driving down the main road, only to find it blocked by debris from buildings destroyed by the tsunami'* (Chung, 1995). The damage in Hokkaido was compounded due to a tsunami. Similarly, after the 1995 Kobe earthquake, the 50-mile long Kobe elevated expressway fell onto the surface street at many locations, resulting in the closure of two of the city's main arteries. Subsequent fire fighting efforts were largely ineffective due to '*disrupted water supplies and blocked roads*' (Chung, 1996).

Even when roads are undamaged, fire fighters may have trouble reaching burning buildings due to traffic jams that often follows an earthquake. After the 2001 Nisqually, Washington State earthquake, city streets and highways were jammed with anxious commuters driving home to make sure their property was undamaged. This phenomenon was also observed after the 1999 earthquake in Kocaeli, Turkey. Table 10-1 lists the number of roadways closed after the Loma Prieta and Northridge earthquakes grouped by hazard types and roadway types (i.e., freeways and highways vs. streets). The data from these two California earthquakes indicate that the impact of direct versus indirect hazards for freeways and streets was significantly different. In general functionality of freeways are impacted mainly by earthquake hazards such as shaking, landslides, or liquefaction. On the other hand, streets have been affected by indirect hazards such as building damage and water main breaks in addition to ground shaking and deformation. However, redundancy of the local streets yield fewer impassible sections compared to less redundant freeways.

Table 10-1 Summary of Street and Freeway closures for the Loma Prieta and Northridge Earthquakes (source: Perkins et al., 1997)

	Hazard Type	Loma Prieta	Northridge	Loma Prieta And Northridge		
		Freeways, Highways & Streets	Freeways, Highways & Streets	Freeways & Highways	Streets	Total
Direct Hazards	Shaking	30	31	49	12	61
	Landslides	23	22	16	29	45
	Liquefaction	17	10	7	20	27
	Fault Rupture	0	0	0	0	0
	Total	**65**	**63**	**68**	**60**	**128**
Indirect Hazards	Building Damage	43	15	0	58	58
	Freeway Hazard	15	13	0	28	28
	Natural Gas Release	0	7	0	7	7
	HAZMAT release	0	3	0	3	3
	Water Main Breaks	17	18	0	35	35
	Other	3	13	0	16	16
	Traffic Control	18	10	10	18	28
	Total	**82**	**77**	**10**	**149**	**159**
Public Safety	Public Safety	109	104	42	171	213
	Impassable	33	36	36	33	69
	Total Closures	**142**	**140**	**78**	**204**	**282**

10.4.2 Past Earthquake Performance with Secondary Impacts on other Lifelines

The 'co-location' of roads and other utilities has sometimes resulted in fires without the means of extinguishing them. Most utilities are located adjacent to or under roads and are carried on bridges. Earthquake damage to roads and bridges may cause fires from broken gas lines with no means of extinguishing them due to broken water mains. An interesting example of this was during the 1994 Northridge earthquake. In Granada Hills, a 22-inch gas line under Balboa Boulevard ruptured, Cover Photo. Figure 10-4 show the excavation in the tension zone exposing the damaged pipes. According to eyewitnesses, the gas was ignited when a motorist tried to start his truck, which had stalled on the flooded street due to a break in a water main under the road (Hall, 1994).

Figure 10-4 Excavation under Balboa Blvd (1994 Northridge earthquake) resulted in several homes burning down, one of which can be seen in the background. (Lund)

In fact, fire fighting was hampered due to the lack of water pressure caused by a number of breaks to water mains under city streets. At the nearby Balboa Blvd Undercrossing on I-118, a broken water main at the bridge abutment washed out the embankment and closed the bridge.

In the 2001 Peru earthquake, subsidence behind a bridge abutment on the Pan American Highway broke an optical fiber cable buried adjacent to the highway that was one of the main communication channels between Lima and southern Peru.

Fires following earthquakes in rural areas require a different set of emergency response operations since the environmental factors are more prominent and redundancy in the system is usually limited. After the 1992 Cape Mendocino (Petrolia) earthquake, the Petrolia fire department called Ferndale for mutual aid, as they couldn't get their firehouse doors open. However, the road between Ferndale and Petrolia was closed due to a rockslide, making Petrolia inaccessible for the crews from Ferndale. After the 1989 Loma Prieta earthquake, building inspectors and medical personnel had to be flown into locations due to disruption of roads around Watsonville.

10.5 Mitigation Strategies

Well-handled efforts towards mitigation can improve communication across agencies, reduce loss of connectivity of the roadway system, and help improve the efficacy of fire fighting efforts. Such efforts include the following:

- Establish protocols for coordination and communication across agencies;
- Enhance post-event response capability through use of technology;
- Minimize damage to roadway system and its components by:
    - locating roads away from earthquake hazards (to the extent possible), and
    - designing and retrofitting roads and bridges to better resist earthquakes;
- Use techniques for fast repair of bridge abutment settlements;
- Optimize the co-location of roads and other utilities;
- Reduce the adverse impacts of co-location of lifelines (e.g., flexible joints are required on pipelines at each end of the bridge to accommodate offsets between the ground and the structure);
- Minimize impact of damaged components on system functionality by:
    - providing redundancy to the highway system,
    - identifying lifeline and emergency vehicle routes for highly likely earthquake scenarios.

10.6 Key Issues Related to Fire Following Earthquake

- Damage to roadway systems limits the ability of fire service and lifeline crews from responding to suppress fires or restore critical services.

- Congestion on roads limits the ability of fire service and lifeline crews from responding to suppress fires or restore critical services.

- Damage to bridges can be particularly disruptive because alternative routes may be limited so that they create bottlenecks.

- Damage to bridges can be disruptive to water and communication systems that use them as conveyance over water bodies.

10.7 Research Needs and Educational materials on FFE

10.7.1 Seismic Risk Assessment to Roadway System and its Components, and Mitigation

Engineering research over the past three decades has led to much better understanding of seismic risk to roadway system and its components (Werner et al., 1999; Williams et al., 2002). While the performance (survivability) of critical components can be predicted with reasonable confidence, the disruption to the network as a whole is still difficult to predict. Collateral affects, such as lifeline interactions and dependency are, in general, difficult to quantify and are the subject of more recent research (Brunsdon, 2000; Bausch et al., 2000; Eidinger et al., 2000).

Research and implementation in the following areas are recommended:

- Establishing a GIS-based roadway and bridge inventory

- Assessment of component vulnerability

- Evaluation of the impacts of lifeline interaction and dependency of fire suppression

- Development of tools for traffic congestion management to determine bottlenecks and system reliability; which can be used for pre-earthquake planning (e.g., component retrofitting, designation of emergency vehicle lanes with stiff fines for violators)

- Resource allocation for fire following earthquakes in order to determine available and necessary resources and their logistics

- Reduction of the adverse impacts of co-location of lifelines (e.g., flexible joints are required on pipelines at each end of the bridge to accommodate offsets between the ground and the structure)

- Application of technology to mitigate physical damage for both new and existing construction

10.7.2 Use of Technology for Emergency Response Operations

As for any other TIM operation, information is key for emergency response operations and suppression of fires. Rapidity in fire suppression is essential. The ability to share information between different agencies, to obtain accurate and timely information would facilitate better decision making under emergency situations. Research and implementation in the following areas are recommended:

- Establishing inventory and resource planning
- Development of a GIS-based decision support system
    - to collect and disseminate data on (e.g., damage to bridges, road closures, as well as areas with reduced bridge load capacity for heavy fire trucks and transportation of heavy equipment) using technology and tools such as, handheld devices, GPS, satellite imaging
    - to quickly determine major and minor incident locations after a disaster through a seamless link to the advanced transportation management system used by state agencies
- Development of tools for traffic congestion management after an earthquake; such a tool would provide emergency detours in near-real time by enabling modifications of the modeled system to reflect post-event changes on the infrastructure and the traffic operations, and would facilitate information dissemination via intranet and Internet

10.8 References

AASHTO. "Proceedings of the National Conference on Traffic Incident Management: A Road Map to the Future", jointly sponsored the American Association of State Highway and Transportation Officials, ITS America, and the Federal Highway Administration Irvine, CA, 107 pp., June, 2002.

Bausch, D.; et al. "GIS-based hazard mapping and loss estimation in the safety element of the general plan for Riverside County, California," Sixth International Conference on Seismic Zonation: Managing Earthquake Risk in the 21st Century, Proceedings [computer file], Earthquake Engineering Research Inst., Oakland, California, 2000

Brunsdon, D. R. "A decade of lifelines engineering in New Zealand," 12th World Conference on Earthquake Engineering [Proceedings] [computer file], New Zealand Society for Earthquake Engineering, Upper Hutt, New Zealand, 2000, Paper No. 2798

Chung, R. M. "Hokkaido-Nansei-oki earthquake and tsunami of July 12, 1993: reconnaissance report," Earthquake Spectra, 11, Suppl. A, Apr. 1995, 166 pages, EERI Publication 95-01

Chung, R. M. "The January 17, 1995 Hyogoken-Nanbu (Kobe) Earthquake," NIST Special Publication 901, July 1996

Eidinger, J. M., Collins, F., and Connor, M. E., "Seismic assessment of the San Diego water system," Sixth International Conference on Seismic Zonation: Managing Earthquake Risk in the 21st Century, Proceedings [computer file], Earthquake Engineering Research Inst., Oakland, California, 2000.

Hall, J.F., "Northridge Earthquake January 17, 1994, Preliminary Reconnaissance Report," EERI, March 1994.

National Academy of Sciences (NAS), "The Great Alaska Earthquake," Publication #1603, Washington D. C., 1973

Perkins, J. B., Chuaqui B., and Wyatt E. "Riding Out Future Quakes - Pre-Earthquake Planning for Post-Earthquake Transportation System Recovery in the San Francisco Bay Region" Association of Bay Area Governments, CAT. NO. P97002EQK, 198 pp, October 1997.

Turner, Fred, "Performance of roll-up garage doors," Lessons Learned Over Time Earthquake Engineering Research Institute, Oakland, California, 1999.

Wang, Z., Personal Communication, Caltrans Traffic Operations, District 3, January, 2003.

Werner, S.D., Taylor, C.E., and Moore, J.E. II. "Seismic Risk Analysis of Highway Systems: New Developments and Future Directions," Proceedings of Fifth U.S. Conference on Lifeline Earthquake Engineering, Seattle WA, August, 1999.

Williams, M., Basöz, N., and Kiremidjian, A. "A GIS-based Emergency Response System for Transportation Networks," Proceedings of the Fourth US-China-Japan Trilateral Symposium on Lifeline Earthquake Engineering, Qingdao China, October, 2002.

Yashinsky, Mark, "The Loma Prieta, California Earthquake of October 17, 1989 Highway Systems," USGS Professional Paper 1552-B, Washington DC, 1998

11 Methods for Mitigating Fires Following Earthquakes

This section discusses mitigation of the fire following earthquake problem. Fortunately, there are numerous opportunities for mitigation of the fire following earthquake problem although, unfortunately, their application is sparse. A framework for the discussion of mitigation opportunities is Figure 4-1, which has been enhanced in Figure 11-1 to show opportunities for intervention in the fire following earthquake process. As indicated in the diagram, following the earthquake damage in the most general sense (even if its only the overturning of a candle) is required for an ignition. Thus,

- *Reduction of damage* is the first opportunity for mitigation of fires following earthquake. This ignition might be extinguished without human intervention, if

- *Automatic sprinklers* or other automatic suppression systems (e.g., Halon) are present, and if automatic system is post-earthquake functional, and if the water for example for the sprinklers is available. If the ignition is not automatically extinguished, then the fire remains to be discovered, and the

- *Citizens* so discovering it may be able to extinguish it, <u>if</u> water is available. If those persons are not able to extinguish the fire, then they must report it to the fire department. For the citizens to report the fire, functional

- *Communications* (i.e., telephone) are required (unless, perhaps, the fire department may observe the smoke column, and self-report).For the fire department to respond, they must have available

- *Fire department personnel*, and their assets (i.e., apparatus) must be able to travel to the fire – that is,

- *Transportation*, in the form of roads etc, must be functional. Upon arrival, the fire department will be able to extinguish the fire, <u>if</u>

- *Water* is available and <u>if</u> the fire is controllable given the fire department resources reporting to the scene. If not, the fire will continue to grow (admittedly, perhaps at a smaller rate given the fire department's efforts), until it meets a firebreak. The provision of firebreaks, via the urban planning process, is the last opportunity for mitigation of fires following earthquake.

We discuss each of these aspects in turn.

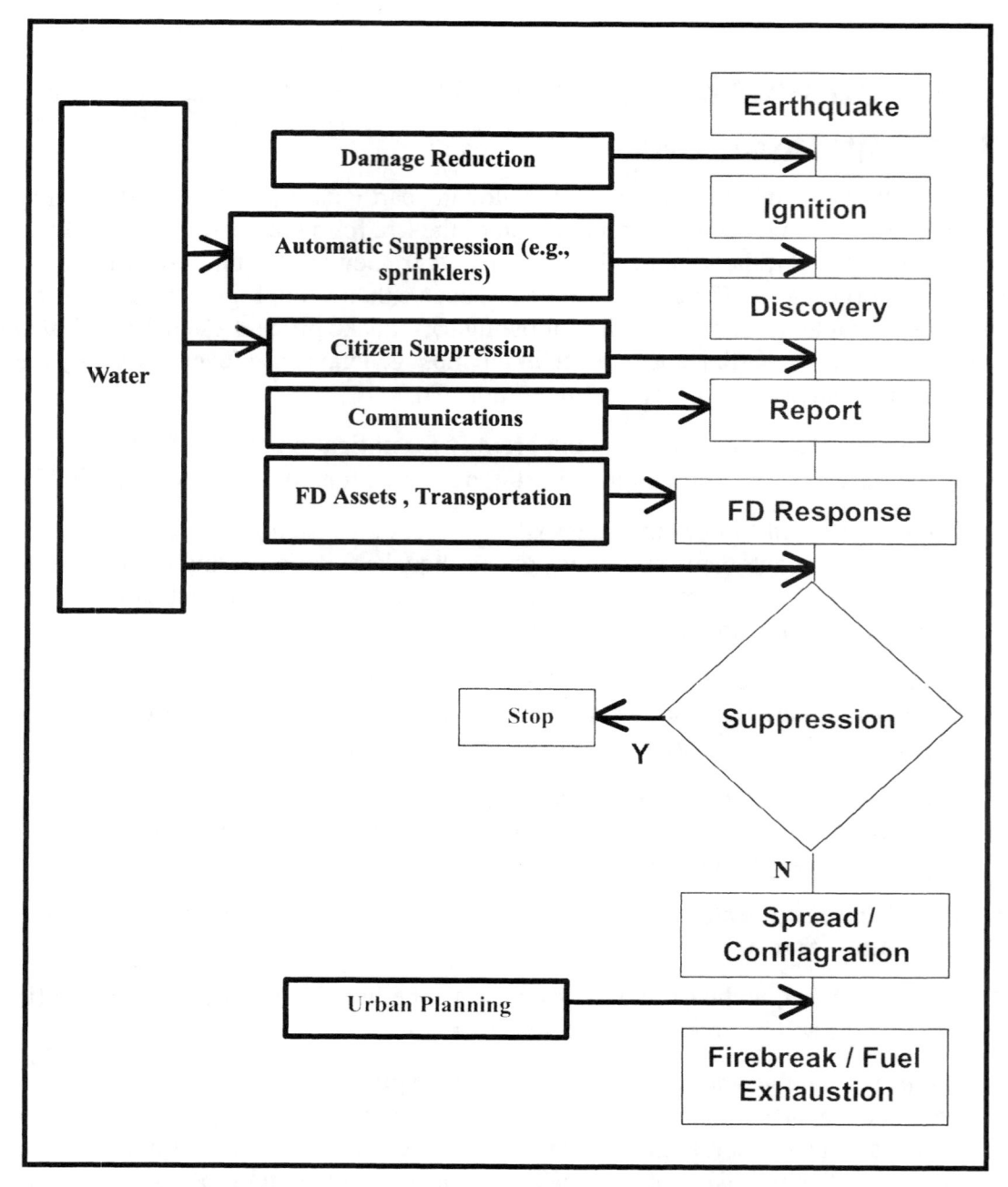

Figure 11-1 Fire following earthquake process, with mitigation opportunities in bold

11.1 Reduction of Damage

In a general sense, reducing earthquake damage or effects due to shaking reduces the potential for ignitions following an earthquake. Obvious examples include restraining sources of ignition such as water heaters, lamps and other heat sources, and containers holding potentially reactive chemicals. Many sources of information and products exist for bracing and anchoring of simple household and office goods (e.g., www.abag.org) – a simple search of the Internet will uncover many.

Gas appliances in the home are a concern to many persons, since overturned water heaters and/or broken gas piping (due to excessive displacements of a seismically weak house) can result in gas leaks which, in a few minutes, can produce an life-threatening explosive mixture of gas and air (although this rarely occurs on a per-household basis, it might occur a few times given many thousands of gas leaks). Figure 11-2 shows how to restrain a gas water heater. Mroz and Soong (1997) evaluated the method depicted as well as other methods of restraint, and found varying effectiveness – using only plumber's tape provided resistance against falling away from the wall, but not towards the wall – thus use of rigid tubing is recommended.

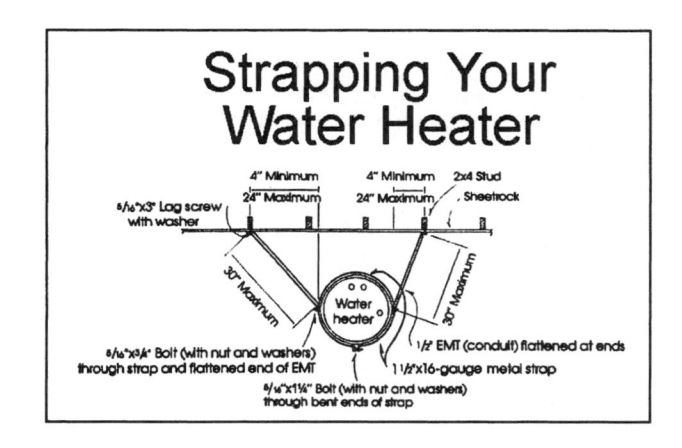

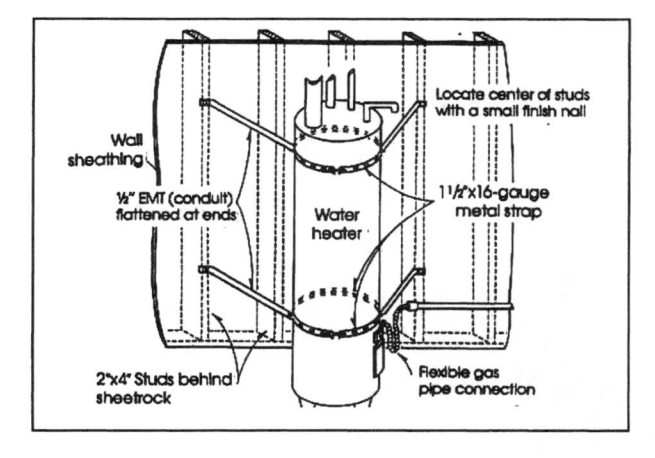

Figure 11-2 Strapping water heater (a) plan view, showing 'plumber's tape' strap around water heater and attachment to framing using rigid tubing; (b) elevation view; (c) photo of actual water heater – note 'plumber's tape' strap around 'belly' of the water heater

Similarly, Figure 11-3 illustrates the gas valve controlling service to a home, together with a 'gas wrench', affixed by the homeowner to the gas valve so as to be always

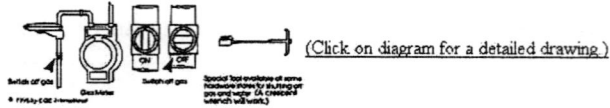

available in the event of an emergency.

Figure 11-3 (a) How to shut off a gas valve, and (b) actual gas service to a home, showing pressure regulator (flat round plate-like object) and valve, with 'gas wrench' wired to pipe, and writing on side of house showing location

Because persons may not be home when an earthquake occurs, and thus unable to turn off the gas service in the event of a gas leak following an earthquake, automatic gas shut-off valves have been invented. Automatic gas shut-off valves are of two types: (a) excess flow valves, actuated by a change in gas pressure, and (b) seismic valves, actuated by motion or movement. Both are intended to actuate when subjected at the point of installation to an earthquake severe enough to cause damage to gas systems and appliances within the structures that they protect. Figure 11-4 shows an example of an automatic gas shut-off valve installation.

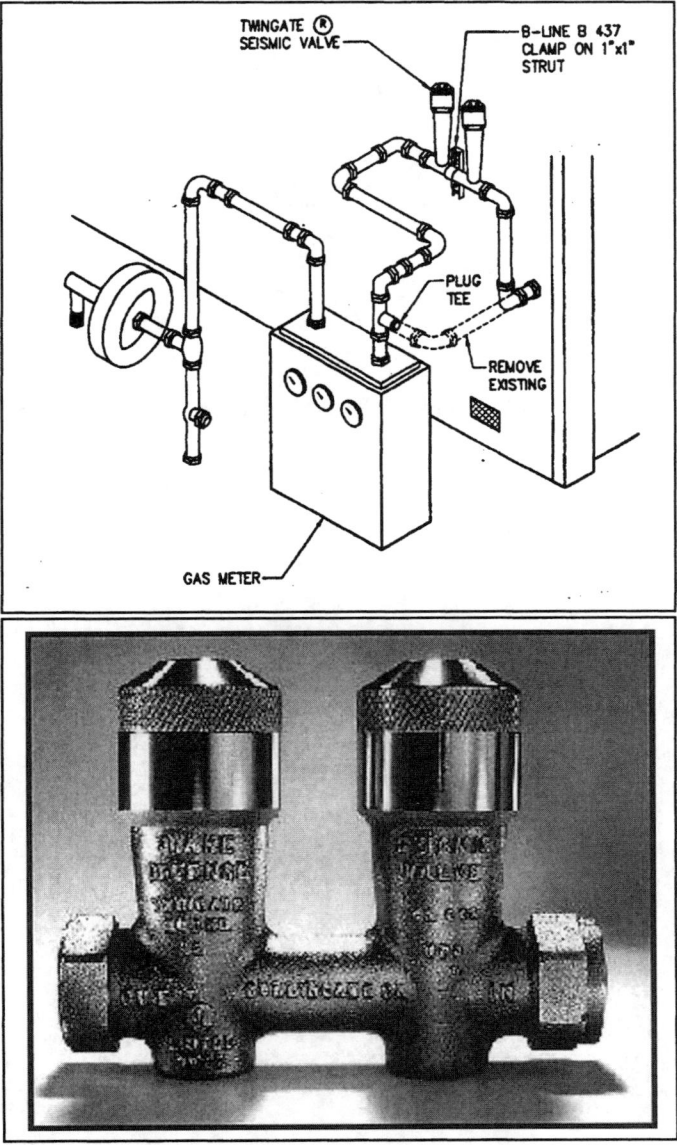

Figure 11-4 Example of automatic gas shut-off valve (a) typical installation, (b) close-up of valve

The above precautions are appropriate assuming the structure is relatively safe – however, should the house slide off its foundation, then the gas service line will be sheared off, with pressurized gas escaping to the atmosphere – one spark produces a gas 'torch' of major proportions, which will quickly ignite the home. In such cases, if one can get to the pipe before the gas ignites, simply plugging the broken pipe with a tapered piece of wood effectively turns the gas off. Older homes with unbolted foundations and/or unbraced cripple walls are particularly common in the western U.S. and especially prone to sliding off their foundations. Figure 11-5 shows an example of how to bolt and brace cripple walls: Note steel straps nailed through plywood into studs, and horizontally bolted into concrete foundation. Figure 11-6 shows houses fallen off their foundations in the 1989 Loma Prieta earthquake.

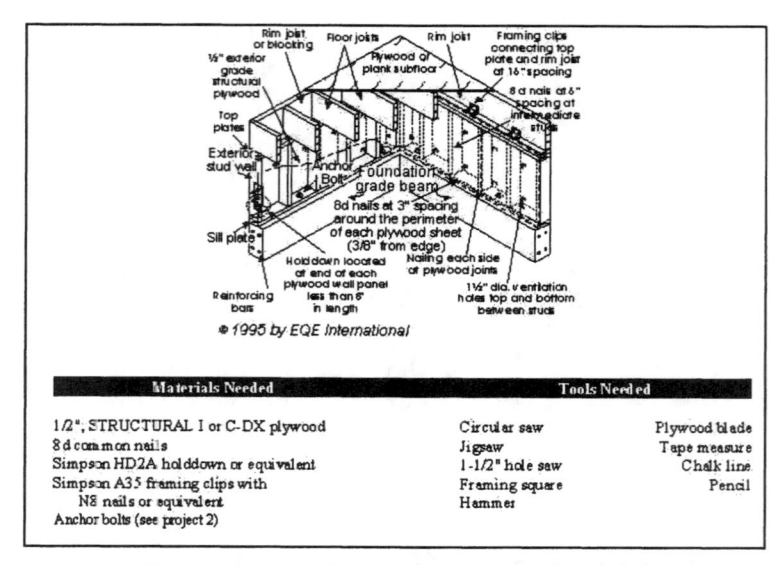

Figure 11-5 (top) diagram, and (bottom) example of cripple wall bracing using plywood

Figure 11-6 Watsonville in 1989 Loma Prieta earthquake, examples of houses fallen off their foundations

Another source of post-earthquake ignition is electricity, which can cause ignitions via a number of mechanisms, including wiring shorts, appliance damage, overhead line arcing, etc. A particularly egregious example of the latter is found in the CBD of Vancouver, B.C., which is an enormous concentration of value, yet is the only major North American city with overhead electric transmission. Wood pole mounted transformers abound in the CBD, in many cases only inches from commercial buildings, Figure 11-7.

Lastly, another major source of post-earthquake fires are chemical reactions. Chemicals in glass and other containers can easily fall off shelves or tip over in an earthquake, spilling their contents and reacting with other spilled chemicals, or other materials such as flooring or furniture. The heat given off by the reaction (termed *exothermic* reactions) can be sufficient to ignite cloth or other materials, thus starting a fire. High-school chemistry laboratories are a source of fires in almost all major earthquakes. In California and other high seismicity regions, it has become relatively common practice to put lips on shelves in laboratories, restrain chemical containers, put door latches on laboratory cabinets and take other precautions to prevent spillage of chemicals.

Figure 11-7 Vancouver, B.C. Central Business District, with overhead electric lines – an obvious fire hazard, especially following an earthquake

11.2 Automatic Suppression

The fire hazard in non-earthquake conditions is such that automatic suppression systems have become a standard part of most larger buildings. These also serve the same purpose following an earthquake, and can be very effective in reducing the overall post-earthquake fire problem, by stopping many ignitions at their initiation. Thus, an item of critical importance for the fire following earthquake problem is the seismic performance of automatic sprinklers. Design and installation of automatic sprinklers in the U.S. is generally governed by NFPA 13[27], which includes provisions for seismic design of sprinklers. This design is relatively simple, and consists of lateral and longitudinal bracing for sprinkler mains at specified distances. These provisions have generally proven to be adequate in most earthquakes (Fleming, 1998), although a flaw in the design is the general overlooking of situations where the sprinklers are restrained by the bracing on either side of a building temperature joint. Large buildings are often built with joints, intended to limit stresses due to normal heating and cooling of the buildings during the day. Where sprinklers cross these joints, flexible couplings have to be provided – otherwise, in an earthquake, the two parts of the building on either side of the joint (effectively, that is, two buildings) will move differentially, thus rupturing the sprinkler. Differential movement of piping supports in general is a seismic design issue often overlooked. One or a few failures of sprinkler mains in this manner can drain the sprinkler system and prevent functional post-earthquake fire suppression.

There are other automatic suppression systems in addition to water-based sprinklers – these include halogenated agents (e.g., HALON), CO_2, dry chemical, foam and other systems, each of which can similarly be of great benefit in the post-earthquake environment, and each of which should be reviewed to assure post-earthquake functionality.

In addition to automatic suppression systems, the post-earthquake functionality of fire detection and alarm systems is vital following an earthquake, if for no other reason than to provide warning for evacuation. The author was involved in the development of an entire methodology for the assessment of such systems, to assure post-earthquake functionality (Johnson et al, 1999), which is discussed in another chapter of this Handbook.

Lastly, the application of *intumescent* paint, which is special paint for walls and other surfaces and which is highly resistive to radiant heat, is emerging as a promising form of passive fire protection in general, which may have special application for fire following earthquake risk reduction. The resistance of intumescent paint is largely gained via chemicals in the paint which utilize the heat gain in expansion, resulting in a swelling or 'blistering' of the paint rather than ignition, and the creation of a fire-retardant foam. Ignition is thus much delayed and, in certain cases, prevented, Figure 11-8. In the test shown in Figure 11-8, both 'buildings' had crib fires set at the same time. Building on the

---

[27] NFPA 13, Standard for the Installation of Sprinkler Systems (most recent edition), National Fire Protection Association.

left is of bare wood, and has flashed over and is fully involved at slightly less than 5 minutes following ignition, while building on the right, painted with intumescent paint, still has only a small fire on the interior, not yet flashed over . San Francisco has approved intumescent coatings for application in a number of historic buildings as an alternative or adjunct to other fire protection systems, and is investigating its possible benefits with regard to fire following earthquake (Kobayashi and Kornfield, 2002).

Figure 11-8 Test of intumescent paint

11.3 Citizen Suppression

Since the essence of the fire following earthquake problem is the multiple simultaneous ignitions following an earthquake, which overwhelm fire department resources when they are most in demand, one approach to mitigating the problem is the use of additional personnel, in the form of citizen volunteers. The concept of volunteers in fire departments is not new – the first fire departments were all volunteers, the U.S. today still has large number of volunteer fire departments outside the major metropolitan areas, and many fire department are part paid and part volunteer. In this regard, several of the larger fire departments in the US have developed specialized training of citizen volunteers, to be mobilized in the event of major disasters[28]. San Francisco for example has trained 10,000 citizens in elementary firefighting and search and rescue, Figure 11-9.

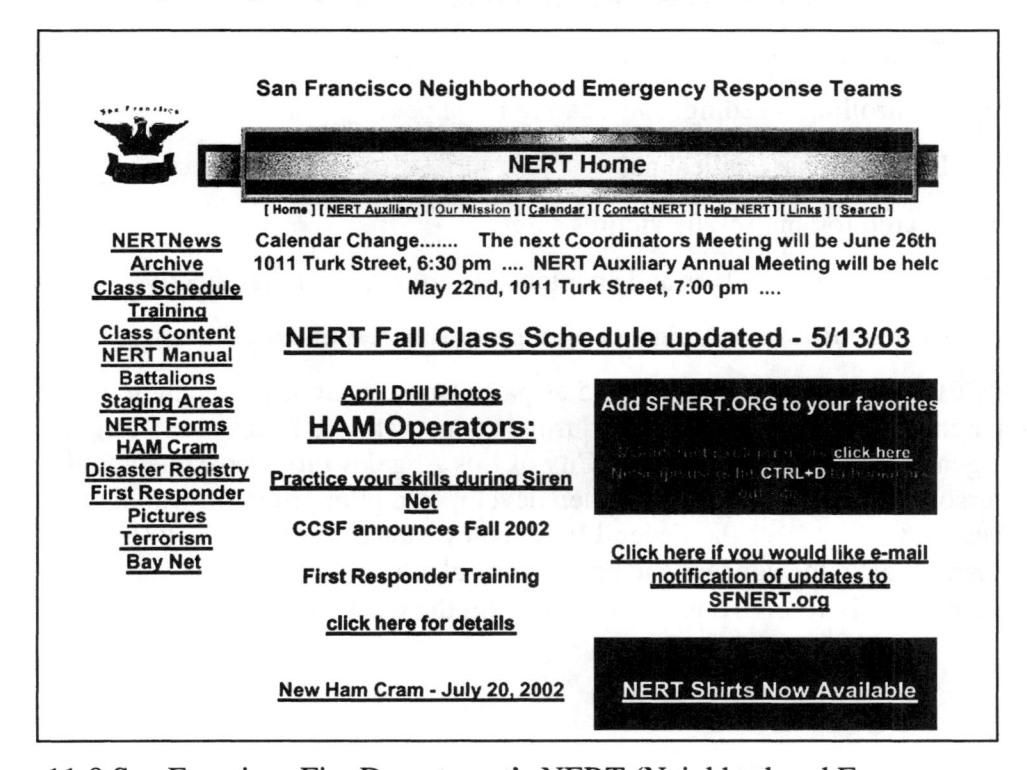

Figure 11-9 San Francisco Fire Department's NERT (Neighborhood Emergency Response Team home page (Source: http://www.sfnert.org/)

---

[28] For example, the San Francisco FD Neighborhood Emergency Response Teams (NERT), (http://www.sfnert.org/), and the Los Angeles region's Community Emergency Response Teams (http://www.cert-la.com/).

To quote the Los Angeles CERT (Community Emergency Response Team) website:

Local government prepares for everyday emergencies. However, during a disaster, the number and scope of incidents can overwhelm conventional emergency services. The Community Emergency Response Team (CERT) program is an all-risk, all-hazard training. This valuable course is designed to help you protect yourself, your family, your neighbors and your neighborhood in an emergency situation. CERT is a positive and realistic approach to emergency and disaster situations where citizens may initially be on their own and their actions can make a difference. While people will respond to others in need without the training, one goal of the CERT program is to help them do so effectively and efficiently without placing themselves in unnecessary danger. In the CERT training, citizens learn to:

- Manage utilities and put out small fires
- Treat the three medical killers by opening airways
- Controlling bleeding, and treating for shock
- Provide basic medical aid
- Search for and rescue victims safely
- Organize themselves and spontaneous volunteers to be effective
- Collect disaster intelligence to support first responder efforts

The CERT concept emerged in 1985 as part of observations on Japanese neighborhood preparedness made by the US side during a cooperative US-Japan exchange of emergency response officials. The City of Los Angeles Fire Department under the leadership of Chief Frank Borden then developed a pilot program, which was so successful that by 1993 the Federal Emergency Management Agency (FEMA) had decided to make the concept and program available to communities nationwide. By 2002, 38 states and six foreign countries were using the CERT training.

11.4 Communications

Discussion above has emphasized the need for rapid reporting of fires following an earthquake, and the fact that this is problematic due to saturation and possible damage to the telephone system, as well as overloading of the emergency dispatch ('911') system. Fire department emergency dispatch centers have contingency plans for coping with a large number of calls in an emergency, but can still be overwhelmed at some point. An additional issue is damage to the emergency dispatch center itself, which should not only be structurally sound, but also have assured backup power and other resources.

11.5 Fire Department Assets

The fire service is the first line of defense against fires following earthquakes. However, the fire service is both vulnerable to earthquakes itself, and is sometimes not as well organized for this specific issue as it might be. This section discusses actions the fire service itself can take to improve its preparedness for fires following earthquakes.

Fire Stations: As noted above (in the discussion of the 1994 Northridge earthquake, where 35 fire stations sustained damage, Sepponen et al, 1997), fire stations can sustain damage themselves in an earthquake, potentially inuring and rendering the assets they house (i.e., firefighters and their equipment) useless in an earthquake. Many larger fire departments in high seismic areas have reviewed their fire stations' structural seismic adequacy, and strengthened inadequate stations. In the late 1980s San Francisco for example undertook an analysis which found that of 55 department facilities, eleven were collapse hazards, and another thirteen would sustain extensive damage in a major earthquake (EQE, 1989). The city embarked on a major program to strengthen and replace fire stations, which has since been completed. Figure 11-10 for example shows a 1913 vintage unreinforced masonry single bay fire station, which was analyzed by a team headed by the author and found to be a collapse hazard. A structural reinforcement scheme, also shown in the figure, was developed and constructed during the 1990s.

In the Vancouver study discussed above, many fire departments were found to have quite new fire stations, or have reviewed and as required strengthened, their fire stations. However, there still remained somewhere between 14 to 30 fire stations (i.e., 18% to 38% of the fire stations in the region), that were still seismically questionable. Thus, a review of fire stations and other emergency facilities, for seismic structural adequacy and to assure post-earthquake functionality, should be undertaken by all fire departments in seismic areas.

Mutual Aid: Fire departments regularly practice mutual aid, and are quite confident of their ability to respond to emergencies, based on their daily experience. however, a major earthquake is a very rare occurrence, which few senior fire officers experience in their lifetimes. Earthquake planning for fire departments is thus mandated. All fire departments in a region should cooperate in the development of a regional earthquake plan with a detailed post-earthquake fire element, which should include procedures for rapid location and reporting of post-earthquake fires, mutual aid response to larger fires, identification of alternative water supplies, and related measures.

Post-earthquake Damage Surveys: Most fire services have as part of their earthquake plan a damage survey by fire companies. This is very unrealistic, in that fire companies self-dispatch or are dispatched immediately following the earthquake. An alternative to damage surveys by fire companies needs to be developed. One possibility is that the police department should perform damage surveys, since they are not equipped to deal with fires or building collapses, yet know their neighborhoods, are mobile and reliable observers, and have reliable communications. Whether the damage survey is performed by police or others can be discussed, but tasking fire companies with this is unrealistic.

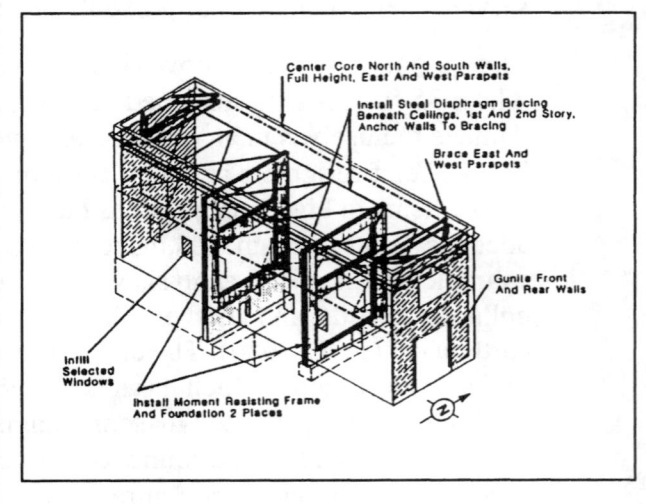

Figure 11-10 (a) Example of single bay unreinforced masonry fire station, built in 1913, and (b) strengthened in 1990s per indicated structural reinforcement scheme.

11.6 Water

Water for firefighting is obviously a vital aspect of the fire following earthquake problem. Earthquakes damage water system, and thus post-earthquake reliability of water supply is a necessary element of post-earthquake fire mitigation. Thus, water supply systems in seismic regions should be assessed for their post-earthquake functionality. Several methodologies for this exist and the main points are highlighted in Chapter 6. If the municipal water supply system cannot be made post-earthquake reliable, then alternative water supplies are required.

Alternative water supplies can be provided for in a number of ways, including:

- Identification of alternative water sources, such as rivers, lakes, swimming pools or other supplies. This is a normal part of fire department planning, which is increasingly being overlooked as the fire department increasingly takes on emergency medical treatment (EMT) as part of its central mission. Los Angeles Fire Department in the response to the No. Balboa Blvd. fire, discussed above, made excellent use of backyard swimming pools as an alternative source of firefighting water; similar small sources of water can often be useful when the fire is not too large. Thus, fire officers need to be aware that the hydrant cannot be relied upon following a major earthquake, and need to identify, and regularly exercise using, alternative sources of water supply. It is not remarkable that San Francisco burnt down in 1906, despite being surrounded on three sides by the largest body of water on earth, because this alternate source of water is not readily available unless pre-planning is adopted to access this water. As part of its AWSS, San Francisco has installed salt-water hydrants (i.e., special connections to by Bay, for drafting) along its shorelines. It is remarkable that other similarly situated cities (e.g., Los Angeles, Oakland, Tokyo, Osaka, Manila, Istanbul...) have not done so.

- On-site secondary water supplies. In the U.S., highrise buildings in higher seismicity zones are required to have secondary water supplies, in recognition of the unreliability of municipal water supplies in earthquakes. These requirements usually translate to a minimum of 15,000 gallons[29]. Other high seismicity regions should adopt this requirement.

- Consideration can be given to development of special auxiliary high pressure water systems, similar to San Francisco's AWSS or Vancouver's DFPS (discussed in Chapter 12). Such high-pressure systems can only be justified for major metropolitan areas in high seismic regions, but there is no shortage of these. It is noteworthy that Vancouver undertook construction of the DFPS in the 1990s.

---

[29] "**903.3.5.2 Secondary Water Supply**. A secondary on-site water supply equal to the hydraulically calculated sprinkler demand, including the hose stream requirement, shall be provided for in highrise buildings in Seismic Design Category C,D,E or F …. The secondary water supply shall have a duration of not less than 30 minutes." (*International Building Code*, 2000 – note that the 1997 UBC has similar requirements).

- Acquisition of a Portable Water Supply System (PWSS), consisting of specialized portable pumps, and 5" to 6" diameter hose, similar to the system pioneered in San Francisco or the *HydroSub* system recently acquired by the Vancouver, Oakland, Vallejo and Berkeley. The PWSS is discussed in some detail in the next section. The *SuperAqueduct* 12" diameter flex hose applications are discussed in Chapter 13.

11.7 Portable Water Supply System (PWSS)

A PWSS is worth discussion in some detail, as it has been adopted by a number of fire departments (e.g., San Francisco, Oakland, Berkeley, Vallejo, all in California, and Vancouver, B.C.) as well as selected water departments (discussed in Chapter 13). The PWSS was conceived of early in the 20th C. by Chief Dennis T. Sullivan of the San Francisco Fire Department (Postel, 1992). However, it was not until the early 1980s that it was fully implemented under the leadership of Chief Frank T. Blackburn of the San Francisco Fire Department.

A PWSS generally consists of the following components:

- A hose tender, typically carrying approximately a mile of large diameter hose (LDH, often 5 inch hose, sometimes as large as 12 inch), portable hydrants, pressure reducing valves, wyes and other fittings, and hose ramps to permit vehicle passage over the LDH. Figure 11-12 shows a PWSS hose tender, Figure 11-13 shows an in-line portable hydrant with Gleeson pressure-reducing valve attached, and

- Figure 11-14 shows LDH (5 inch) and hose ramps. While larger diameter hose is available (see Chapter 13), 5 inch hose is about the maximum size that can be feasibly handled by firefighters; whereas 12 inch hose can be deployed with special systems, possibly at a slower pace.

- A portable pump, capable of drafting or pumping water from ponds or other bodies of water. Because water can only be practicably drafted a vertical distance of about 26 feet (8 m), if a bridge, seawall, pier or other access point is more than this distance above the water, then a fire engine cannot access the water – it must get closer, which is often difficult. To overcome this problem, several systems, such as the Dutch-manufactured *HydroSub,* are designed to lower a powered pumphead to the water, and *push* rather than *draft* the water through the hose. For example, a powered pumphead can have a capacity of 1,200 gpm; the pumphead system would consist of a diesel driven hydraulic pump, which via long hoses powers a hydraulically driven centrifugal pump in a separate portable pumphead, which can be carried up to several hundred feet across a flat topography from the unit. LDH is connected to the portable pumphead, and can then convey the water up to a mile away. In this manner, the *HydroSub* for example can access water more than 100 ft (33 m) below its location, or several hundred feet laterally, and from that source supply a location up to a mile away. Figure 11-11 shows a trailer-mounted *HydroSub* unit.

San Francisco Fire Department PWSS units were employed in the Marina in the 1989 earthquake (this is discussed in Chapter 2) and in the 1991 East Bay Hills Fire (see Chapter 3), as well as in many other urban and wildland incidents. These and other incidents revealed that an essential element of a PWSS system are special procedures and techniques, which should be regularly reinforced via training and exercises.

Figure 11-11 PWSS system, consisting of Hose Tender towing HydroSub portable pumping unit

Figure 11-12 Oakland FD Portable Water Supply System Hose Tender, showing portable hydrants above rear axle), Gleeson pressure-reducing valves (to right of wheel) and hose ramps strapped underneath chassis

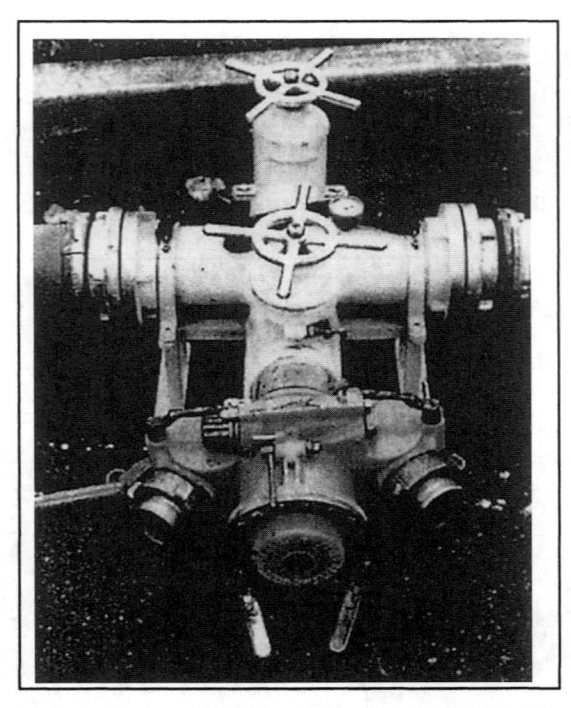

Figure 11-13 In-line portable hydrant, with Gleeson pressure-reducing valve attached

Figure 11-14 PWSS hose ramps over 5-inch LDH, permitting traffic to freely pass over the hose, which is otherwise an obstacle

11.8 Credits

Figure 11-2, 11-3, 11-5. EQE International and Scawthorn. Figure 11-4. Quake Defense, Inc. Figure 11-6, 11-10b. EQE International. Figure 11-7, 11-8, 11-10a, 11-13, 11-14. Scawthorn. Figure 11-11. Scawthorn, courtesy Vallejo Fire Department. Figure 11-12. Scawthorn, courtesy Oakland Fire Department.

11.9 References

EQE Engineering. 1990. Emergency Planning Considerations, City of Vancouver Water System Master Plan, prepared under subcontract to CH2M-Hill, for City Engineering Department, City of Vancouver, B.C.

EQE. 1989. San Francisco Fire Department Facilities, Summary Report, prepared by EQE/AGS under contract to Department of Public Works, City and County of San Francisco, for the San Francisco Fire Commission.

Kobayashi, M. and L.M. Kornfield. 2002. Strategies for reducing post-earthquake fire damage in historic wood structures, Proc. 7th National Conference on Earthquake Engineering, Boston, Earthquake Engineering Research Institute, Oakland CA

Los Angeles City Fire Department, 1987. Earthquake Emergency Operational Plan, 1987.

Mroz, M.P and T. T. Soong. 1997. Fire Hazards and Mitigation Measures Associated with Seismic Damage of Water Heaters, Report NIST GCR 97-732, Building and Fire Research Laboratory, National Institute of Standards and Technology Gaithersburg, Maryland 20899

ICLR. 2001. Assessment of Risk due to Fire Following Earthquake Lower Mainland, British Columbia, report prepared for the Institute for Catastrophic Loss Reduction, Toronto, by C. Scawthorn, and F. Waisman, EQE International, Oakland, CA.

Schussler, H., 1906. The Water Supply of San Francisco, California, Spring Valley Water Company.

Sepponen, C. 1997. Fire Department Emergency Response, Ch. 16 in Schiff (1997).

12 High Pressure Water Supply Systems

12.1 Introduction

In the early 20th century, as a result of major urban conflagrations, a number of US cities built special 'high-pressure' water supply systems to protect their central business districts. The 1871 Chicago, 1904 Baltimore and 1906 San Francisco fires are only the best known of many large urban conflagrations that plagued the U.S. at that time. There were a number of contributing factors to these conflagrations, but lack of reliable water supplies were a key factor that lead to the construction of the high pressure systems. For various reasons having to do with improved building stock and fire service capability, most of these cities have since abandoned or de-emphasized these systems, with San Francisco being the one major exception. San Francisco maintains its Auxiliary Water Supply System (AWSS) due to its high seismic risk and the trauma of the 1906 fire. Recently, considering the benefits of seismically resistant redundant water supplies, several cities at high risk due to earthquake have again considered the benefits of high-pressure systems. This Chapter discusses several of these systems:

- The City of San Francisco's AWSS system.

- The City of Vancouver, B.C., as a result of analysis of its water supply system's seismic reliability (or lack thereof), embarked on the design and construction of a system similar to San Francisco's, which Vancouver called its *Dedicated Fire Protection System* (DFPS). Today, the DFPS is a fully functioning seismically resistant redundant water supply protecting the CBD and other high-density districts.

- The City of Berkeley, CA designed a saltwater system, but construction has not occurred. While not constructed, the analysis and design of the Berkeley system provides valuable lessons and insights.

- Kyoto, Japan is in the early stages of considering an auxiliary system, due to its high seismic risk and the lessons to be learned from the 1995 Kobe earthquake. While only at a very early conceptual stage, discussion of the problems and possible solutions for Kyoto also offers some valuable lessons and insights.

Prior to discussing these three systems, however, we first briefly review the 'parent' of these systems – the San Francisco Auxiliary Water Supply System.

12.2 San Francisco Auxiliary Water Supply System (AWSS)

As noted above, early 20th C. US cities were very conflagration prone, and each eventually had 'its fire'. This situation was especially true of San Francisco, which had grown very rapidly following the Gold Rush, and which had been built with the plentiful inexpensive wood from the forests of California. As a result the National Board of Fire Underwriter's in their Report on San Francisco in 1905 was able to write:

"...In fact, San Francisco has violated all underwriting traditions and precedent by not burning up. That it has not done so is largely due to the vigilance of the fire department, which cannot be relied upon indefinitely to stave off the inevitable."

In recognition of the high fire danger in San Francisco, an Auxiliary Water Supply System (AWSS) had been proposed for the City in 1905 by Chief of Department Dennis T. Sullivan (Postel, 1992) – that is, *prior to* the 1906 earthquake and fire.

San Francisco in 1906 suffered a great (M7.9) earthquake which, however, was known for several generations as the San Francisco Fire, because San Franciscans understood the fire had been the real problem. The earthquake caused multiple simultaneous ignitions in a large city almost entirely built of wood while at the same time breaking the water supply pipes in many locations. That the City had burned due in large part to lack of water was particularly ironic in that San Francisco is surrounded on three sides by the largest body of water on earth – the Pacific Ocean.

As a result of the 1906 earthquake and fire, Marsden Manson (San Francisco City Engineer) in 1908 proposed the AWSS, which was developed with a $5.2 million bond issue approved by the people of San Francisco in that year. Today, San Francisco maintains its Auxiliary Water Supply System due to its high seismic risk and the trauma of the 1906 fire. *A key aspect of San Francisco's ability to maintain and even extend this unique system is that fact that it is, by city charter, owned and operated by the fire department.* This is a vital point, in that it has served to protect the AWSS a number of times over the years, from budget cuts as well as diversion to other uses, such as irrigation, which would eventually impair the system's seismic reliability. The San Francisco AWSS was described in Chapter 2, and only selected aspects will be summarized here:

- The AWSS is intended as an *auxiliary* system, to supplement the use of the municipal water supply system (MWSS, the ordinary potable water supply system) for fighting large fires, under non-earthquake as well as earthquake conditions. This is an important point – it does not sit around for decades, waiting for an earthquake. Rather, the department uses it at most greater alarm incidents, thereby gaining valuable experience, confirming its continued functionality and reliability, and justifying the system's existence.

- It has multiple sources of supply, including over 175 cisterns distributed throughout San Francisco. These cisterns, most having 72,000 gallons or about one hour pumper supply, are completely isolated from all piping, and thus highly reliable. The fresh water supply also includes a 10 million gallon reservoir at Twin Peaks, the highest point in the City, partitioned into two halves. The supply also includes two pumping stations, founded on rock and each capable of pumping 10,000 gpm of salt water from San Francisco Bay, as well as a number of manifolds distributed around the bayshore, into which the City's two fireboats (*Phoenix*, and *Guardian*) can pump 9,600 and 24,000 gpm respectively, from San Francisco Bay, as well as thirty-six water suction connections distributed along the City's waterfront to allow fire engines to draft saltwater from the Bay of San

Francisco. Thus, there are multiple redundant sources of supply. The Twin Peaks reservoir is intended for large non-earthquake fires, while the pump stations and fireboat manifolds represent an unlimited supply of water.

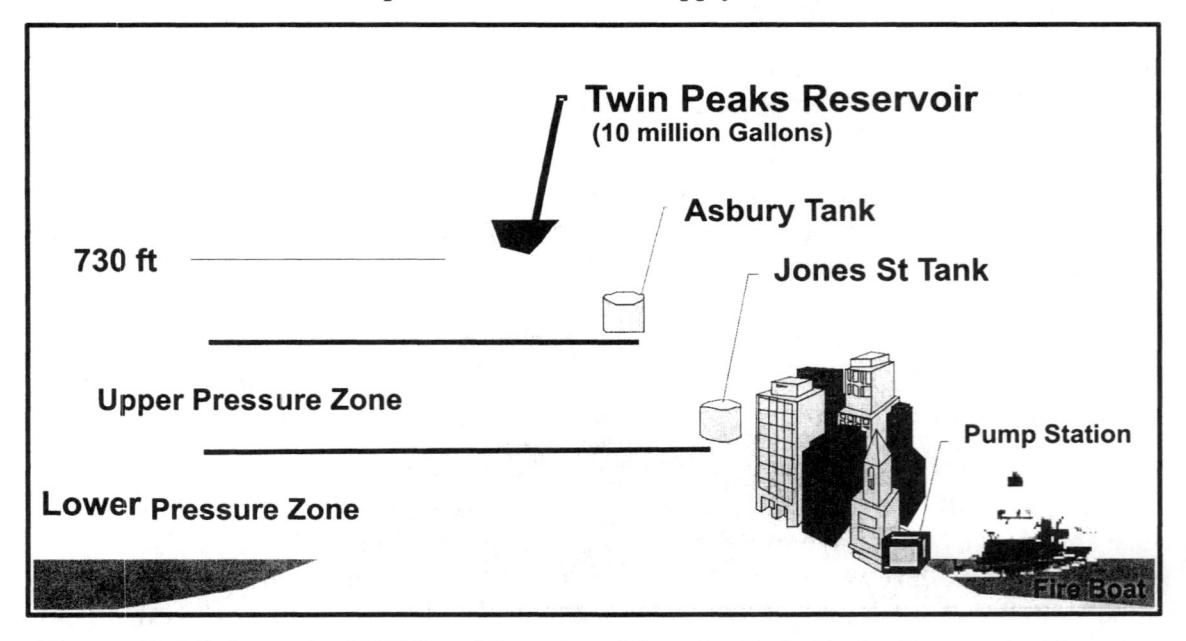

Figure 12-1 Schematic AWSS and Sources of Supply: Twin Peaks Reservoir, Cisterns, Pump Station and Fireboat(s)

- The underground piping system was designed from the beginning to be highly earthquake resistant – the piping is extra heavy walled, and has restrained joints at numerous points. San Francisco has several areas of highly liquefiable soils – these were observed to fail in 1906 and to correlate with damage to underground piping. These 'infirm zones' were all mapped, and the system designed so that, while AWSS pipe passes through these zones, the system can be quickly isolated should pipes in those zones fail. In modern times, the gate valves isolating the infirm zones have been motorized and can be remotely controlled via radio.

- As a result of the elevation of the Twin Peaks reservoir, and the capacity of the pumping stations and the fireboats, very high pressures, in excess of 300 psi, can be sustained in the AWSS. This pressure assures a high volume supply, but is too high for many applications, and can be reduced via Gleeson valves – a patented pressure reduction valve invented in the San Francisco Fire department shops. The Gleeson valve permits a firefighter to attach one or several handlines to an AWSS hydrant, and apply firestreams as if from a fire engine. Thus, the AWSS reduces the need for fire engines, and permits a continuous water curtain to be sprayed from a line of hydrants along a defensive line.

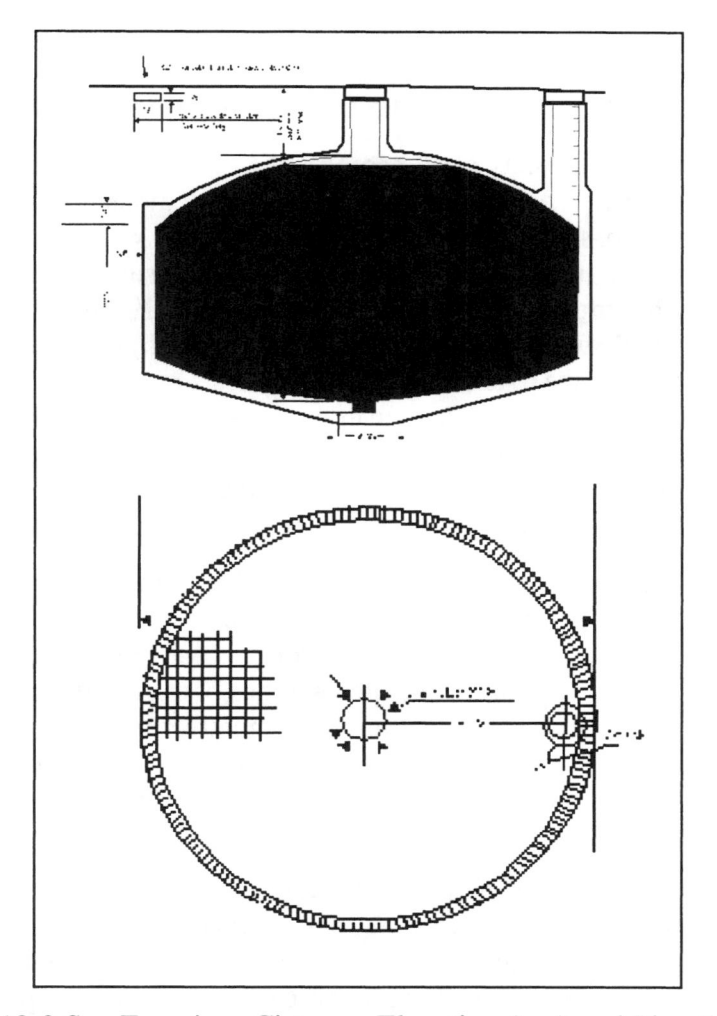

Figure 12-2 San Francisco Cistern – Elevation (top) and Plan (bottom)

Designed almost a century ago with great foresight and skill, the San Francisco AWSS was intended to be a seismically reliable water supply system for fire protection. Most of the original pipeline was extra heavy cast iron pipe with more recent installations using thick-walled ductile iron pipe with restrained joints at high thrust locations. It has been maintained for almost a century, and embodies the key attributes of redundancy in supply and layout, reliability via layout and seismic design of components, flexibility in application, economy via reducing the need for fire engines and apparatus, and integration in the fire department's day-to-day operations. Even so, the 1989 Loma Prieta earthquake damaged a few components of the AWSS (see Section 2.4.2), which, coupled with human inaction, prevented the system from supplying water to the Marina fires; thus demonstrating that there is room for improvement.

12.3 Vancouver, B.C. Dedicated Fire Protection System (DFPS)

As discussed in Chapter 4, the City of Vancouver, B.C. is a large metropolitan area at great risk due to earthquake. Analysis (ICLR, 2001) has demonstrated that the City has the potential for a large loss due to fire following earthquake, which is ironic in that similarly to San Francisco, the City of Vancouver is literally surrounded by the largest body of water on earth – the Pacific Ocean. Analysis (EQE, 1990) also demonstrated that the City's existing potable water supply system, used for firefighting, was inadequate in the event of a large earthquake. As a result of the analysis of its water supply system's seismic reliability (or lack thereof), Vancouver embarked on the design and construction of a system similar to San Francisco's, which Vancouver called its Dedicated Fire Protection System (DFPS). This section briefly describes Vancouver's DFPS as it was finally designed and constructed.

The City of Vancouver's primary water supply is via the 1st and 2nd Narrows Crossings, which both cross liquefiable deposits and were identified a decade ago as seismically vulnerable (EQE, 1990). The only in-city water storage is at Little Mountain reservoir in the center of the city – the reservoir is operated and has been recently strengthened. Fully 77% of the domestic water distribution system in Vancouver is seismically vulnerable cast iron pipe more than 30 years old, while 30% of the larger steel trunk mains in the city are more than 70 years old and of highly vulnerable riveted construction (MTR, 1991).

The DFPS is a major asset for protection of the high-value district of Vancouver, and consists of two seismically reliable pump stations (at Coal Harbor, and at False Creek) each capable of 10,000 gpm, Figure 12-3. Due to the large tide range in Vancouver, the pump stations required a deep wet well, with vertical axis turbine pumps, which are powered by direct-drive diesel engines via a right angle gear drive, Figure 12-4.

The two stations are connected by a loop around the CBD of 20 inch diameter welded steel pipe, Figure 12-5(b). Special large diameter hydrants connect to this loop, from which firefighters can receive high volume high pressure flows. VFR is acquiring special large diameter hose which, together with the DFPS, can be used to protect a 1000 ft. radius from any hydrant off of this loop, thus 'blanketing' the CBD.

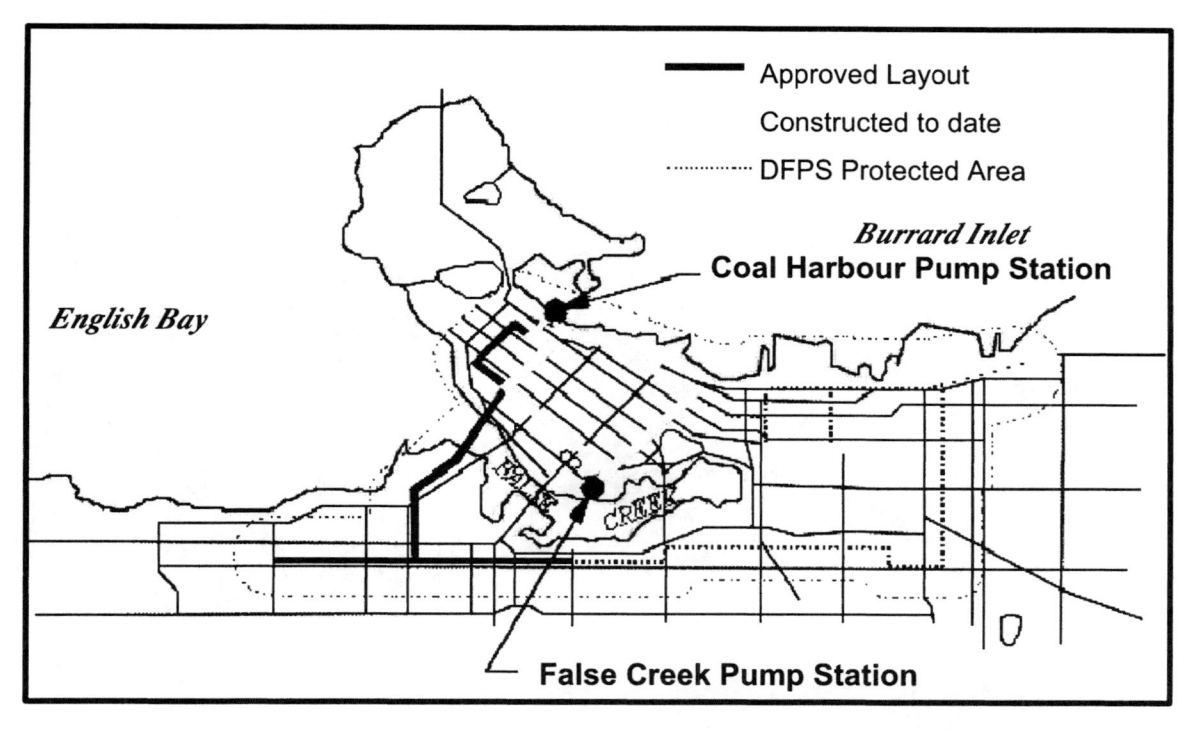

Figure 12-3 Dedicated Fire Protection System (DFPS), City of Vancouver, B.C.

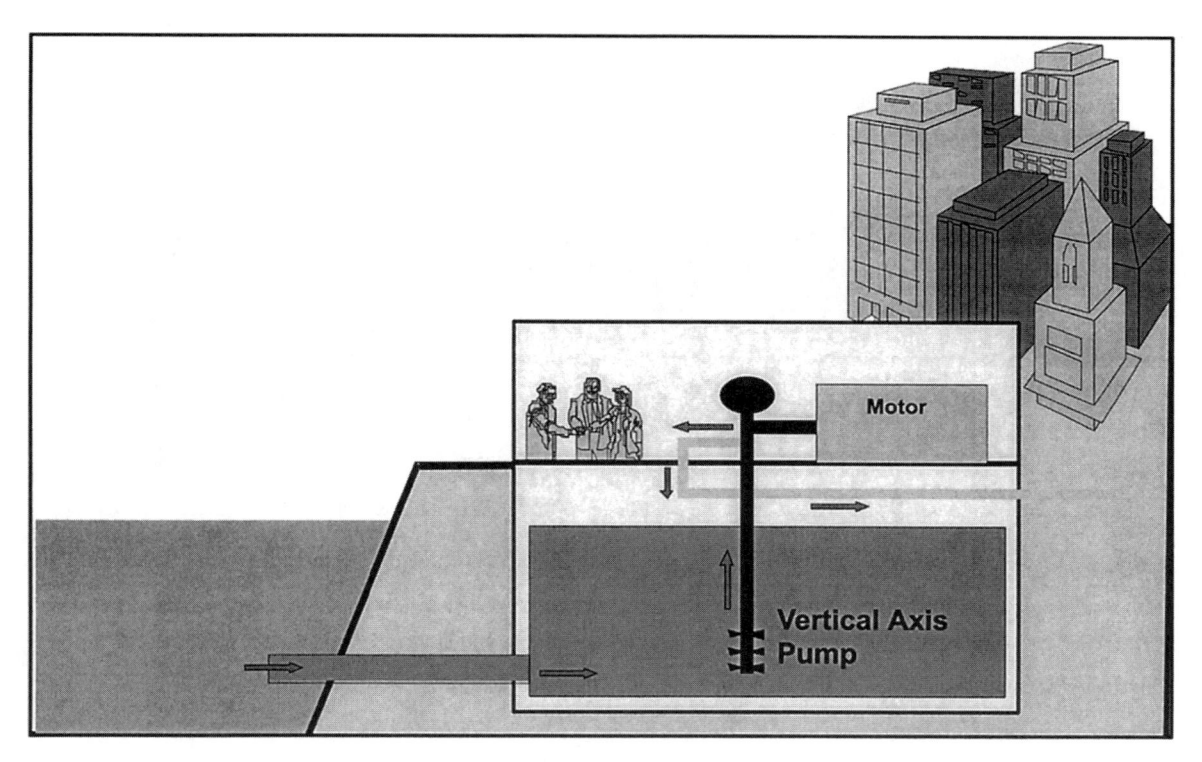

Figure 12-4 DFPS Pump Station Schematic

Figure 12-5 Vancouver, B.C. DFPS (a) False Creek pump station (foreground, nearing completion in 1995); (b) welded steel pipe under construction, 1995; (c) False Creek station proof test, 1995.

12.4 Berkeley Saltwater Fire System (SFS)

This section examines the concepts being used for design of the City of Berkeley Saltwater Fire System (SFS).

12.4.1 System Performance Goals

The SFS was designed with three performance goals:

- The SFS should be able to deliver 20,000 gpm following a large earthquake.

- The SFS should be able to deliver 20,000 gpm at multiple concurrent fire locations throughout the service area at a minimum residual pressure of 150 psi along the buried pipe system and 20 psi throughout the remainder of the service area.

- The SFS should be able to deliver 20,000 gpm as a concentrated fire flow at a location within the service area at a minimum residual pressure of 150 psi along the buried pipe system and 20 psi throughout the rest of the service area.

These performance goals were established in a multi-step process, as follows:

The number of fire ignitions likely to occur within the City of Berkeley was estimated using equation 4-3. For the City of Berkeley, the number of serious fires requiring fire department response, in the first day or so after a magnitude 7 earthquake on the Hayward fault, is about 19 to 26.

The demand for water for fire fighting purposes was then estimated assuming: a moderately slow time for fire department response; some staging of fires (not all fires occur at the same time); limited spread of fires by the time the fire department arrives; fire flow durations of 2, 3 or 4 hours for single family, multi family or commercial/industrial fires, respectively. The types of fires were estimated based upon the local building stock (single family, multi-family, commercial, industrial). The available flows from the local potable water distribution system should supply some but not all these flows, owing to concurrent damage to the local water distribution system. The total fire flow demand less the available flows from the local water distribution system was estimated to be 20,000 gpm. The time duration to provide this flow rate was assumed to be 2 to 3 days, recognizing that there would be a chance that fire flows will be required for more than 24 to 48 hours after the earthquake.

The SFS was designed to provide these flows under three types of fires:

- Fires following earthquake. These could be multiple small fires at diverse locations, or one large fire.

- Urban interface fires. The City of Berkeley has several times before been subjected to urban interface fires. In 1923, over 600 structures were burned due to a fire that started in the urban-interface region northeast of Berkeley. High winds spread the fire into residential neighborhoods. The fire continued to burn through the neighborhoods until the fire reached a location with a large supply of water from the local distribution system. In 1970, 40 structures were burned in the Claremont Canyon fire. In 1980, 5 structures burned near the ridgeline in Berkeley. In 1991, over 3,000 structures burned (Oakland plus Berkeley) due to a fire that started in a high fuel load area in the Oakland - Berkeley Hills. The fire continued to burn through neighborhoods until the fire reached locations with large supplies of water from the local distribution system. The SFS is thus designed to provide high flow rates for essentially unlimited time for such a repeat of these types of fires, which could occur with or without earthquakes.

- Normal fires.

12.4.2 System Layout

The layout of the SFS has the following main features (see Figure 12-6).

The intake structure would be sited in the closest deep water (over 25 foot depth) location in the San Francisco Bay. Other bay locations within 3 miles of the Berkeley shoreline are about 6 feet depth, at low tide. The intake pipeline to the salt-water pump station was designed to accommodate any level of ground movements possible in the young alluvials and bay muds at the bay shorelines. This was accomplished though pipeline material selection and depth of burial.

The salt-water pump station was designed with multiple diesel engine driven pumps. The design flow was 20,000 gpm with one pump out of service. Backups systems were provided to ensure reliable automatic operation of the pump station under remote operational control or local manual controls.

The salt-water pump station could draw upon salt water, or potable water from the local domestic water distribution system, if it is available. Under post-earthquake operations, water from the local domestic system was assumed to be unavailable.

The salt-water pump station would boost water pressure to 300 psi at the operating flow rate of 20,000 gpm.

A 26,000 foot long welded steel buried distribution system would extend from the salt water pump station eastwards through the flatlands of Berkeley to the central business district. At its highest elevation and points furthest from the pump station, the minimum pipeline pressure would be 150 psi, when flowing at 20,000 gpm. The steel pipeline system was designed to accommodate liquefaction induced settlements which could occur anywhere up to one inch, and in areas closer to the bay, up to three inches. Localized settlements at stream crossings could exceed these amounts.

The welded steel distribution system includes one loop. The function of the loop is to provide a redundant hydraulic path to allow flows throughout the remaining undamaged part of the system, should there be an unexpected break in the other leg of the loop.

A total of 20,000 feet of 12" diameter ultra large diameter hose (ULDH) would provide the ability to draw high flows of water from the steel pipeline system. The ULDH would be deployed from special 12" manifold outlets off the main steel pipe, and then deployed for distances up to 7,000 feet away from the steel pipe, flowing at rates up to 5,000 gpm.

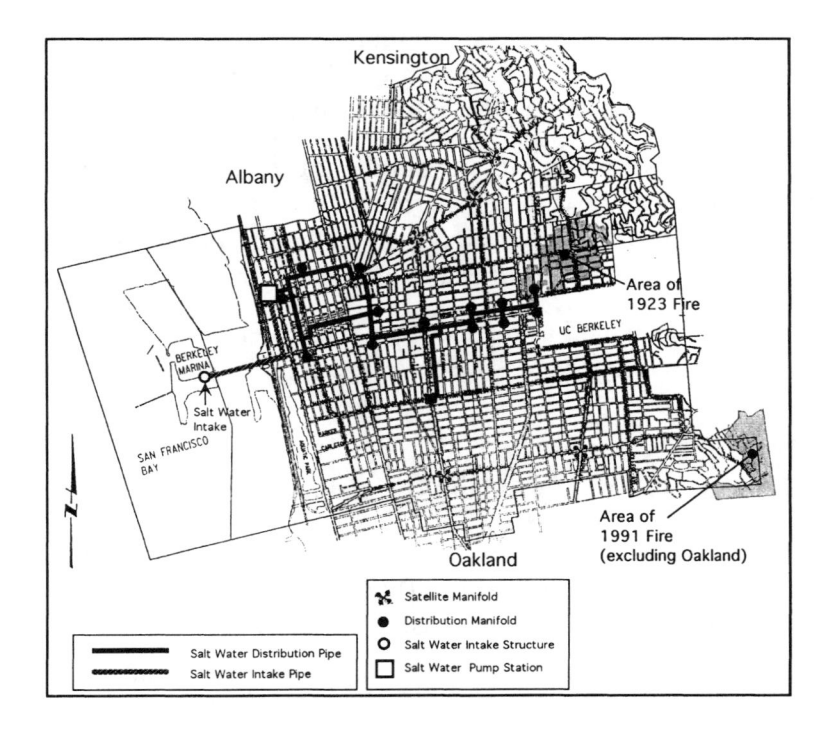

Figure 12-6. City of Berkeley Proposed Salt Water Fire System

A total of 14 manifold outlets are provided in the system, to allow the fire department to choose between several different outlet points from the steel pipe system for hose deployment to a particular fire location. For any location served by the SFS, there are an average of 7 different manifold outlets that could be used. This provides flexibility in laying the ULDH along streets.

At particular busy street intersections, buried steel pipe and suitable manifolds would be installed to allow deployment of ULDH throughout the coverage areas, without having to tie up critical traffic corridors.

Should a break occur in the unlooped portion of the steel distribution pipe, the ULDH could be used to "bridge" the break, by hooking into 12" manifolds either side of the break. If used in this mode, the distribution system performance on the side of the break away from the salt-water pump station would operate at lower pressures, but still be able to deliver water.

12.4.3 Coverage of the System

For locations near the pipeline, fire fighters would be able to draw water from hydrants, without the use of pumper trucks, using normal handline hoses carried on most fire trucks. A pressure adapter may be needed to reduce pressures from the pipeline into the hose. This is called Type I service.

For moderate distances from the pipeline, ULDH hose would be used to connect to the steel pipeline. 5" hose and smaller handlines would be used at the end of the ULDH hose to deliver water to the fire. No pumpers would be required to boost pressures. This is called Type II service.

For long distances from the pipeline, ULDH hose would be used to connect to the steel pipeline. The end of the ULDH hose would then connected to pumper trucks, for boosting water pressure. 5" hose and smaller handlines would then be used from the pumper truck to deliver water to the fire. This is called Type III service.

The coverage ratio represents the average number of manifolds that the fire department could get water from for any particular fire location within the area served by the SFS. Higher coverage ratios provide more flexibility to the fire department. A coverage ratio of 1.00 means that only one manifold can be used to fight any individual fire.

The coverage ratio is calculated as the total coverage area, summed over all manifolds, divided by the actual coverage area. The coverage ratio is calculated as follows:

$$Area_i = \pi r_i^2 \qquad \text{(equation 12-1)}$$

Where $Area_i$ is the area served by manifold i, and πr_i^2 is the area with radius r.

The next step in the overlap calculation is to add up the total area served by all manifolds, assuming that each manifold is operational (gets water). This would be the case under the normal (non-earthquake) condition, or a case where there is no damage in the post-earthquake case.

$$TotalArea = \sum_{i=1}^{number\ of\ manifolds} Area_i \qquad \text{(equation 12-2)}$$

where the summation is from $i=1$ to number of manifolds.

The TotalArea calculated by equation (12-2) will double count some areas of Berkeley. For example, if two (or more) manifolds can each serve a particular location, the TotalArea value in equation (12-2) will overstate the actual area that can be served by any manifold. The following is done to exclude this double counting:

- Develop a mesh covering the area of Berkeley and adjacent areas potentially served by the SFS. Each grid element of the mesh is 20' x 20'. The total mesh has 960,000 elements, covering a grid with 1,200 x 800 elements, totaling 24,000 feet (east to west) by 16,000 feet (north to south).

- The selected mesh covers 5,383 acres of the land area of Berkeley. The model also includes 2,357 acres in Oakland and Albany.

- For each manifold, calculate the potential area served. For example: for manifold 1, calculate $Area_1$; for manifold 2, calculate $Area_2$; etc.

- Calculate if a grid element can be served by $Area_1$, $Area_2$, etc.

- Add up the number of grid elements that can be served from any manifold. Multiply this by 400 (each grid box represents 400 square feet), and divide by 43,560 (number of square feet per acre). This is the actual area served. The actual area served is calculated twice: once for land area within the City of Berkeley and once for land area outside the City of Berkeley. Note that although some manifolds can provide service to areas over water (in the Berkeley Marina area), these areas are excluded from the calculations.

- $$ActualArea = \frac{400}{43,560} \sum_{k=1}^{960,000} \langle 1 \text{ (if served)} | 0 \text{ (if not served)} \rangle \quad \text{(equation 12-3)}$$
  where $k =$ grid elements

- where summation is from $k =$ grid elements

- The coverage ratio is: $CR = \dfrac{TotalArea}{ActualArea}$. (equation 12-4)

- The coverage ratio is calculated using only the area within the City of Berkeley.

Five different choices as to the number of manifolds used in the pipeline system were considered:

- 13 manifold System. This analysis considers what is "lost" by saving the cost of one manifold.

- 14 manifold system. This is the baseline system, shown in Figure 12-6.

- 19 manifold system. This adds 5 manifolds to the baseline system.

- 21 manifold system. This adds 7 manifolds to the baseline system.

- 23 manifold system. This adds 9 manifolds to the baseline system.

An analysis of the coverage ratios is shown in Table 12-1 highlighting that:

- The more manifolds included in the system, the more Type I coverage is provided. Type I service is limited to a short corridor nearest the pipeline.

- Type II coverage area increases from 2,226 acres to 2,395 acres, as the number of manifolds is increased from 13 to 23.

- Type III coverage area increases from 4,052 acres to 4,060 acres, as the number of manifolds is increased from 13 to 23.

Table 12-1 System Service Areas and Coverage Ratios

Service Type, Coverage	13 Manifolds	14 Manifolds	19 Manifolds	21 Manifolds	23 Manifolds
Type I - Berkeley	436 acres	462 acres	579 acres	642 acres	678 acres
Type I - Albany / Oakland	0 acres	0 acres	0 acres	0 acres	0 acres
Coverage Ratio	1.18	1.20	1.30	1.30	1.34
Type II - Berkeley	2,226 acres	2,248 acres	2,276 acres	2,277 acres	2,395 acres
Type II - Albany / Oakland	83 acres	83 acres	89 acres	89 acres	91 acres
Coverage Ratio	2.25	2.40	3.22	3.56	3.86
Type III - Berkeley	4,052 acres	4,052 acres	4,060 acres	4,060 acres	4,060 acres
Type III - Albany / Oakland	1,259 acres	1,259 acres	1,277 acres	1,277 acres	1,277 acres
Coverage Ratio	6.55	7.06	9.10	10.33	11.03

A factor to be considered in the number of manifolds is the potential need to supply very high volumes of water to a single fire location using ULDH. If a large fire requires flows reaching 20,000 gpm, it may be possible to lay multiple sets of ULDH from several manifolds, drawing sufficient flows from each ULDH to achieve the desired flow rate. With an average coverage ratio of 7, a 14 manifold system appeared adequate.

12.4.4 Earthquake Hazard

Table 12-2 lists the scenario earthquakes that were selected for evaluation purposes. The four most likely events are the Hayward North M 7, San Andreas Peninsula M 7, the Hayward South M 7, and the Rogers Creek M 7; the probabilities for these events were developed by the USGS (1990).

Table 12-2 Scenario Earthquakes

Fault	Magnitude	Closest Distance to Salt Water System (km)	Probability of Occurrence through the Year 2020
Hayward North	7.0	1	28 %
Hayward North	7.25	1	< 3 %
Hayward North	7.5	1	< 1 %
San Andreas North Coast	8.0	26	2 %
San Andreas Peninsula Segment	7.0	36	23 %
Calaveras	7.0	24	< 10 %
Concord	6.5	23	< 10 %
Loma Prieta	7.0	79	not applicable
Hayward South	7.0	14	23 %
Rogers Creek	7.0	27	22 %

Given the occurrence of any of the ten scenario earthquakes, there will be ground shaking throughout Berkeley. The level of shaking at each pipeline location of the SFS was calculated. Table 12-3 provides the median level peak ground accelerations at the surface, at two locations:

- Salt Water Pump Station. This is the westernmost location of the pipeline distribution system.
- Oxford and Berkeley Way. This is the easternmost location of the SFS pipeline distribution system. Expected ground motions at other locations would be at intermediate levels.

Table 12-3 Ground Shaking Levels (Median Levels)

Fault	Magnitude	Salt Water Pump Station (PGA - g)	Oxford and Berkeley Way (PGA - g)
Hayward North	7.0	0.46	0.54
Hayward North	7.25	0.47	0.55
Hayward North	7.5	0.48	0.55
San Andreas North Coast	8.0	0.26	0.24
San Andreas Peninsula Segment	7.0	0.12	0.12
Calaveras	7.0	0.16	0.18
Concord	6.5	0.13	0.14
Loma Prieta	7.0	0.08	0.08
Hayward South	7.0	0.24	0.26
Rogers Creek	7.0	0.16	0.15

Figure 12-7 provides an engineering geologic map of the Berkeley area. Various geologic units are identified. These are briefly described below, along with their associated seismic hazards:

Unit 1 - Artificial Fill (Qaf). There are areas along the Berkeley waterfront where fill has been placed over the last 60 years. The fill was placed directly on soft tidal marsh deposits along the shoreline. These areas are highly susceptible to liquefaction, with potential for settlement everywhere and lateral spreading near open shorelines.

Unit 2 - Bay Mud (Qbm). Bay Mud refers to the soft marsh deposits occurring in the tidal flats as well as the mud accumulating on the bay bottom. Bay Mud usually consists of unconsolidated, water-saturated, dark, plastic carbonaceous clay and silty clay, with minor lenses of sand, shells, and organic material. These deposits are located beneath the San Francisco Bay. These areas are highly susceptible to liquefaction, with potential for settlement everywhere and lateral spreading near open shorelines.

Unit 3A, 3B - Young Alluvial, Fluvial and Basin Deposits (Qya). These sediments have been deposited in the last 10,000 years. Fluvial deposits (3A) are composed of fine grained sand, silt, and silty clay. Basin deposits (3B) are composed primarily of clay and silty clay. These deposits are generally located west of San Pablo Avenue and east of the Southern Pacific Railroad alignment (darker areas - 3A, lighter areas - 3B). Additional deposits are expected to occur in the vicinity of historic creek alignments throughout Berkeley. As creek alignments may have migrated over the past several thousand years, these soils can occur locally in many places. These soils are moderately susceptible to liquefaction, primarily manifested by settlement.

Unit 4 - Old Alluvial Soils (Qoa). Older alluvial soils consist of weakly consolidated, slightly weathered, poorly sorted deposits of clayey gravel, sandy silty clay, and sand-silt-clay mixtures. These deposits are located east of San Pablo Avenue, extending eastward to the Berkeley Hills. These deposits are also located at depth beneath the younger alluvium west of San Pablo Avenue and the Bay Muds. These soils have low susceptibility to liquefaction.

Unit 5 - Undifferentiated Rock. Berkeley Hill areas east of the Hayward fault are generally rock, overlain by a veneer of residual soil formed by in-place weathering of rock. There are also some rock outcrops located west of the Hayward fault (hatched areas in Figure 12-7). Rock areas east of the Hayward fault are shown in white in Figure 12-7. Occasionally, the hill areas are overlain with colluvial soils that accumulate on the lower reaches of slopes as a result of downslope movement of soil and rock waste. These areas are susceptible to landslide movements.

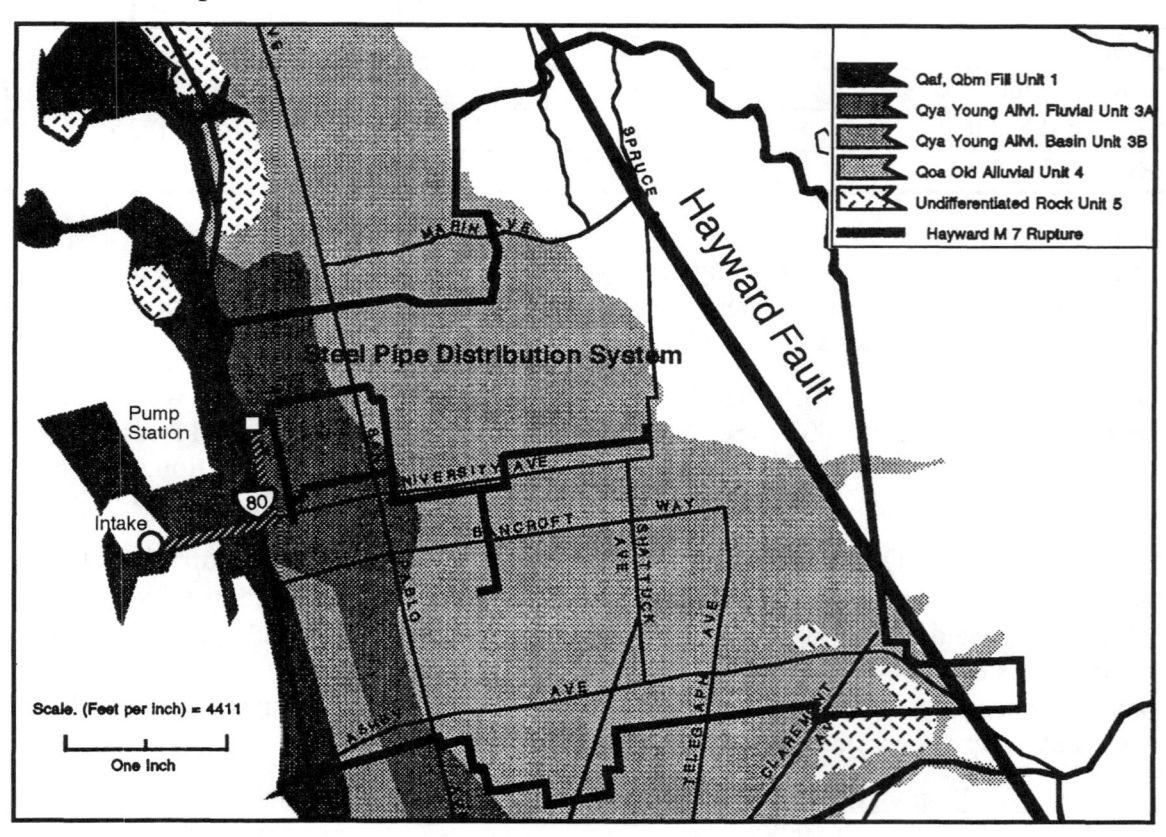

Figure 12-7. City of Berkeley – Surface Geology

The main liquefaction hazard for the SFS steel pipelines occurs in geologic Unit 3A. Expected settlement in this area is between 1 and 3 inches for a Hayward magnitude 7 event. There may also be some liquefaction hazard, on the order of 3 to 4 inches, at some stream crossings. The liquefaction hazard for the SFS intake pipeline is more severe.

12.4.5 Pipeline Design

12.4.5.1 Performance of Damaged Pipeline System

The SFS was designed to be able to sustain some level of damage while still reliably delivering the target 20,000 gpm flow, following a major earthquake. The main location of vulnerability to the pipeline system is in the soils susceptible to liquefaction. These soils mainly occur west of San Pablo Avenue. To counter this hazard, the pipeline system includes a loop with in-line valves in this area. If one branch of the loop is significantly damaged, water can escape from the damaged pipe. Without in-line valves, if a sufficient volume of water leaks at the damaged location, then the entire SFS system will be depressurized, preventing flows from reaching manifolds at undamaged locations.

12.4.5.2 Construction Issues

The selection of the buried steel pipeline diameter considers the constructibility of field girth joints.

If a full penetration butt weld design is selected, then the cost of construction will increase. A well-made full penetration butt weld joint provides extra seismic capacity as compared to a single lap joint. A double lap welded joint is somewhat more expensive to construct than a single lap welded joint. Overall, the reliability of a double lap welded pipe will be substantially better than for a single lap welded pipe.

12.4.5.3 Pipeline Coating

The primary function of the exterior pipeline coating is corrosion protection. For earthquake purposes, it is desirable to reduce the friction between the pipeline exterior and the soil backfill. The lower the friction, the easier the pipeline can accommodate differential settlements (liquefaction) of the soil. Coal tar enamel, tape, and plastic coatings will generally produce a somewhat lower level of friction than mortar coating.

12.4.5.4 Pipeline Appurtenances

The SFS pipelines would include a number of different types of appurtenances. The main seismic design issue related to these appurtenances is how much "anchorage" they provide to the pipeline. By "anchorage", it is meant that the appurtenance will provide localized restraint to prevent the pipeline from slipping axially through the soil. Such slippage is desirable for pipelines that must accommodate permanent ground deformations (PGDs). The more anchors on the pipeline, the more difficult it is for the pipeline to distribute strain induced by PGDs. PGDs can arise from differential settlement (liquefaction), lateral spreading (at stream crossings), and other ground movements.

12.4.5.5 In-line Valves

The SFS pipeline includes a number of in-line valves. The function of these valves is to isolate the pipeline for two purposes:

- Normal maintenance
- Post-earthquake isolation of damaged sections of pipeline

In-line valves can be manual gate valves, butterfly valves, or motor- or hydraulic-operated (highest cost) valves. Bypass pipes and valves (nominal 6" diameter) are used to simplify the opening and closing of large valves. The more valves installed into the pipeline, the more flexibility is provided in terms of maintenance and ability to isolate smaller portions of the pipeline. However, adding more valves into the pipeline has some disadvantages:

- Large diameter valves are relatively expensive. The more valves added to the pipeline, the higher the cost of the pipeline

- Large diameter valves require a maintenance program. They should to be turned (exercised) at least once every six months

- Buried large diameter valves will increase anchorage in the pipeline system

- Bypass pipes will increase anchorage in the pipeline system

12.4.5.6 *Main Branch Connections and Pipe Bends*

There would be many "T" and 90-degree turns of the main SFS pipeline in the configuration shown in Figure 12-6. Each "T" connection or 90 degree turn would effectively create an anchor in the pipeline. To limit the effects of these anchors, particularly in the soils susceptible to liquefaction, the next "anchor" placed in the pipeline would be located preferably at least 150 pipe diameters away. The anchor spacing criteria is based on the pipe length required to distribute accumulated axial strains due to pipe settlement.

12.4.5.7 *Manifold Branch Connections*

Manifold branch connections would create "T" connections, with the manifold pipeline branching off the main pipeline. The lateral may provide significant anchorage to the main pipeline. To reduce the effects of such anchorage, the manifold "T" connections would be placed no closer than the spacing above (150 pipe diameters). Where this would not be practical, the branch line would include suitable expansion joints / flex couplings near the main pipeline, or would be housed in a larger pipe to provide an annulus between the pressurized pipe and the outer pipe. The space between the two pipes should not be filled with grout or other solid materials.

12.4.5.8 *Hydrant Laterals*

The design of these laterals would incorporate the same seismic issues as for manifold branch connections.

12.4.5.9 *Air Valves*

A number of air valves would be included in the pipeline. These act to introduce air into or vent air from the pipeline. The diameter and length of these branch lines can be significant in providing restraint to the main pipeline, as well as increasing stress at the branch connection points.

Experience has shown that many air valves are "snapped off" due to shearing forces of the main pipeline slipping through the soil, especially in areas prone to PGDs; or can be damaged due to inertial forces when placed in a vault. Failure of an air valve can result in soil erosion due to the release of high-pressure water into the ground, which may undermine the main pipeline and have other secondary impacts.

The length of the hard pipe from the main pipeline to the air valve should be kept as short as practical, to limit anchorage. Spacing the air valves to prevent anchorage within the lengths described above will limit their impact. Alternately, air valves could be installed in vaults that allow free axial movement of the main pipe, while allowing for inertial loading on the stem pipes allowing for accumulated corrosion.

12.4.5.10   Blow Offs

A number of blow offs would be included in the pipeline. These act to allow dewatering and purging of the pipeline. The diameter and length of these branch lines can be significant in providing restraint to the main pipeline, as well as increasing stress at the branch connection points. The seismic design issues for blow offs would be similar as for air valves.

12.4.5.11   Manholes

A number of manholes would be included in the pipeline. These are buried or installed in vaults. While the diameter of the manholes is large, the length of the stubs is usually small, thereby limiting the effective restraint they add to the pipeline.

12.4.5.12   Couplings, Flanges, and Other Mechanical Joints

A number of couplings and flanges might be included in the pipeline. Couplings are installed to provide flexibility between two parts of the pipeline and to allow removal of valves or other pipe appurtenances. Flanges are included to allow fit up of valves and to provide insulation joints. These types of fittings will provide limited anchorage to the pipeline.

12.4.6 Reliability Analysis

The Berkeley Fire Department wished to be able to operate the SFS with very high reliability following earthquakes. A constraint in the design process was therefore to achieve the highest possible reliability for the system, while minimizing initial construction costs.

System reliability was defined as the percentage of the SFS coverage area that is expected to remain in service following the earthquake. System reliability was calculated as the area of Berkeley that can receive water from any manifold following the earthquake, divided by the area of Berkeley that can receive water from any manifold prior to the earthquake. Since the use of 12" ULDH allows that several manifolds from the pipeline system can be used to reach a particular fire, there is built-in redundancy to the system.

The reliability of the steel pipeline part of the SFS system was calculated as follows:

Step 1. A SERA model of the SFS system was developed (Section 4.6.3). The model included the following attributes: location of the pipes; pipeline design; location of in-line valves; location of manifolds; soil conditions at each pipe location; location of the earthquake faults.

Step 2. A Monte Carlo simulation technique was used to evaluate the response of the SFS under a given scenario earthquake. For each simulation, the following steps were taken:

- At each pipe location, the ground motion was calculated. This includes soil-surface PGA and PGV.

- For pipe locations in potentially liquefiable soils, the probability that the ground will liquefy was calculated. A Monte Carlo simulation was performed to determine whether liquefaction occurs at the pipe location. If no, then there was no PGD. If yes, then a randomized amount of soil movement was calculated.

- Given the PGD and PGV for a pipe, the number of pipe repairs was projected. A pipe was considered functional if it leaks, but not functional if it breaks. Given all the pipe damage, it was assumed that broken pipes were automatically or manually valved out.

Step 3. The reliability of the SFS was calculated. A system analysis was performed to determine whether water can reach the manifold location from the pump station, once broken pipes (if any) are valved out. If there were no unbroken loops available, then the manifold gets no water. If there was one (or more) unbroken loops, then the manifold gets water. Results were tabulated using four reliability measures:

- The number of manifolds that could get water was summed up over all simulations. This is called "Manifold Reliability" in the following tables.

- The "Type III Service Reliability" was calculated. The loss of water supply to any individual manifold does not necessarily mean that there would be no water available for a particular fire. This is because the fire fighters can use ULDH from another nearby manifold that does have water. The Type III Service Reliability percentage would always be higher than the Manifold Reliability percentage, as there is always some redundancy in the system. For the baseline case (14 manifold system), the Type III service area is 4,052 acres in Berkeley.

- The average number of repairs to the pipeline system was tabulated. This is the average over a number of simulations. For example, 1.33 repairs means that given the scenario earthquake, there is likely to be 1 or 2 locations in the SFS pipeline that suffers some sort of damage requiring repair following the earthquake. This includes damage to pipeline joints, air valves, blow offs, hydrant laterals, and manifold laterals which may result in leaks in the pressure boundary.

- The percentage of times that less than 40% of all the manifolds have no water supply is calculated. If less than 40% of all the manifolds can deliver water, this means that a significant part of the SFS is out of service. For example, 3% means that, 3 times in 100, at least 60% of the manifolds will have no water.

Step 4. Steps 1, 2 and 3 above were repeated 500 times for each scenario earthquake, and the results were tabulated. Steps 1, 2, 3 and 4 were then repeated 10 more times, once for each scenario earthquake, and the results are tabulated in the following tables.

Table 12-4 summarizes the results for the baseline design. The baseline design reflects a "standard" steel pipeline design using double lap welded joints, with burial depths of at least 5 feet. The change in reliability for the Hayward North Magnitude 7, 7.25 and 7.5 earthquakes was modest.

Table 12-4. System Reliability. Baseline Case

Scenario Earthquake	Manifold Reliability	Type III Service Reliability	Average Number of Repairs	Simulations with Under 40% Total Manifold Reliability
Hayward North 7 to 7.5	93.0 %	95.9 %	1.02	3.2 %
San Andreas North Coast 8	97.9	99.1	0.31	0.4
S.A. Peninsula Segment 7	99.9	100.0	0.04	0.0
Calaveras 7	99.4	99.7	0.10	0.4
Concord 6.5	99.8	99.9	0.04	0.0
Loma Prieta 7	100.0	100.0	0.01	0.0
Hayward South 7	98.2	99.0	0.30	0.2
Rogers Creek 7	99.8	99.9	0.08	0.0

The reliability of the system was also examined for changes under several alternate design conditions. Table 12-5 lists the alternatives explored.

Table 12-5. Alignment Alternatives

Alignment Alternative	Length of Pipe (Feet)	Number of Manifolds	Number of In-Line Valves	Length of Pipe using Single Lap Welds (feet)
1. Baseline Alignment	26,000	14	19	0
2. Improved design for laterals	26,000	14	19	0
3. Single lap welded joints in best soils	26,000	14	19	13,900
4. Single lap welded joints throughout	26,000	14	19	26,000
5. Single lap welded joints throughout, shallow burial, improved laterals	26,000	14	19	26,000
6. Shallow burial with improved laterals.	26,000	14	19	0
7. Shallow burial with 21 in-line valves	26,000	14	21	0
8. Worst liquefaction areas with 21 in-line valves	26,000	14	21	0
9. Baseline Alignment, but with no loop	24,000	14	16	0
10. Improved design for laterals and 23 manifolds	26,000	23	19	0
11. Fewer in-line valves	26,000	14	5	0

Table 12-6 lists the system reliability for each of the 11 alignment alternatives, for the major scenario earthquakes. The results for a repeat of the 1989 Loma Prieta earthquake are provided for comparison. The results for the other scenario earthquakes (Concord M 6.5, Calaveras M 7, Rogers Creek M 7, San Andreas Peninsula M 7) are all over 99%.

Table 12-6. Alignment Reliabilities

Alignment Alternative	Hayward North M 7 to M 7.5	San Andreas M 8	Hayward South M 7	Loma Prieta M 7
1. Baseline Alignment	95.9 %	99.1 %	99.0 %	100 %
2. Improved design for laterals	97.3	99.3	99.3	100
3. Single lap welded joints in best soils	97.0	99.3	99.3	100
4. Single lap welded joints throughout	94.2	98.7	98.6	99.9
5. Single lap welded joints throughout, shallow burial, improved laterals	93.1	98.3	98.1	99.9
6. Shallow burial with improved laterals.	95.3	98.7	98.8	100
7. Shallow burial with 21 in-line valves	95.3	98.7	98.8	100
8. Worst liquefaction areas with 21 in-line valves	93.6	98.2	98.3	100
9. Baseline Alignment, but with no loop	91.2	96.7	97.8	100
10. Improved design for laterals and 23 manifolds	97.5	99.3	99.3	100
11. Fewer in-line valves	93.8	97.6	98.0	100

The reliability analysis suggests the following for the design of the main steel pipeline.

- The loop provides substantial extra reliability to the overall system.

- Use 14 manifolds. The cost for extra manifolds provides little increase in reliability.

- Use 19 in-line valves. The cost savings by reducing the number of in-line valves does appear worthwhile. The increase in reliability by adding more valves is very modest.

- Use 30" diameter double lap welded 0.25" (minimum) thickness steel pipeline in all areas prone to moderate liquefaction susceptibility; and the pipe from the pump station to the loop.

- In areas prone to low liquefaction susceptibility, use of 24" single lap welded pipe is acceptable.

- Use an improved design for all laterals off the main pipeline that are closely located, in soils prone to settlement.

12.5 Kyoto

This section presents a conceptual design for a high-pressure system for Kyoto, Japan. Examples of innovative concepts are highlighted by which a high-pressure system might be inexpensively constructed in a large modern, congested city, where normal concepts and construction costs would be prohibitively expensive, precluding post-earthquake fire protection.

Kyoto, Japan is a city of over 1 million population, located in south central Honshu just north of Osaka, and not far from Kobe. Figure 12-8 shows Kyoto, which is bounded on the north, east and west by mountains, and bisected by several rivers. Crossing the City east-southwest is the main Tokaido railroad line and, connecting with the main train station is a recently constructed underground subway line. Currently the line has two lines – a north-south line connecting with the main railroad station, and an east-west line. A second east-west line is under construction, and other extensions are planned. Kyoto's water supply is via aqueduct (canal and pipe) from Lake Biwa, crossing the East mountains. Scattered throughout the city are natural and manmade ponds and other bodies of water. Kyoto was the home of the Emperor and capital of Japan for over one thousand years, and was one of the few cities not heavily damaged in World War 2. It thus preserves a great many historical buildings and art treasures, and is a cultural treasure unparalleled in Japan, and in few places in the world – in recognition of this, Kyoto has been designated a World Heritage Site by the United Nations.

Kyoto is also at great earthquake risk, with several major faults in or near the city, and a long history of major earthquakes. The Hanaore fault is in the northeast part of the City, and other faults also exist in the region that could cause large ground motions in Kyoto. Table 12-7 shows selected larger earthquakes which have damaged Kyoto during the last one thousand years, from which it can be seen that these events have occurred on average about every 200 years, with the last occurring approximately 250 years ago. Based on this and other seismological evidence, it is widely believed that Kyoto could at any time sustain an event comparable to the earthquake that devastated Kobe in 1995. The resulting large ground motions would result in major damage to Kyoto's potable water supply system, which would leave the City without an adequate water supply for fire protection following a large earthquake. As in Tokyo in 1923, Kobe in 1995 and virtually all Japanese urban earthquakes, fire following earthquake is a major concern in Kyoto.

Table 12-7 Selected historic earthquakes affecting Kyoto, Japan

Date	Magnitude
938	7
1185	7.4
1317	6.5+
1350	6
1425	6
1751	6-

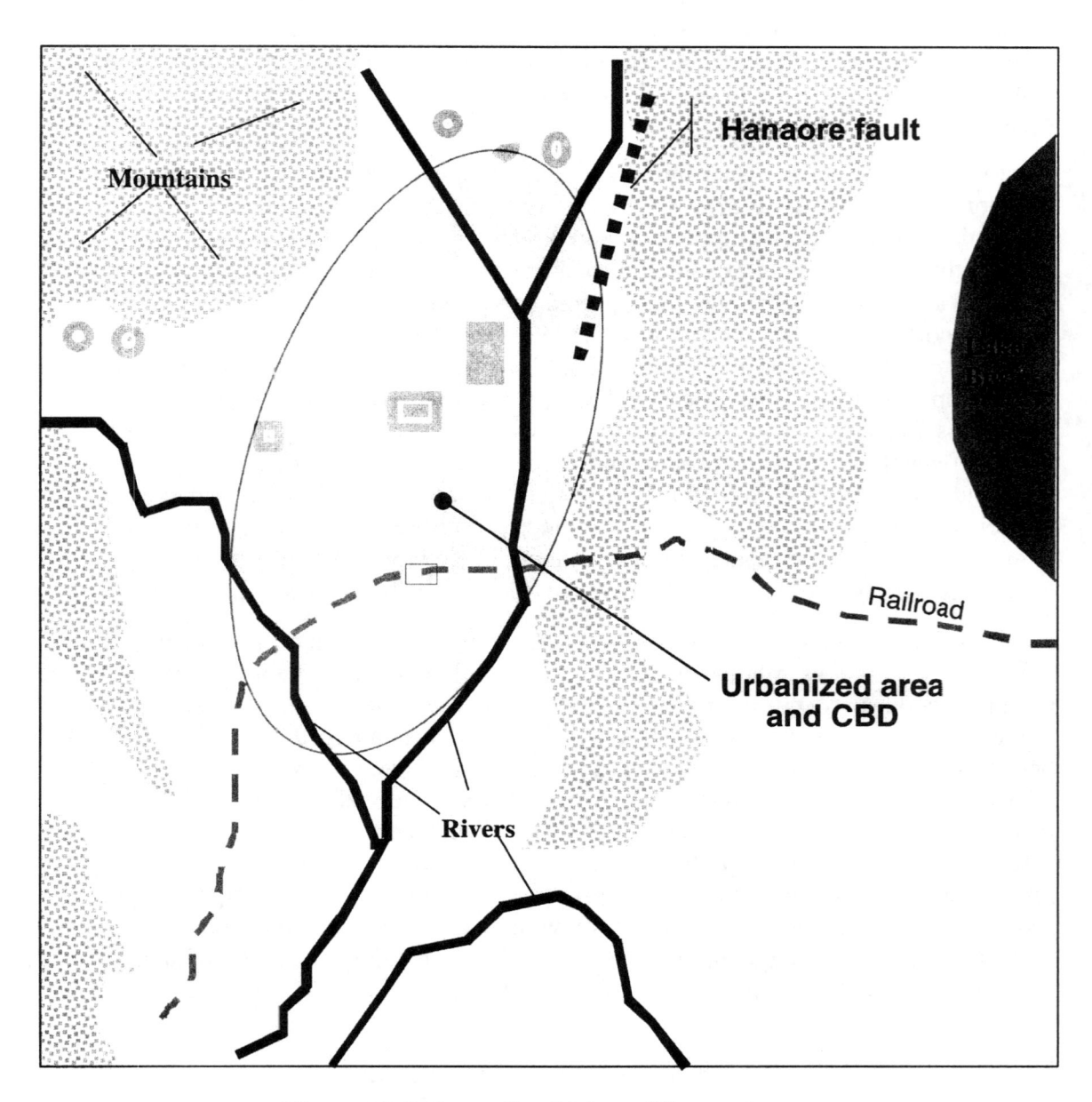

Figure 12-8 Generalized Map of Kyoto, Japan

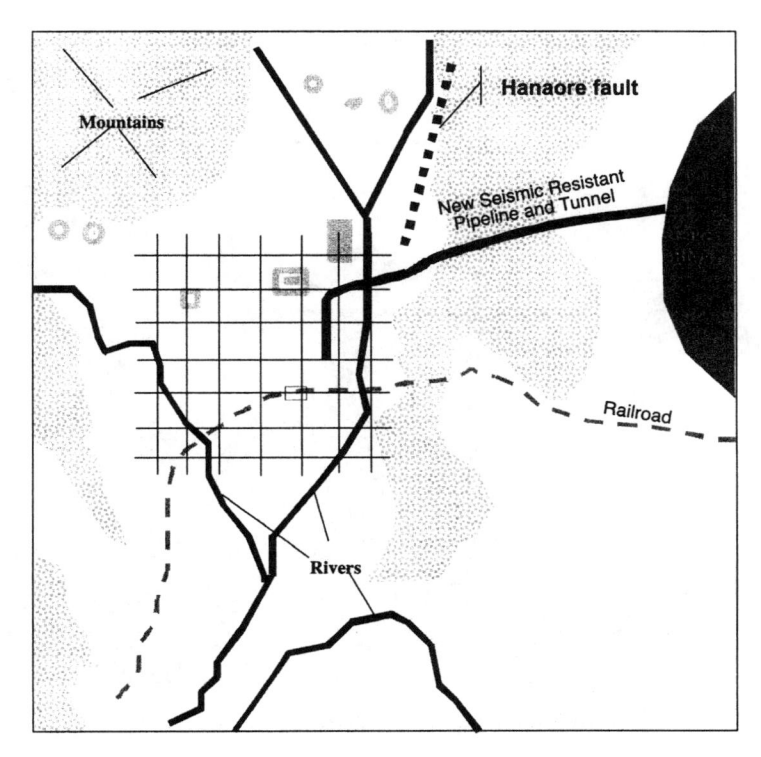

Figure 12-9 Kyoto. Upgrade option using new transmission and local grid pipeline network

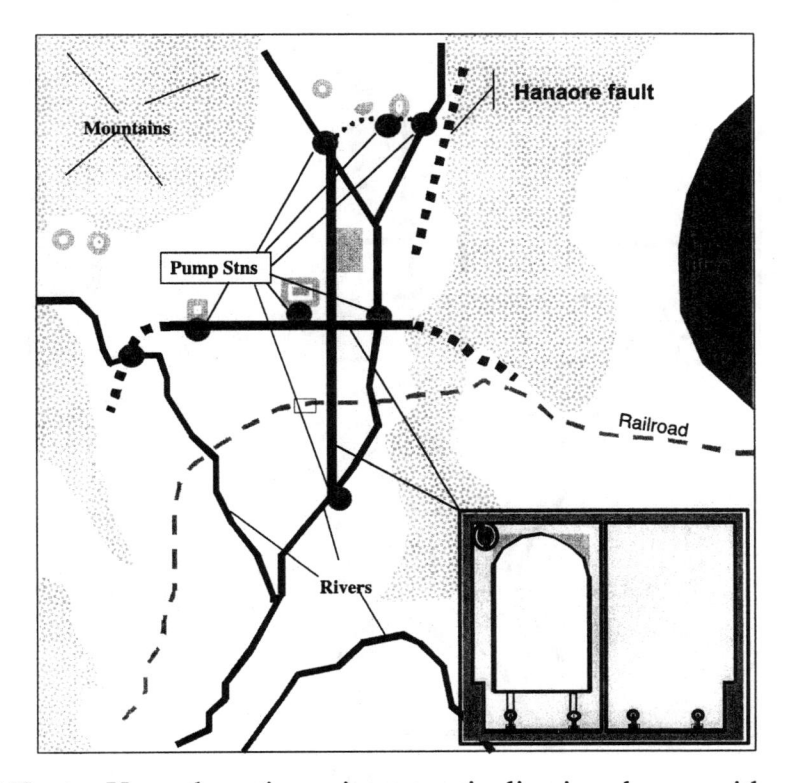

Figure 12-10 Kyoto. Upgrade option using new pipeline in subway, with pump stations drawing water from local water supplies

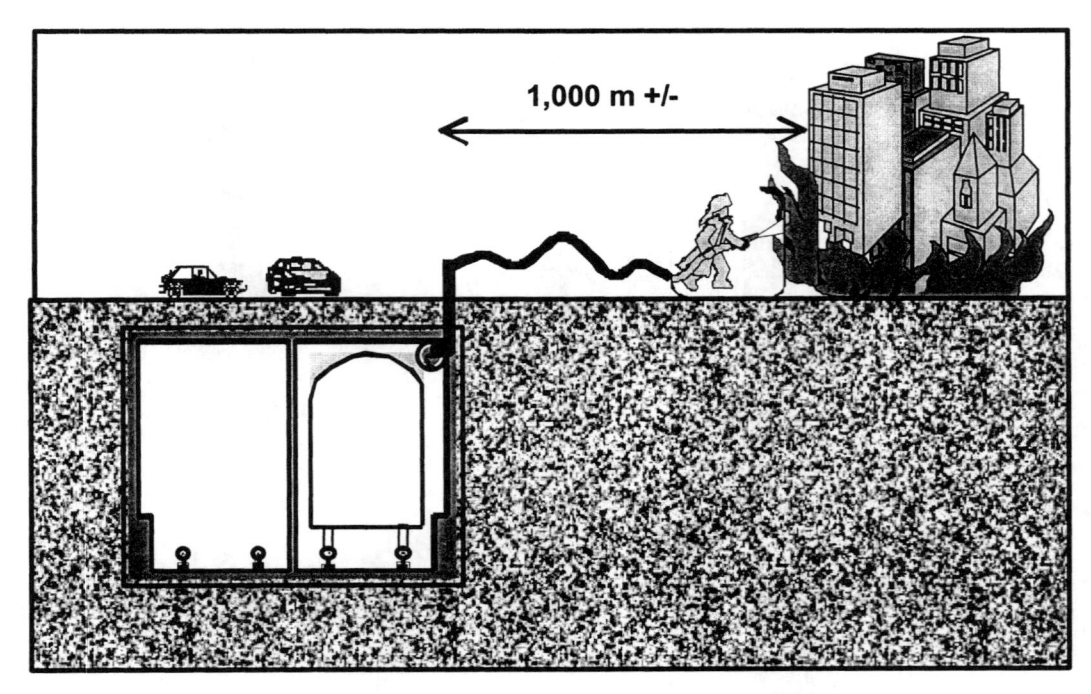

Figure 12-11 Kyoto. The proposed system could feed handlines wielded by firefighters over perhaps as much as a 1,000 m radius, without need of fire engines

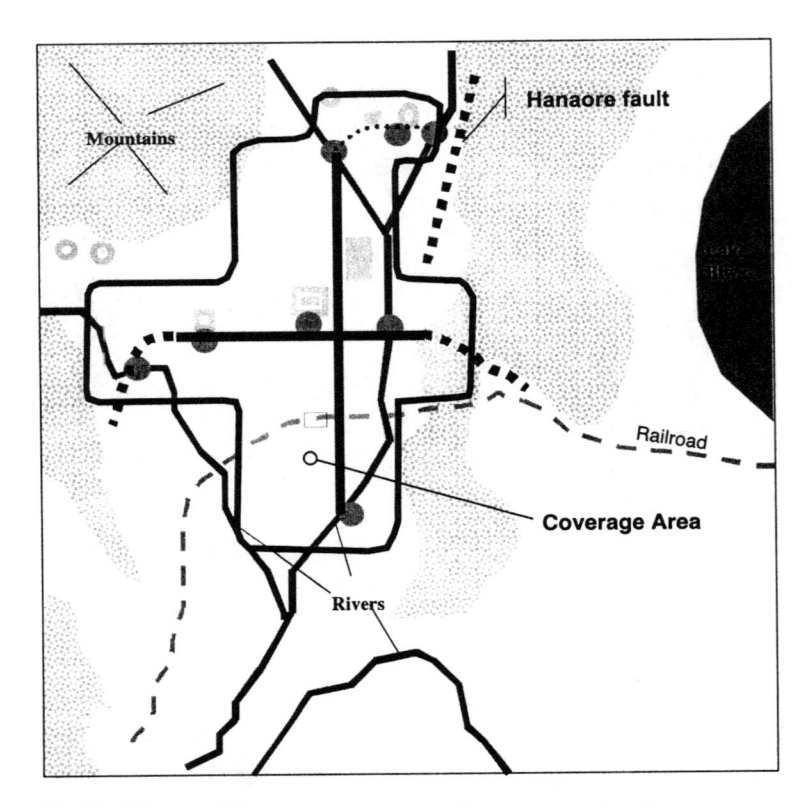

Figure 12-12 Kyoto. Fire coverage area by new redundant water system

Two options for accessing alternative water supplies are highlighted in Figures 12-9 through 12-12.

One option to provide an alternative water supply would be to construct a high volume seismically resistant pipeline / tunnel from Lake Biwa, to feed a distribution system (shown in Figure 12-9 as a gridwork). This solution would be relatively expensive, especially constructing an underground network in the highly congested portions of Kyoto.

However, scattered throughout the city are a number of natural and manmade ponds and other bodies of water. This provides the basis for an alternative water supply system, which would be to utilize a recently constructed subway system, shown in Figure 12-10 as the heavy black line, running north-south and east-west, with planned extensions shown as heavy dashed lines. This subway line was only recently constructed, to high seismic standards, and is highly likely to withstand any earthquake without significant damage. The existing route of the subway intersects or is very close to a number of water supplies, such as the rivers and various ponds. A welded steel or other pipe could be placed within the existing subway tunnel, and fed via short connectors linked to self-contained modularized pump stations installed at the rivers and various ponds, to provide alternative water supplies. Similar to other high pressure systems, the proposed system could feed handlines (or perhaps ultra large flex hoses connected to handlines) wielded by firefighters over perhaps as much as a 1,000 m radius, Figure 12-11, without need of fire engines. The proposed system utilizing the existing earthquake-resistant Kyoto subway structure, in combination with handlines feeding off the high-pressure system, would provide protection to a large part of the urbanized part of Kyoto, Figure 12-12.

While conceptual, the proposed system for Kyoto is an example by which a high-pressure system might be inexpensively constructed in a large modern, congested City, where normal concepts and construction costs would be prohibitively expensive, precluding post-earthquake fire protection.

12.6 Credits

Figure 12-5. Scawthorn. Figure 12-6. Eidinger

12.7 References

Eidinger, J. M., Dong, W., "Fire Following Earthquake," in <u>Development of a Standardized Earthquake Loss Estimation Methodology</u>, prepared for National Institute of Building Sciences by RMS Inc., February, 1995.

Eidinger, J. M., Goettel, K., Lee, D., Fire and Economic Impacts of Earthquakes, Proceedings, 4th U. S. Conference on Lifeline Earthquake Engineering, ASCE, San Francisco, 1995.

EQE Engineering. 1990. Emergency Planning Considerations, City of Vancouver Water System Master Plan, prepared under subcontract to CH2M-Hill, for City Engineering Department, City of Vancouver, B.C.

ICLR. 2001 Assessment of Risk due to Fire Following Earthquake Lower Mainland, British Columbia, report prepared for the Institute for Catastrophic Loss Reduction, Toronto, by C. Scawthorn, and F. Waisman, EQE International, Oakland, CA.

MTR. 1991. Pre-Design Report for Saltwater Pumping Stations and Dedicated Distribution Systems, prepared by MTR Consultants Ltd., and James M. Montgomery Consulting Engineers, Vancouver, B.C. for the Waterworks Engineer, City of Vancouver, B.C.

Postel, F.F. 1992. *The Early Years of the Fire Department*, address Delivered at the Palace Hotel during 1992 Ceremonies Marking the 125th Anniversary of the San Francisco Fire Department, By Frederick F. Postel, Chief of Department (available online at http://www.sfmuseum.net/hist1/hist2.html)

Sadigh, K., Chang, C.Y., Egan, J.A., Makdisi, F., Youngs, R.R, "Attenuation Relationships for Shallow Crustal Earthquakes Based on California Strong Motion Data", Seismological Research Letters, v. 68, No. 1, p. 180, 1997.

Working Group, Probabilities of Large Earthquakes in the San Francisco Bay Region, California, USGS Circular 1053, 1990.

13 Seismic Retrofit Strategies for Water System Operators

13.1 Introduction

At various locations throughout this report, a variety of mitigation options have been suggested for adoption by the fire service community. Chapters 11 and 12 provide specific recommendations as to how to reliably design a brand new water system for purpose of fighting fires. However, under most circumstances, a fire department or water utility will not want to build a brand new water system solely to fight fires following earthquakes. Instead, the water utility will wish to rely, to the greatest extent, on its own existing water system.

It has been well established in this report, that water systems are prone to damage in earthquakes. Chapter 12 presents a several dedicated water system concepts that have been put into place or have been considered by various fire departments. These concepts are geared towards providing water for fighting fires within the first few hours to a day after the earthquake, but not for providing potable water service to customers.

Water utilities have other priorities in restoring water serving, including restoring water to critical care facilities, maintaining adequate water quality, providing a suitable flow of water for minimum use (excluding irrigation), etc. A comprehensive treatment of post-earthquake service reliability guidelines is provided in (Eidinger and Avila, 1999).

In the following sections, we provide examples of actual mitigations adopted by water utilities. Observations are made as to the potential of each type of mitigation as to addressing the fire following earthquake issue.

13.2 Pipe Replacement or Pipe Bypass?

It is well established that the primary type of damage to water systems in earthquakes is failure of buried pipelines. Other types of damage also occurs, such as failure to tanks, damage to buildings, loss of electric power that hampers pumping, damage to emergency generators, landslides into reservoirs that impact water quality, toppling of outlet towers, possible failure of dams, etc. However, the most prevalent kind of damage has been, will likely continue to be for the foreseeable future, leaks and breaks to underground pipes.

In a large water system of say 1,000 to 4,000 miles of distribution pipelines, having a few pipe breaks per week is not unusual. However, a large earthquake could generate about one pipe repair per mile of pipeline, should the pipeline inventory have the following features:

- Poor materials of construction. Cast iron pipe is one of the weakest styles of pipeline construction. Chapter 6 covers the issues with other common pipeline materials. Overall, it would be fair to say that pipes designed to modern AWWA standards, are for the most part, not reliable under earthquake conditions. While pipe material and construction techniques are available to provide highly reliable designs, most modern water utilities do not install seismically-designed pipelines, even in high seismic risk areas.

- The pipelines traverse areas that are prone to ground failure. These areas include: fault crossing locations; landslide zones; liquefaction zones. It is not uncommon for about 5% of the entire pipeline inventory of a water utility in a large urban area to traverse these zones. Perhaps 70% or more of all pipeline damage in a water system will occur in these 5% of all areas. Depending on the hydraulic characteristics of the water system, heavy damage to these 5% of pipelines might result in a complete loss of water supply to 80% or more of the water system.

Given this problem, water utilities are faced with the following two extreme choices:

- Choice 1. Do nothing. This costs nothing up front, but can result in $ billions of losses after earthquakes.

- Choice 2. Replace all pipelines in soils prone to ground failure. Assuming that a large water utility has 250 miles of pipeline traversing poor soil areas, and allowing that the pipe diameters range from 6" to 36", averaging about 8" diameter, and allowing that special seismically designed pipeline can be installed in urban areas for about $20 per inch foot, then a complete seismic replacement program might cost: 250 miles * 5280 feet / mile * 8 inches * $30 per inch-foot = $317,000,000. This is a rather large capital cost, even for a moderately large utility. The utility might wish to do an even more comprehensive job, like replacing all of its aging cast iron pipe, all or its 4" small diameter pipe, etc. that are located in areas with corrosive soils. This might involve replacing about 1,000 miles of pipe with new "standard" design pipe (often rubber gasketed PVC or ductile iron) pipe, which should perform reasonably well in soils not prone to permanent ground failure. This cost would be: 1,000 miles * 5,280 feet / mile * 8 inches * $20 per inch foot = $845,000,000. For this example, the total pipeline retrofit program would be about: $317 million + $845 million = $1,162 million.

Clearly, choice 2 is preferable over choice 1 with regards to post-earthquake performance of a water system. But, somebody has to pay the $1.2 billion capital cost for choice 2. Using the benefit cost model in Section 14, it will usually be found that a current capital cost of $1.2 Billion will not be worth the avoided losses ("benefits") from future earthquakes, if one uses a reasonable time value for money (say 4% to 7%).

Given this, some water utilities have tried to implement some sort of "middle ground" when dealing with the potential of damage to pipelines. This "middle ground" approach goes something like this:

- Identify all "backbone" pipelines in the system. Backbone pipelines are usually those pipes 12" and larger in diameter, that bring water from sources (water treatment plants, pump stations) to destinations (tanks, critical customers).

- Map out the hazards on these backbone pipelines. If there are no significant hazards (i.e.., the risk of failure of the pipeline is less than a few percent, given a major earthquake), then that pipeline might be adequate "as-is". If there are significant hazards (fault crossing, active landslide, substantial liquefaction) such that the existing pipeline cannot reliably accommodate the hazard, then this backbone pipeline is "not-reliable".

- Consider if the destinations served by these unreliable pipelines can be served by alternative reliable sources, including local wells, other pipelines, etc. If no other reliable alternative source exists, then consider building a new reliable pipeline to that destination, if hydraulic demands (increasing population, etc.) so require another pipeline over the 20 to 30 year planning horizon.

- If there is no need for new pipeline for hydraulic purposes, then consider two style of backbone pipeline retrofit:
  o Option 1. Replace the pipeline in place with a new pipeline with better seismic design features, such that the new pipeline can reliably accommodate the hazard without failure; or
  o Option 2. Install a bypass system to allow the rapid restoration of water service using above ground flexible hose.

For fire fighting purposes, the first option (pipe replacement) is the best choice in terms of keeping water available immediately after an earthquake. For restoration of customer service, the second option (usually much cheaper to install) might be the most cost effective for purpose of rapid (under 24 hour) restoration of water to large areas in a water distribution system. For a 24" diameter pipe with a hazard crossing of length 1,200 feet, the pipeline replacement option might cost: 1,200 feet * 24 inch diameter * $30 / inch-foot, plus valves = $864,000 + valves, say $1,000,000. The installation of a bypass system might cost: valves + appurtenances = $200,000, plus some length of above ground flexible hose.

For a large water utility with perhaps 50 larger diameter pipes crossing high hazard zones, there might be situations where the pipe replacement option is best, or situations where the bypass option is best. When considering how much hose to have on hand, one also has to consider that not all pipes will fail at all hazards, and so the total hose needed to have on hand should be just sufficient to accommodate the total expected pipe failures.

13.3 Above Ground Ultra Large Diameter Hose Pipe Bypass

Figure 13-1 shows the actual deployment of above ground flexible hoses uses to bypass a 24" diameter pipeline that crosses the Hayward fault. In the foreground of this figure are three pipe elbows. Attached to the elbow on the left is a 12" diameter blue-colored "super aqueduct" hose manufactured by Angus. We call this 12" ultra large diameter hose, ULDH, in distinction to 5" large diameter hose, LDH, commonly used by fire departments. Attached to the middle elbow is another hose, but unpressurized. The third elbow is currently shown with no attached hose. The original 24" diameter steel pipeline is located in the road to the left of the hoses. The Hayward fault traverses across this street, a few hundred feet away from the elbows seen in the foreground. A similar set of three elbows is located at the other side of the fault, about 900 feet from the elbows seen in the foreground.

Figure 13-1. Flex Hose Deployment in EBMUD System

Figure 13-2 shows the same three elbows seen in the foreground of Figure 13-1, from a side view. The hose is attached to the elbows using victaulic couplings. The three elbows are in turn connected to a buried 12" diameter pipe which laterals off the main 24" diameter transmission backbone pipe. The street repair (new pavement) seen in Figure 13-2 shows where the new 12" lateral has been installed.

Figure 13-2. Flex Hose Deployment in EBMUD System

In actual practice, the flex hose can be deployed by a minimum of 2 men, and connected up in about 1 hour. In a post-earthquake environment, this might take longer, but with a reasonable emergency response plan, a crew of 8 people could reasonably deploy about 6 such installations in 8 hours. It would take these same 8 people about 15 to 20 days to repair the same damaged pipes. In this way, restoration of water service is spec up from 15 to 20 days, to perhaps 8 hours.

Should fires be occurring within the first 8 hours, then the flex hose installation technique will not be as ideal as the pipe replacement technique. This should be factored into the decision as to what type of pipe replacement option to undertake.

The flex hose option has some other drawbacks. First, the utility will be deploying above ground hose. At a pressure of perhaps 100 to 150 psi, the force within a 12" diameter hose is about 11 to 17 kips. If the hose should break for any reason, then the hose will move at great force, impacting cars and nearby passersby, and even structures. The likelihood of a hose break can be minimized by selecting suitable non-traffic crossings, or by closing the street to traffic. The quality of ULDH manufacture can be quite high, and the hose shown in Figure 13-2 has a nominal burst pressure of 400 psi; connection fittings are stronger than the hose with regards to burst pressure. However, any motorist with a SUV could still try to straddle a 12" hose, and possible a hot muffler will get caught on the hose. The author has witnessed taxi cab drivers driving over 5" hose with a similar outcome; and it should be understood that it is hard to restrain citizens from driving through street barriers and over hose. A 12" diameter pressurized hose looks like quite a formidable obstacle, but citizens with Hummers, taxi cabs and the appetite for unrestricted access by the news service (that led to failure of the PWSS in the 1991 Oakland Hills fire) must be considered.

Large diameter flex hoses can also be hooked up to fire hydrants, as for example in Figure 13-3. The hose is attached to a special fitting placed under the hydrant, in order to accommodate the change in diameter. A different fitting could be used to attach the hose directly to the screwed 4.5" diameter pumper outlet fitting on the hydrant. The small pipe outlets seen in Figure 13-1 to Figure 13-3 provide an easy way to empty water from the large diameter flex hose when it is time to roll it up; and also to flush the hose should that be necessary.

Figure 13-3. Flex Hose Attachment To Fire Hydrant

Figure 13-4 shows a "flaking" truck being hauled by a heavy pickup truck. The large diameter hose (up to 600 foot length) is stored in the flaking box, and then driven to the site, and then deployed. Figure 13-5 shows a "hose reel" as an alternative way to store and deploy large diameter hose (in practice, the storage of hose in flaking boxes, Figure 13-4, might be most cost effective).

Figure 13-4. Deployment of Flex Hose

Figure 13-5. Alternate Deployment of Flex Hose

13.4 Above Ground Large Diameter Flexible Hose Pipe Bypass

Depending upon the length of hazard to be bypassed, and the required flow rates, it will sometimes be feasible to use large diameter (5" to 6") above ground flex hose (LDH) to do, more-or-less, the same function as the 12" ultra large diameter flex hose. Fire departments are well versed with the use of large diameter flex hose. Figure 13-7 shows just one such deployment.

Figure 13-7. Deployment of Medium Diameter Flex Hose

When using 5" diameter flex hose, it might be feasible to also install ramps to allow traffic to go over the hose. One such example is seen in Figure 13-8. These ramps will require additional time / manpower to set up the hose. Even with these ramps, there is no guarantee that vehicles will use them – as was evidenced when a taxicab got stuck on a 5" hose.

13.5 Hydraulics

The use of flex hose to bypass existing underground pipelines must consider the required flow rates and allowable pressure drops.

The use of a single 5" or 12" diameter flex hose over a moderate distance (say 950 feet) will result in the following pressure drops for typical flow rates:

- 500 gpm = 0.3 psi pressure drop (12") =   18 psi pressure drop (5")
- 1,000 gpm = 0.9 psi pressure drop (12") =   65 psi pressure drop (5")
- 2,000 gpm = 3   psi pressure drop (12") = 233 psi pressure drop (5")
- 5,000 gpm = 18 psi pressure drop (12")

Given these flow rates and pressure drops, it is apparent that a single 12" diameter flow rate can probably be sufficient to flow at 2,000 to 3,000 gpm rate over a 950 foot length, with acceptable pressure drops. If the target flow rate is to meet winter time demands, plus possibly a fire, then a single 12" diameter hose might be sufficient to bypass a 24" to 30" diameter pipe.

Smaller diameter flex hose (5" diameter) cannot handle flow rates much above 500 gpm over lengths of about 1,000 feet unless booster pumps are used. While a 5" diameter flex hose is often adequate for delivering water from a hydrant to a fire pumper truck, over a length of no more than 250 to 300 feet, it is not very useful for modest flow rates over the moderately long distances usually involved with pipelines that traverse hazard zones.

Figure 13-8. Use of Ramp Over A 5" Diameter Flex Hose

13.6 Credits

Figures 13-1 to 13-8, J. Eidinger.

13.7 References

Eidinger, J. M., Avila, E. Editors, Guidelines for the Seismic Upgrade of Water Transmission Facilities, ASCE Technical Council on Lifeline Earthquake Engineering, Monograph No. 15, January 1999.

14 Benefits and Costs of Mitigation

14.1 Introduction

Seismic upgrades of water systems to limit the potential for losses from fire following earthquake, as well as from other causes, are often considered too expensive, despite the significant benefits. Unfortunately, decisions about whether or not a prospective seismic upgrade is "worth it" are sometimes made on subjective grounds. The purpose of this chapter is to review benefit-cost analysis and other quantitative economic methods for investigating whether or not a prospective seismic upgrade of a water system is "worth it."

The most direct benefits of seismic upgrades of water systems are a reduction of system damages and service outages in future earthquakes. Reducing service outages also reduces the sometimes very large economic losses experienced by customers impacted by interruptions of water service. Seismic upgrades may also have other benefits including enhanced post-earthquake fire suppression capability and improved system performance under non-earthquake conditions, including possible reductions in operation and maintenance costs or improved water quality. Seismic upgrades may also reduce life safety risks to employees and customers due to collapse of buildings or large system components and/or to failures that result in releases of large volumes of water.

Seismic upgrades are any measures designed to reduce system damages and/or service outages in future earthquakes. Most common seismic upgrades are "hard" upgrades such as retrofitting reservoirs, pumping plants and other system components, replacing vulnerable components, and anchoring control and communications equipment. Augmenting system capacity or redundancy may also be considered as seismic upgrades, at least in part, if such measures improve post-earthquake performance. "Soft" upgrades, such as enhancements in emergency planning, mutual aid agreements, operational improvements, and improved stockpiling of supplies and equipment necessary for seismic repair, can also be considered seismic upgrades because they are designed to reduce service outages in future earthquakes. Although this review focuses primarily on "hard" upgrades the principles are also applicable to other types of seismic upgrades.

This review focuses primarily on benefit-cost analysis, a powerful tool for determining whether or not a seismic upgrade is "worth it." Although benefit-cost analysis provides a wealth of information for decision-makers, there are limitations that arise because benefit-cost analysis is an expected value calculation which uses best estimates and a probabilistic approach. More complete information is provided by considering not only mean or expected value calculations but also the dispersion or uncertainty in such calculations. Furthermore, especially in the context of whether or not system performance is "acceptable," deterministic scenario studies also add useful information to the strictly probabilistic decision-making approaches. Finally, the "expected value" approach may be limited if the water system owner (either the utility or the rate payer) is "risk-averse": in other words, if the decision maker is risk-averse, then the "expected utility" will be greater than the "expected value", and the "expected utility" should be used in the economic decision making analysis.

In evaluating seismic upgrades it is important to note that benefits accrue both to the operators of the water system and to the customers. In some cases, there may be a mismatch between those paying for upgrades and those benefiting from the upgrades. Thus, a major factor in determining whether or not a prospective seismic upgrade is "worth it" is whether the benefits to be counted are those which accrue to operators, to customers or to both. In some cases, the benefits to customers (e.g., reductions in economic losses from water outages and from fires following earthquakes) may be substantially larger than the benefits to the operators of the water system (in the case of public utilities, the "operators" and "customers" are the same). The methodology discussed below includes benefits that accrue both to operators and to customers (i.e., for publicly owned water systems). However, all results are presented in disaggregated form, so that users of this methodology may count whichever subsets of benefits are deemed appropriate for their specific case.

Sections 14.2 to 14.7 provide the technical basis of the benefit cost analyses for water system with regards to fire following earthquake and other losses. Section 14. 98 provides some examples of how this methodology was applied for two water utilities: the East Bay Municipal Utility District and the City of San Diego Water Department.

14.2 Benefit-Cost Analysis

The benefits of a seismic upgrade are the reduction in damages and losses in future earthquakes. More specifically, benefits are the net present value of the avoided damages and losses over the useful lifetime of the upgrade project. Future benefits are discounted to net present value and compared to the costs of the upgrade. Benefits must be estimated probabilistically because they are the reduction in damages and other economic losses from future earthquakes and neither the timing nor the severity of future earthquakes are known a priori.

The costs of seismic upgrades can be reasonably forecasted, being estimated by the engineering design of the upgrade and the usual cost-estimating methods. If the upgrade requires an increase in maintenance costs to ensure its effectiveness, then the net present value of these costs over the project lifetime must also be included in the upgrade cost.

This chapter on benefit-cost analysis draws heavily on the principles outlined in benefit-cost software and user's manuals developed for the Federal Emergency Management Agency for evaluation of the seismic upgrade of buildings [FEMA, Goettel]. However, this review focuses on evaluation of seismic upgrades of water systems that requires counting of somewhat different types of benefits than in the FEMA benefit-cost programs.

Benefit-cost analysis of seismic upgrades may be considered in four steps:

- Seismic Hazard. The seismic hazard must be specified for the full range of damaging earthquakes affecting the upgrade project site.

- Seismic Vulnerability Before Upgrade. The seismic vulnerability of the water system facility or component must be estimated for the before upgrade as-is condition.

- Seismic Vulnerability After Upgrade. The seismic vulnerability of the water system facility or component must be estimated for the after upgrade condition.

- Benefit-Cost Calculation. Benefits (i.e., the net present value of avoided future damages and losses) are estimated from the above three sets of information, along with the seismic upgrade projects' useful life and the discount rate.

14.3 Seismic Hazard Curves

The benefits of seismic upgrades accrue for all future earthquakes that are large enough to have caused any damages and fire ignitions and ensuing losses for the as-is system. When the probabilities of various levels of ground shaking are factored in, the benefits of avoiding the smaller but more frequent damages and losses from low to moderate levels of ground shaking may be as important or more important statistically than the higher levels of damages and losses experienced at higher levels of ground shaking which occur much less frequently. Therefore, to evaluate the benefits of upgrades, the probabilities for the full range of damaging ground motions must be considered. This approach differs from the typical engineering design approach that focuses primarily on the design basis for ground motions (i.e., a single level of ground shaking).

Damaging ground motions can be expressed in several different ways including peak ground acceleration, velocity and displacement as well as in spectral terms. All of these ground motion measures are valid and each may be appropriate for some components of water systems. However, to illustrate the underlying concepts of benefit-cost analysis, we use peak ground acceleration (PGA). If desired, any of the other measures of ground motion could be used equally well in benefit-cost analysis. We recognize that actual damage patterns depend in a complex manner on the level of ground shaking, the duration of ground shaking, the spectral content of the ground motions, and the various forms of permanent ground deformations (liquefaction, landslide, fault offset).

Table 14-1. Expected Annual Frequency of Earthquakes

PGA Range (% g)	Modified Mercalli Intensity	Expected Annual Frequency of Earthquakes
4 - 8	VI	0.0794
8 - 16	VII	0.0614
16 - 32	VIII	0.0315
32 - 55	IX	0.0101
55 - 80	X	0.0022
80 - 100	XI	0.0004
>100	XII	0.0002

For computational purposes, the seismic information shown in the hazard curve is converted to discrete ranges (bins) of PGA and the expected annual frequency (annualized number) of earthquakes in each bin is calculated by the difference from the upper and lower limits of the PGA ranges. In the FEMA benefit-cost methodology, the PGA bins shown in Table 14-1 are used. For reference, and because much of the older seismic damage literature is given in terms of Modified Mercalli Intensity (MMI) the approximate MMI levels corresponding to these PGA levels are also shown. We recognize that there is no universally agreed upon conversion from PGA to MMI. However, as discussed in Section 14.4, we recommend that seismic vulnerability assessments be made relative to PGA or another quantitative scale, rather than in MMI, so that the MMI/PGA conversion does not affect numerical benefit-cost results.

For benefit-cost analysis of an upgrade at any specific site, the site characteristics must be taken into account. Therefore, amplification or deamplification depending on site characteristics (e.g., soft soil sites), spectral content vs. frequency characteristics, the potential for liquefaction or permanent ground displacements must be properly accounted for either in the seismic hazard curve or in the seismic vulnerability assessment. For spatially distributed systems, such as water systems, the spatial variation of site characteristics may have to be considered.

14.4 Seismic Performance of Existing Systems

The seismic performance of the existing as-is water system or component is evaluated in a four-step process:

- Direct physical damages to the system or component
- Restoration curve and service outages
- Fire losses and Economic and losses to customers from service outages
- Casualties

For each of these four steps, damages and losses are estimated on a scenario (i.e., per earthquake event) basis for the seven ranges of PGA discussed above in Section 14.3. Then, these scenario damages and losses are multiplied by the annual frequency of earthquake events in each PGA range to obtain the long-term probabilistically annualized damages and losses (i.e., the expected annual damages and losses).

14.4.1 Physical Damages to Facility or Component

The first step in evaluating the seismic performance of the existing water transmission system or component is to estimate the direct seismic damages to the facility or component under evaluation as a function of PGA (or whatever quantitative measure of ground motions is used), for the full range of damaging ground motions. Then, the impact of these physical damages on service outages, the economic impact of the outages on customers, and casualties must be estimated. First, we deal with the physical damages estimates and then we consider the impacts that arise from the physical damages.

Benefit-cost analysis requires an explicit, quantitative estimate of the seismic vulnerability of the existing (before upgrade) facility or system component under evaluation. Physical damages are estimated in dollars or, equivalently, in percentages of damage relative to the replacement value of whatever system component is being evaluated. Qualitative evaluations, such as "does not meet current code", are insufficient: in order to determine whether or not the benefits of upgrades are worth the cost it is absolutely necessary to make quantitative estimates of the seismic performance of the as-is facility.

Quantitative seismic vulnerability estimates can be made in several ways. Seismic performance at various PGA levels can be estimated by the historical damage patterns of similar facilities in past earthquakes. However, the most useful damage-estimating format is the fragility curve. Fragility curves define damage states (e.g., slight, moderate, extensive, complete) in terms of the median PGA at which 50% of similar facilities are expected to reach or exceed the defined damage state, along with log-normal standard deviations (betas) which indicate the dispersion in the damage state estimates. Sample fragility curves for water treatment plants, pumping plants, transmission pipelines and other water system components and descriptions of the damage states are given in Chapter 5 of this report. Comprehensive catalogs of fragility curves for water system components (pipelines, tanks, canals, etc.) are provided in (Eidinger, 2001). Fragility curve representations of the seismic vulnerability of each facility or component would have a similar form as the example shown in Table 14-2.

Table 14-2. Example Fragility Curve for a Water Treatment Plant

Damage State	Median PGA (%g)	Beta
Slight	.22	.40
Moderate	.58	.40
Extensive	.87	.45
Complete	1.57	.45

From such fragility curve estimates the mean damage percentage at any desired PGA level can be calculated, along with estimates of the dispersion (statistical uncertainty) in the damage estimate. Damage estimates, however made, must include the best available information about the seismic characteristics of the facility under review as well as information about site characteristics. For example, variations in site amplification, spectral characteristics, or expected ground displacements may strongly affect seismic vulnerability estimates. A limitation of the fragility curve in Table 14-2 is that it represents damage to a complex facility; while this type of simplification may be crudely sufficient for initial "order of magnitude" loss estimation analysis, it will not provide the detail needed by the water utility owner to assess the efficacy of particular retrofit alternatives, and component specific fragility curves will generally be required. For purposes of this chapter, the example fragility in Table 14-2 is used for illustrative purposes, but in actual practice the analysis should be done at a more granular level.

For benefit-cost analysis two types of damages are considered. Scenario damages indicate how severe the damages are expected to be per earthquake event. Scenario damages depend only on the vulnerability of the facility or component, not on the probabilities of damaging earthquakes. Expected annual damages combine the scenario damage estimates with the seismic hazard curve to show the annualized damages estimates over a long time period. Thus, the information contained in the scenario and expected annual damage estimates is complementary. Seismic damage estimates are presented in the format shown in Table 14-3.

Table 14-3. Damage Estimates Before Upgrade

PGA Range (% g)	Scenario Damages (% of replacement value)	Scenario Damages (dollars)	Expected Annual Frequency of Earthquakes	Expected Annual Damages (dollars)
4 - 8	1	$10,000	0.0794	$794
8 - 16	3	$30,000	0.0614	$1,842
16 - 32	10	$100,000	0.0315	$3,150
32 - 55	40	$400,000	0.0101	$4,040
55 - 80	60	$600,000	0.0022	$1,320
80 - 100	100	$1,000,000	0.0004	$400
>100	100	$1,000,000	0.0002	$200
TOTAL	N/A	N/A	0.1852	$11,746

In Table 14-3, scenario damages summed over all PGAs is not a meaningful number. However, total expected annual damages ($11,746 in the above example) represents the total expected annualized damages for the before-upgrade state of the facility and is thus one of the key decision making parameters. The sum of the expected annual frequency of earthquakes indicates the total expected annual frequency (number) of earthquake events with PGAs ≥ 0.04 g. As always, "expected annual" means the long-term statistical average per year, not that this number of events or damages occurs every year.

14.4.2 Restoration Times and Service Outages

The second step in the evaluation of seismic vulnerability is to estimate the restoration time to repair the damages estimated previously. In some cases, a facility may be only operable or inoperable, so that restoration time is simply the time to make sufficient repairs to resume operations. In other cases, partial operations may be possible. In this case, a service restoration curve is the estimate of the percent functionality vs. time. Restoration times depend on the extent of damages, on the numbers of personnel and equipment required to make repairs, on the availability of replacement equipment, on mutual aid, and on many other factors. However, utilities can make reasonably good estimates of how long repairs are likely to take, by adding up the total damage expected to the water system, and then seeing how long it will take to repair that damage assuming a certain number of in-house and (possibly) mutual aid repair crews.

Two sample restoration time examples are given in Table 14-4. Case 1 is the simplest case where the facility is either 100% operable or completely inoperable. Case 2 assumes a linear restoration of service over the time to restore service 100%. More realistic cases may have non-linear restoration curves but such cases can be estimated based on the operating characteristics of whatever facility is under evaluation. In the present context, a "system day" is a measure of the percentage loss of service to customers. One system day could be 100% loss of service for 1 day, or 50% loss of service for 2 days or 5% loss of service for 20 days.

Table 14-4. Restoration Time Estimate

PGA Range (% g)	Scenario Damage (% of replacement value)	Restoration Time to 100% Service (days)	System Days Lost (Case 1)	System Days Lost (Case 2)
4 - 8	1	0.0	0.0	0
8 - 16	3	0.5	0.5	0.25
16 - 32	10	3.0	3.0	1.5
32 - 55	40	30.0	30.0	15
55 - 80	60	60.0	60.0	30
80 - 100	100	120.0	120.0	60
>100	100	120.0	120.0	60

14.4.3 Economic Losses to Customers From Service Outages

The third step in the evaluation of seismic vulnerability is to estimate the economic impacts of the interruption of water service. The previous two sections have discussed the estimation of physical damages to water system facilities or components and the estimation of service restoration times. Making both of these estimates requires experience and good engineering judgment / analysis, but the procedure is relatively straightforward.

However, estimating the economic impacts of the resulting estimated service outages is often outside the realm of most engineers' professional experience. We consider three types of economic impacts: revenue losses to the utility, economic losses to customers, and fire following earthquake.

Revenue losses to the utility are straightforward to estimate, based on the system service days lost, the average daily volume of water, and the average revenue per unit volume of water. However, in estimating revenue losses, the amount of redundancy and excess capacity in the system must be considered. For example, the loss of a 100 MGD facility does not necessarily result in revenue loss if the lost capacity is covered by excess capacity elsewhere in the system.

Economic losses to customers because of water service interruptions can, in some cases, be the largest negative impact of water system damage, dwarfing the repair costs and the lost revenues. However, counting economic losses to customers should be done if and only if the perspective of the benefit-cost analysis is to assess the overall impacts of the seismic upgrade. If the economic perspective is only that of the utility, then economic losses and benefits to customers do not factor into the calculation.

Estimating the losses to customers is a somewhat difficult task. A simple approach was adopted by (Eidinger, Goettel and Lee, 1995) who estimated the economic impact to customers of an urban water district from earthquake water service outages in a three-step process. First, for the affected region, an estimate of the gross regional product by industry sector was obtained from the California Statistical Abstract (similar economic data are available for other geographic regions). Second, the impact of loss of water services on gross regional product by industry sector was estimated using the water importance factors given in ATC-25 (1991). Third, each industry sector was assumed to bear the same fractional loss of water service (i.e., relative geographic variability in service outages and industry sector locations was not considered). Thus, economic losses were estimated as the aggregation by economic sector of days of business interruption due to loss of water supply, considering the importance of water supply to production in each sector. Double counting of economic disruptions due to building damages, power outages, and water outages can be avoided by reducing the effective system days lost to account for lost power and by adjusting post-earthquake water demand downwards to account for building damages. This approach for estimating the economic impact on customers of interruptions of water service is illustrated in Table 14-5, with an example for the East Bay Municipal Utility District (EBMUD) service area in the Oakland, California area.

This simple method of estimating the economic impact of water outages on customers probably overestimates the economic impact. In economic sectors with excess capacity, losses of water supply may not reduce total production over time because lost production can be made up by producers unaffected by water outages or by affected producers after water supplies are restored. Thus, a more accurate analysis would have to take into account these factors.

For a more detailed discussion of methods to evaluate indirect or secondary economic impacts of earthquake damages (such as the impact of water system damage on customers' economic production), see Chapter 16 of the National Institute of Building Sciences earthquake loss estimation methodology [NIBS].

Fires following earthquakes may, in some cases, contribute substantially to the total economic losses from the earthquake. Eidinger, Goettel, and Lee (Eidinger et al, 1995) presented a fire following earthquake model, which includes the impacts of reduced availability of water for fire suppression. This fire following earthquake model includes methodologies for modeling fire ignition rates, fire spread rates, and fire suppression capabilities (including water supplies). Reduced water supplies for fire suppression may be due primarily to distribution system damages. In a case study for the EBMUD service area, total fire following earthquake losses (after a Hayward magnitude 7 event) were estimated at $120 million, $330 million, and $1,860 million for calm, light and high wind conditions, respectively. At the highest level of seismic upgrades considered, these losses were estimated to be reduced by $85 to $230 million. Thus, augmenting post-earthquake fire suppression capability can be one of the major benefits of seismic upgrades of water systems.

14.4.4 Casualties

Seismic damages to water transmissions systems may also pose life safety risks. Casualties (deaths and injuries) from failures of water transmission system facilities or components can arise in three ways: directly due to failure of buildings or other facilities (affecting primarily employees), directly due to releases of large volumes of water which may cause drownings or collapses of nearby structures (primarily affecting non-employees), or indirectly through reductions in water supply and/or water quality (illness or deaths due to drinking contaminated or improperly treated water).

Casualty estimation is one of the most difficult aspects of benefit-cost analysis of seismic upgrade projects. The absence of recent earthquakes within the United States with large numbers of casualties makes calibration of casualty estimation models difficult.

For buildings and other facilities, casualty estimates can be made by assessing the damage states and the expected modes of failure (e.g., the probable extent of collapse).

Casualty estimates for water system failures which result in the release of large volumes of water would have to be made on a case by case basis, taking into account the probable volumes and rates of water release, the topography, and the population patterns in the inundation area. Similarly, casualty estimates for water system failures which are likely to result in contamination of water supplies and thus to result in illness or deaths would have to be made on a case-by-case basis.

14.4.5 Example: Total Damages and Losses Before Upgrade

As an example of the approach outlined above, we present a hypothetical example in Table 14-6a and Table 14-6b for a 100 MGD facility, showing both the scenario damages and losses and the expected annual damages and losses for each of the damage and loss categories discussed above. The scenario damage and loss estimates are for the before-upgrade, as-is condition of the facility that could be a reservoir, a pumping plant, a transmission line or any other system component. The expected annual damages and losses are calculated from the scenario damage and loss estimates, using the expected annual number of earthquakes shown in Table 14-1, for each PGA bin.

Table 14-5. Estimated Reduction in EBMUD Service Area Gross Regional Product (GRP) per System Day of Water Service Interruption ($000)

Economic Sector	Metropolitan Area GRP	EBMUD Service Area GRP	Water Importance Factor (ATC-25)	EBMUD Service Area GRP Losses
Agriculture	$330	$248	0.70	$174
Mining	$1,286	$964	0.15	$145
Construction	$7,171	$5,378	0.50	$2,689
Manufacturing- Nondurable Goods				
Food products	$2,474	$1,855	0.70	$1,299
Paper & allied	$681	$511	0.60	$307
Printing & publishing	$1,316	$987	0.30	$296
Chemicals & allied	$2,112	$1,584	0.80	$1,267
Petroleum & coal	$4,605	$3,454	0.50	$1,727
Other non-durable (apparel)	$465	$349	0.70	$244
Manufacturing - Durable Goods				
Lumber, wood & furniture	$388	$291	0.50	$146
Stone, clay & glass products	$510	$382	0.50	$191
Primary metal products	$456	$342	0.90	$308
Secondary metal products	$1,158	$868	0.80	$694
Industrial machinery	$2,274	$1,706	0.60	$1,024
Electronic Equipment	$1,635	$1,226	0.60	$736
Transportation equipment	$1,043	$782	0.60	$469
Instruments	$966	$725	0.90	$653
Miscellaneous manufacturing	$179	$134	0.60	$80
Transportation	$4,233	$3,174	0.20	$635
Communications & utilities	$10,487	$7,865	0.40	$3,146
Trade, wholesale and retail	$22,366	$16,775	0.20	$3,355
Finance, insurance & real estate	$29,458	$22,094	0.20	$4,419
Services	$29,491	$22,118	0.50	$11,059
Federal government	$5,886	$4,414	0.20	$883
State & local government	$13,444	$10,083	0.40	$4,033
TOTALS	$144,414	$108,309	0.37	$39,977

Table 14-6 (a). Scenario Damages and Losses: Before Upgrade

PGA Range (% g)	Facility Damage ($K)	System Service Days Lost (Days)	Revenue Lost ($K)	Economic Loss to Customer ($K)	Fire Following Earthquake Loss ($K)	Statistical Value of Deaths ($K)	Total ($K)
4 - 8	$100	0.00	$0	$0	$0	$0	$100
8 - 16	$300	0.25	$250	$1,666	$1,000	$0	$3,216
16 - 32	$1,000	1.50	$1,500	$10,000	$5,000	$0	$17,500
32 - 55	$4,000	15.00	$15,000	$100,000	$50,000	$2,200	$171,200
55 - 80	$6,000	30.00	$30,000	$200,000	$75,000	$8,800	$319,800
80 - 100	$10,000	60.00	$60,000	$600,000	$125,000	$22,000	$817,000

Table 14-6 (b). Expected Annual Damages and Losses: Before Upgrade

PGA Range (% g)	Facility Damage ($K)	System Service Days Lost (Days)	Revenue Lost ($K)	Economic Loss to Customer ($K)	Fire Following Earthquake Loss ($K)	Statistical Value of Deaths ($K)	Total ($K)
4 - 8	$8	0.000	$0	$0	$0	$0	$8
8 - 16	$18	0.015	$15	$102	$61	$0	$198
16 - 32	$32	0.047	$47	$315	$157	$0	$551
32 - 55	$40	0.152	$151	$1,010	$505	$22	$1,729
55 - 80	$13	0.066	$66	$440	$165	$19	$704
80 - 100	$4	0.024	$24	$240	$50	$9	$327
>100	$2	0.012	$12	$120	$30	$4	$168
Totals	$117	0.316	$316	$2,227	$969	$55	$3,685

The example shown in Tables 14-6a,b is hypothetical, but shows many of the common trends in the evaluation of water system upgrades. The physical damages to the facility itself are smaller than the loss of revenue from service outages. The largest economic impact is on customers, the economic losses from disruption of business activity due to loss of water service. Losses from fire following earthquake can also be very large; in this example, they are the second largest economic impact.

For this example, the scenario losses for very high PGA levels are in the hundreds of millions of dollars. However, the probabilities of these events (see Table 14-11) are very low. The expected annual damages and losses present the "annualized" damages and losses, taking into account the probabilities of each level of earthquake. The "annualized" damages of $3,685,000 do not happen every year. Rather, they represent the estimated long-term statistical average damages per year from the cumulative impact of the full range of damaging earthquakes over a long time period.

14.4.6 Seismic Performance of Upgraded System

The seismic performance of the upgraded system is evaluated in the same manner as the seismic performance of the as-is before upgrade system, using the same approach as discussed above. As before, damages and losses after upgrade are considered on both a "scenario" or per earthquake event basis, on an annualized basis, taking into account the annual probabilities of earthquakes. The enhanced seismic performance of the upgraded system is reflected in the reduced damages and losses expected in all or nearly all of the categories of damages and losses considered.

We show an example hypothetical "post-upgrade" compilation of estimated scenario and expected annual damages and losses in Table 14-7a,b. The expected annual damages and losses in this table are calculated in the same manner as in Table 14-6a,b as the product of the scenario damages and losses and the expected annual number (frequency) of earthquakes (Table 14-1).

Table 14-7(a). Scenario Damages and Losses: After Upgrade

PGA Range (% g)	Facility Damage ($K)	System Service Days Lost (Days)	Revenue Lost ($K)	Economic Loss to Customer ($K)	Fire Following Earthquake Loss ($K)	Statistical Value of Deaths ($K)	Total ($K)
4 - 8	$0	0.00	$0	$0	$0	$0	$0
8 - 16	$10	0.00	$0	$0	$500	$0	$510
16 - 32	$100	0.00	$0	$0	$2,500	$0	$2,600
32 - 55	$500	0.50	$500	$3,333	$30,000	$0	$34,333
55 - 80	$1,000	1.50	$1,500	$10,000	$45,000	$0	$57,500
80 - 100	$2,000	5.00	$5,000	$50,000	$75,000	$0	$132,000
>100	$2,000	5.00	$5,000	$50,000	$75,000	$0	$132,000

As a result of the upgrade, the estimated annualized damages and losses are reduced from about $3,685,000 before upgrade to about $665,700 after upgrade, for an annualized reduction in damages and losses of $3 million (difference between Table 14-6a and Table 14-7b). The key question to be answered is whether this reduction in damages and losses from future earthquakes is large enough to justify the up-front capital costs for the upgrade. This question is addressed in the following section.

Table 14-7(b). Expected Annual Damages and Losses: After Upgrade

PGA Range (% g)	Facility Damage ($K)	System Service Days Lost (Days)	Revenue Lost ($K)	Economic Loss to Customer ($K)	Fire Following Earthquake Loss ($K)	Statistical Value of Deaths ($K)	Total ($K)
4 - 8	$0	0.000	$0	$0	$0	$0	$0
8 - 16	$0.6	0.000	$0	$0	$30.7	$0	$31.3
16 - 32	$3.2	0.000	$0	$0	$78.8	$0	$81.9
32 - 55	$5.1	0.005	$5.1	$33.7	$303	$0	$346.8
55 - 80	$2.2	0.003	$3.3	$22.0	$99	$0	$126.5
80 - 100	$0.8	0.002	$2.0	$20.0	$30	$0	$52.8
>100	$0.4	0.001	$1.0	$10.0	$15	$0	$26.4
Totals	$12.2	0.011	$11.4	$85.7	$556	$0	$665.7

14.5 Benefit-Cost Analysis and Results

This section reviews benefit cost results, first on an annualized basis and then on a net present value basis.

14.5.1 Benefits (Avoided Damages and Losses)

Comparisons between the damage and loss estimates in (Before Upgrade) and 14-7 (After Upgrade) show the estimated effectiveness of the prospective upgrade in reducing damages and losses in future earthquakes. The effectiveness is not uniform across the damage and loss categories but varies, in accord with the specifics of each upgrade. In the present example, the fire following earthquake losses are reduced by the upgrade but not by the same percentage as the other damage and losses, because suppression of fires following earthquakes depends not only on water supplies but also on the water distribution system, the availability and deployment of fire suppression personnel and apparatus, and the various other factors described in Chapter 4 or this report.

Comparisons between the before and after upgrade status of the system or component under evaluation can be made explicitly by showing the scenario and expected annual benefits of the upgrade (i.e., the reduction in damages and losses in each category). These scenario and expected annual benefits are calculated by simply subtracting the estimates in each category after the upgrade from those before the upgrade (i.e., Table 14-6 entries minus Table 14-7 entries).

In this example, we note that of the total annualized benefits of approximately $3.0 million, only about $400,000 accrues directly to the utility operators (avoided facility damages and avoided revenue loss). About two thirds of total annualized benefits are avoided economic loss to customers. About 15% of annualized benefits are avoided fire following earthquake, which will accrue predominantly to customers. While these examples are hypothetical, they illustrate the important point that benefits of seismic upgrades of water systems may accrue predominantly to customers.

14.5.2 Net Present Value Calculations

The total expected annual benefits (avoided damages and losses) is a statistical representation of the average annual benefits expected over a long time period, averaging the benefits expected from the full range of damaging earthquakes in proportion to the annualized frequencies of earthquakes. For comparison to the up-front capital costs of the prospective upgrade this stream of future benefits must be converted to a net present value.

The net present value of a stream of annualized benefits depends on two parameters: the useful lifetime of the seismic upgrade project and the discount rate. The useful lifetime of a seismic upgrade is the time period over which the project is judged to be effective in reducing damages and losses from future earthquakes. Project useful lifetimes will vary from a few years for projects relating to equipment which may only have a 3 or 5 or 10 year lifetime to as long as 50 to 100 years for major reservoir or canal projects. For many building or pipeline projects, useful lifetimes of 30 to 50 years are often assumed.

The second factor necessary to convert annualized benefits to net present value is the discount rate, which accounts for the time value of money. For example, a dollar received 30 years from now is worth less today than a dollar received today. Similarly, the benefits of a seismic upgrade, reduced future damages and losses, which are received (statistically) in the future are worth less than benefits which are received today.

The choice of an appropriate discount rate can be a somewhat complex exercise in economic theory, which is beyond the scope of the present review (see Chapter 3 in Volume 2 of the FEMA benefit-cost user's guide (FEMA 1994). Briefly, discount rates can be "real" which means the rate includes the true time value of money, or "nominal" which means that the rate includes the true value of money, plus the expected inflation rate. For benefit-cost analysis, the appropriate discount rate is generally the "real" rate, excluding future inflation. Thus, benefit-cost calculations are done in present dollars, with the discount rate accounting only for the true time value of money.

Future inflation is not germane to the net present value calculation. If estimated future inflation rates were included to adjust the expected future benefits, then a "nominal" discount rate would have to be used and the effects of estimated future inflation on net present value would simply cancel out. By working with real values, rather than nominal, the analyst is not assuming that future inflation will be zero. Instead, by working with real rates the analyst can be agnostic about inflation and can avoid making any particular assumptions about it.

In detail, the choice of an appropriate real discount rate may depend on how the seismic upgrade is paid for and whether the owner is a taxable or tax-exempt entity. However, at the current time, real discount rates in the range of 3% to 5% are probably appropriate for most analysis. In this example, we assume a discount rate of 4%. Discount rates and interest rates are conceptually distinct. However, since most people are more familiar with interest rates, we note that if a bond to finance a capital improvement project carries a 7% nominal interest rate, then the real rate of interest is 4% if the assumed rate of future inflation is 3%.

14.5.3 Benefit-Cost Results

For the hypothetical seismic upgrade presented in Tables 14-6 and 14-7, we assume a 30-year project useful lifetime and a 4% real discount rate. For this combination of project useful lifetime and discount rate, the present value coefficient (the net present value of $1.00 per year in benefits over the project useful lifetime) is 17.29. A table of present value coefficients for a wide range of combinations of discount rates and project useful lifetimes is given in Chapter 10 of the FEMA benefit-cost user's guides (FEMA 1994), while the underlying equations are given in Appendix 1 of that reference.

For the hypothetical upgrade, used to illustrate the principles of benefit-cost analysis, we assume that the prospective seismic upgrade has a cost of $20,000,000. Then, the net benefit results are the difference between

Table 14-6 and Table 14-7. On a net present value basis, the benefits (avoided future damages and losses) exceed the costs by about $32,000,000 and the benefit-cost ratio is 2.61.

Table 14-8. Benefit-Cost Results

Annualized Benefits (Table 14-6b less 14-7b)	$3,018,892.00
Present Value Coefficient	17.29
Net Present Value of Benefits	$52,196,643.00
Seismic Upgrade Project Cost	$20,000,000.00
Benefits Minus Cost	$32,196,643.00
Benefit-Cost Ratio	2.61

In addition to the seismic benefits discussed above, seismic upgrades may also have other benefits, such as reduced maintenance/operations costs under normal conditions and better fire suppression capabilities under non-earthquake conditions. These benefits are additive with the seismic benefits discussed above and could be counted, if appropriate.

However, such benefits are more easily considered directly on an annualized basis, rather than on a scenario earthquake basis, since they do not depend on the frequency or severity of earthquakes. Thus, for example, reductions in annual maintenance costs or reductions in annual fire losses could be considered benefits, with the net present value calculated simply as the annualized benefit times the present value coefficient.

Other benefits which are more difficult to quantify in monetary terms may also be considered if desired. Examples of such benefits would include improvements in water taste or quality or environmental enhancements. Discussion of such benefits is beyond the scope of the present review.

14.5.4 Interpretation of Benefit-Cost Analysis

As discussed at the beginning of this review, benefit-cost analysis is an expected value calculation which means that the analysis is based on estimates of the long-term statistically annualized damages and losses expected from future earthquakes. Thus, the benefit-cost ratio of 2.61 obtained in the example discussed above means that, on average over a 30-year time period, the benefits will exceed the project costs by a factor of 2.61. However, because it is not possible to predict the timing or severity of future earthquakes, the actual benefits over the next 30 specific years will vary from this long-term statistical average. Nevertheless, the expected value calculation provides the best estimate of the long-term benefits of a seismic upgrade project.

Prospective seismic upgrade projects with benefit-cost ratios above 1.0 are economically justified on an expected value, net present value basis. Benefit-cost ratios and net benefits (benefits minus costs) can also be used to prioritize among competing projects. However, in using benefit-cost results in this manner it is very important to consider projects of a similar scale and scope. Thus, a $5 seismic upgrade project with a benefit-cost ratio of 2.0 is not necessarily a "better" project than a $1,000,000 project with a benefit-cost ratio of 1.5 because the net benefits are very different as are the potential negative impacts on both the owner and the customers if the project is not completed.

14.6 Limitations of Benefit-Cost Analysis and Other Decision Making Approaches

14.6.1 Limitations of Benefit-Cost Analysis

As an expected value calculation, benefit-cost results present the "expected" or mean outcome over long time periods. A major limitation is that the dispersion in the mean results (i.e., the uncertainty) is not considered. For example, consider the scenario damages estimate (table 14-6) of $171,200,000 for ground shaking between 32 and 55% of g at the facility under evaluation. These scenario damage estimate represents the best available estimate of the average total damages and losses expected for ground shaking in this range. However, as expressed in the fragility curve for the facility there is a finite probability that the facility will be a complete loss at this level of ground shaking. The mean or expected value damage estimate is the weighted average of all plausible damage states for this facility at this level of ground shaking.

Similarly, the level of ground shaking expected from a particular earthquake at a particular site is not determined exactly but depends on the details of the specific earthquake include the fault break mechanisms, the possible directionality of ground motions and so on. Thus, for any specific earthquake the actual level of ground motions may be significantly higher or lower than the mean or expected value.

These effects combine with the uncertainty about the timing and severity of future earthquakes to produce considerable uncertainty in the actual future damages and losses over a specific time period. Thus, during the hypothetical 30-year useful lifetime of the proposed seismic upgrade, a major earthquake may or may not occur and the damages may be higher or lower than the mean or expected damages for an event with that level of ground shaking.

A risk-tolerant owner may be willing to gamble that the next 30-years will be free of major earthquakes, even if on an expected value basis the benefits of the upgrade exceed the cost. On the other hand, a risk-adverse owner may be unwilling to accept the possibility of a major loss, even though on an expected value basis the benefits of the upgrade are less than the cost.

14.6.2 Deterministic Damage and Loss Estimates: Scenario Studies

Deterministic damage and loss estimates provide a complementary perspective to the expected value perspective of benefit-cost analysis. The scenario damage and loss estimates for the before upgrade condition of the facility (

Table 14-6) provide deterministic estimates of the estimated damages and losses for seven ranges of PGA ground motions. Alternatively, deterministic damage and loss estimates could be made for a specific event such as a particular magnitude earthquake on a particular fault, using the same damage and loss estimation approaches as described for the estimates in

However, purely deterministic damage and loss estimates have limited applicability to decision making. Estimating that there will be, for example, $200,000,000 in damages and losses from a particular earthquake is of little decision-making significance unless the probability of the event is also considered. The inference is quite different if the event occurs every 30 years on average or every 30,000 years on average.

14.6.3 Probabilistic Damage and Loss Estimates: Threshold Studies

Another risk assessment approach which is also complementary to benefit-cost analysis is to examine the probabilities that damages and losses will exceed threshold levels over various time periods. For the hypothetical seismic upgrade considered above, we evaluate the probabilities that various total damage and loss levels will be exceeded in 1 year, in 10 years, and in 30 years, based on the scenario damage and loss estimates in Table 14-6 and the seismic hazard data shown in Table 14-1.

Table 14-9. Probabilistic Damage and Loss Estimates

PGA Range (% G)	Total Scenario Damages and Losses	Probability of Exceeding in Time Period 1 Year	Probability of Exceeding in Time Period 10 Years	Probability of Exceeding in Time Period 30 Years
4 - 8	$100,000	0.1852	0.8710	0.9978
8 - 16	$3,215,667	0.1058	0.6732	0.9651
16 - 32	$17,500,000	0.0444	0.3650	0.7440
32 - 55	$171,200,000	0.0129	0.1218	0.3226
55 - 80	$319,800,000	0.0028	0.0277	0.0807
80 - 100	$817,000,000	0.0006	0.0060	0.0178
>100	$842,000,000	0.0002	0.0020	0.0060

As shown in Table 14-9, over a 30 year time period, there is an estimated 32% probability that total damages and losses from an earthquake will exceed $171 million, an 8% probability that damages will exceed $319 million, and nearly a 2% probability that damages will exceed $817 million. Deciding whether or not these probabilities constitute acceptable risk may be determined on grounds other than the expected value basis of benefit cost analysis. Thus, a risk-adverse owner may decide that a given probability of some level of total damages and losses over some time period is unacceptably high, even if on an expected value basis, the benefits of the upgrade are less than the costs. Decisions of this type are particularly common when life safety is significant issue. The probability of casualties may be too high to be acceptable, separate from benefit-cost results. Thus, in some cases, owners may wish to lower the probability of experiencing such losses by seismic upgrades which go beyond those justified simply on an expected value basis.

14.7 Example Application

The procedure to develop benefit cost analyses presented in Sections 14.2 to 14.7 was applied to the City of San Diego water system (Eidinger, 2000). The San Diego water system serves a population of about 1,200,000 people. Figure 14-1 shows the major faults surrounding the City of San Diego. Using the ignition models presented in chapter 4 of this report, coupled with suitable attenuation models and scenario earthquakes suitable for planning purposes for San Diego, Table 14-10 shows the forecasted median number of fire ignitions for the City of San Diego.

Table 14-10. Fire Ignitions for San Diego after Scenario Earthquakes

Scenario Earthquake	Total Number of Ignitions	Fire Ignitions in Gravity Zones	Fire Ignitions in Pumped Zones
Rose Canyon M 6.5	99	83	16
Silver Strand M 6.5	71	58	13
RC – SS M 7.2	115	94	21
Elsinore M7.4	15	10	5
La Nacion M 6.6	101	76	25

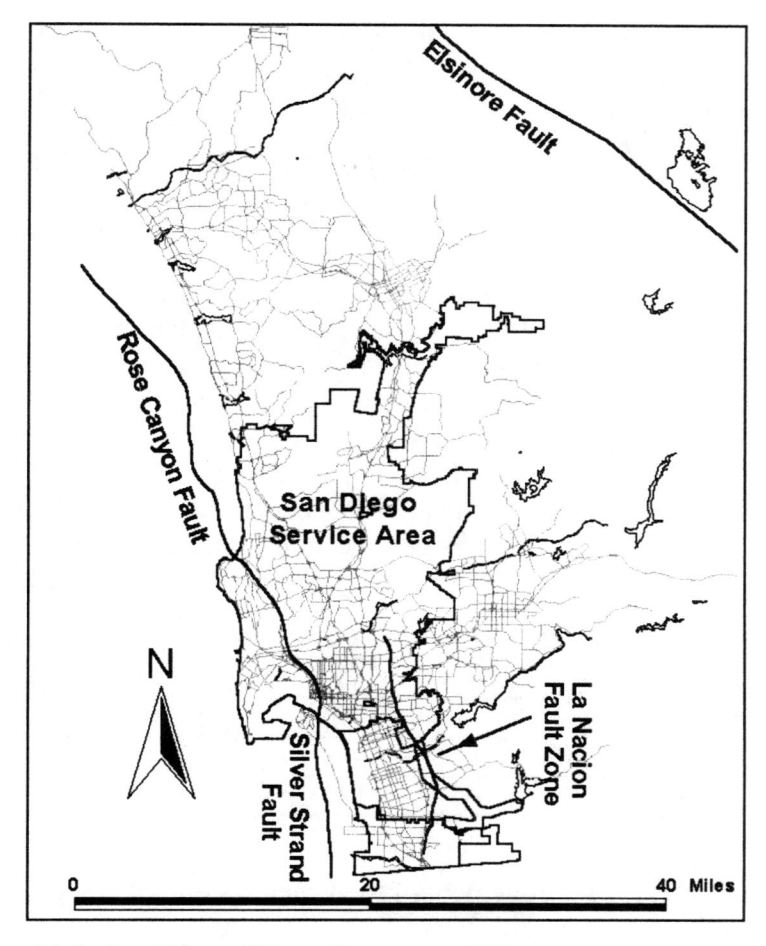

Figure 14-1. San Diego Water System and Nearby Earthquake Faults

Figure 14-2 summarizes the losses to San Diego assuming the water system in its year 2000 configuration (i.e., before any seismic retrofit). For example, a Magnitude 7.2 earthquake on the combined Rose Canyon – Silver Strand faults would generate about a $1.052 billion loss in San Diego. Losses due to fires would be about 25% of the total loss. As can be seen in Figure 14-2, the only type of loss greater than from fire is due to the reduction in gross regional product due to loss of water supply. (All data in Figure 14-2 and Figure 14-3 are in $2000).

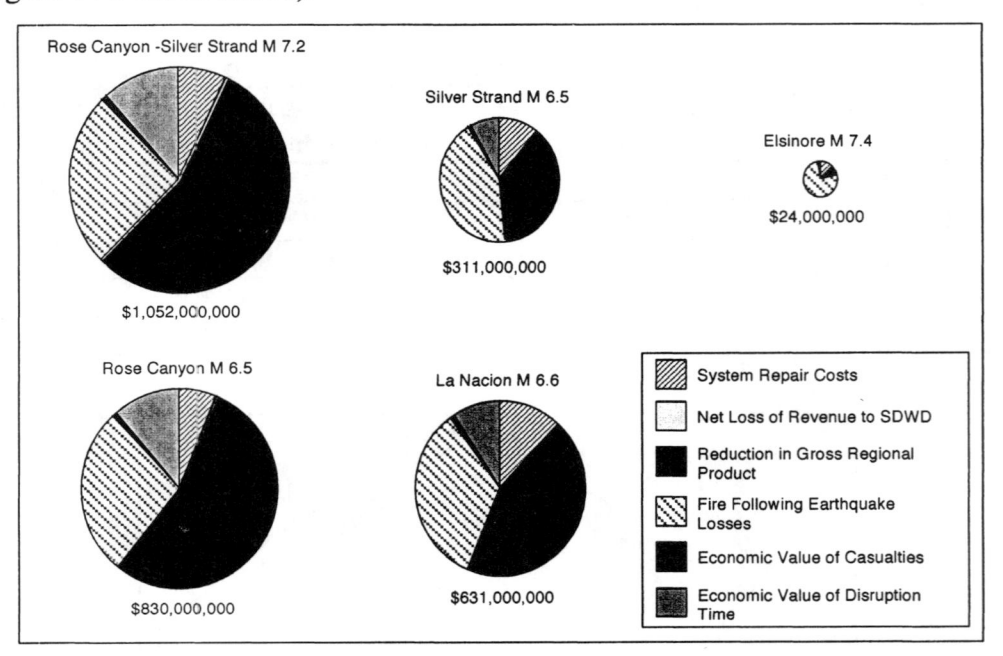

Figure 14-2. Scenario Losses to San Diego – As Is Water System

A series of capital improvements to the San Diego water system was proposed. Four possible levels of seismic retrofit were considered, ranging from $0.7 million to $90 million. Seismic evaluations of the water system were re-run assuming each level of capital upgrade was installed, for the five scenario earthquakes identified in Table 14-10, as well as for many other causative faults and varying magnitudes. Table 14-11 provides some of the results, in terms of forecasted numbers of structures burned, for the water system in its "as-is" and with "recommended" seismic retrofits. The data in Table 14-11 were calculated using models similar to those described in Chapter 4 of this report.

Table 14-11. Structures Burned – Rose Canyon M 6.5 Scenario Earthquake

Wind Conditions	Calm	Light	High
Number of Ignitions	99	99	99
Existing (As Is) Water System	328	793	4,342
Upgraded Water System ($62 million)	124	348	3,970

As seen in Table 14-11, the improvement of the water system makes a substantial reduction in the number of structures burned, should there be calm or light winds at the time of the earthquake. For the "Calm" wind condition case, the initial 99 ignitions would ultimately result in the loss of 328 structures (as-is system) or 124 structures (upgraded system), or a net reduction in fire losses by 63%. Under the ""High" wind condition, the seismic improvements is not very effective (9% reduction). This is not surprising, in that even completely intact water systems have been incapable of preventing fire conflagrations under high wind conditions (see Chapter 3 for various examples).

Figure 14-3 shows the scenario losses for the same five scenario earthquakes listed in Figure 14-2, with the assumption of a $62 million seismic upgrade. As can be seen there is about a 60 to 70% reduction in total losses. The full benefit cost model described previously in Chapter 14 was applied, and the results are listed in Table 14-12. The key points are as follows:

The "Recommended" level of seismic upgrade for the City of San Diego water system is up to $62 million. This would provide about a $60 million "benefit". This level of seismic upgrade has a benefit cost ratio of 0.97 (about equal to 1).

The "Extensive" level of seismic retrofit would cost about $28.2 million more, but would provide only about $2.4 million more in benefits.

Table 14-12. Costs and Benefits for Various Levels of Seismic Upgrade

Level of Seismic Upgrade	Capital Cost of Seismic Retrofit ($million)	Net Present Value of Benefits ($million)	Benefit Cost Ratio
Minimum	$0.7	$2.4	3.38
Limited	$16.3	$28.0	1.71
Recommended	$62.2	$60.3	0.97
Extensive	$90.4	$62.7	0.69

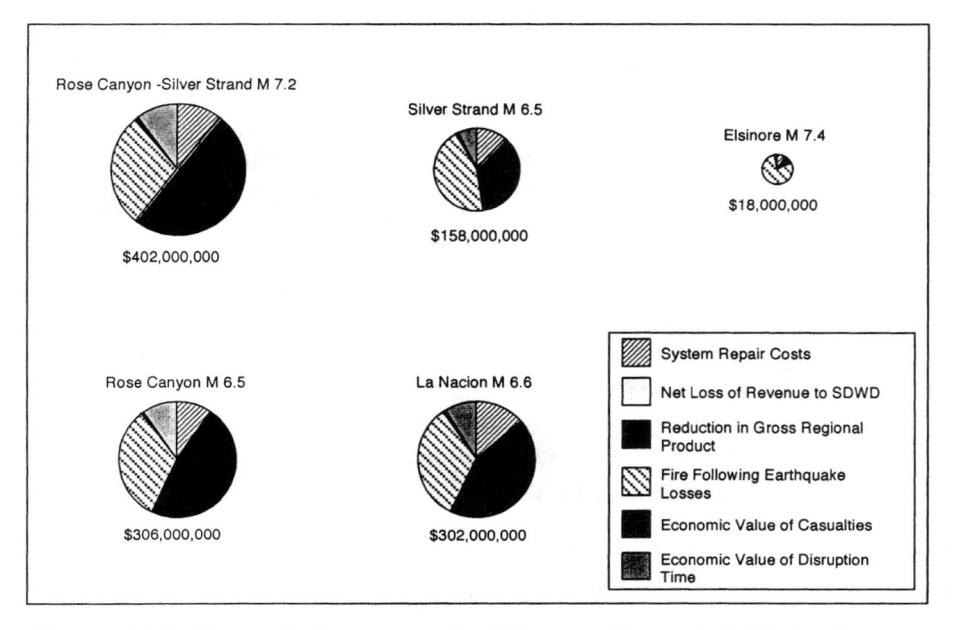

Figure 14-3. Scenario Losses to San Diego – Upgraded Water System

14.8 Conclusions

This Chapter presents a model that can be used to perform benefit-cost analyses for the seismic upgrade of water systems. The examples highlight that:

- Not all levels of seismic upgrades of water systems are cost effective. It may be reasonable to allow some damage to water systems in earthquakes.

- Losses to the community from fires can range from 25% to perhaps 45% of all losses from damage to water systems in earthquakes. When considering whether to upgrade a water system, the economic losses from fires as well as economic impacts should be considered.

- Seismic upgrades for a water system will likely be most effective in controlling the spread of fires under calm to light wind conditions. Under high wind conditions, there is still a reasonable chance of conflagration.

- Cost effective strategies to limit the chance of conflagration might be best pursued by a combination of improvements to the water system, coupled with ways to reduce the number of ignitions (electric power, natural gas) as well as ways to reduce the spread of fire (improved fire service capability, restrictions of fuel load, etc.)

14.9 References

Applied Technology Council, Seismic Vulnerability and Impact of Disruption of Lifelines in the Conterminous United States, ATC-25, 1991.

Eidinger, J. (ed.), Seismic Fragility Formulations for Water Systems, American Lifelines Alliance, Revision 0, April 2001.

Eidinger, J., Collins, F., Conner, M., Seismic Assessment of the San Diego Water System, 6th International Conference on Seismic Zonation, Earthquake Engineering Research Institute, Palm Springs, CA, November 2000.

Eidinger, J. M., Goettel, K. A. and Lee, D. Fire and Economic Impacts of Earthquakes, in M. J. O'Rourke (editor) Lifeline Earthquake Engineering, Proceedings of the Fourth U.S. Conference, 1995, pp. 80-87.

Federal Emergency Management Agency, Seismic Rehabilitation of Federal Buildings: A Benefit/Cost Model. Volume 1 - A User's Manual and Volume 2 - Supporting Documentation. FEMA-255 and FEMA-256, September 1994.

Goettel & Associates Inc., Benefit-Cost Analysis of Hazard Mitigation Projects, Volume 5 Earthquake, User's Guide Version 1.01, prepared for the Federal Emergency Management Agency, October 1995.

National Institute of Building Sciences, <u>Development of a Standardized Earthquake Loss Estimation Methodology</u>, Draft Technical Manual, Chapter 8, Direct Physical Damage to Lifelines-Utility Systems, Chapter 16, Indirect Economic Losses, 1995.

Glossary

Attack time	Elapsed time from the beginning of effective work on a fire and ending when control work is begun.
Attenuation	The rate at which earthquake ground motion decreases with distance
Branding	See firebrands
CBD	Central business district
Citizen alarm	Fire service terminology for a fire report by a citizen in person, to fire service personnel (e.g., running up to the fire station yelling Fire!). Also termed a still alarm.
Conflagration	A large destructive fire – in the fire service, in the urban context, conflagration usually denotes a large fire that spreads across one or more city streets. Conflagration usually connotes such a fire with a moving front – that is, a mass fire, as distinguished from a firestorm. The term is sometimes used for a large fire destroying a complex of buildings.
Control time	Elapsed time from the beginning of effective work on a fire and ending when the fire is controlled.
Diablo winds	See foehn
Discovery time.	Elapsed time from the start of the fire until the time of the first discovery which results directly in subsequent suppression action.
Earthquake cycle	Concept that seismicity follows a cyclical pattern, with repetitions of similar events every several hundred to thousands of years, depending on the region. Clearly demonstrated by repeated similar events where record is long, such as on the Nankai trough offshore Japan.
Epicenter	The projection on the surface of the earth directly above the hypocenter
Fault	A zone of the earth's crust within which the two sides have moved - faults may be hundreds of miles long, from one to over one hundred miles deep, and not readily apparent on the ground surface.
Fire break	A partially or wholly cleaned barrier constructed before a fire occurs and designed to stop or check fires that may occur, or to be used as a line from which to work.
Fire flows	Fire flows are the water used by fire departments to suppress urban fires. Fire flows are usually expressed in terms of flows (gallons per minute) for a duration (hours) at a minimum pressure (generally 20 pounds per square inch).
Fire foam	A term applied to the product of various chemicals which when mixed with water, cause a great increase of volume by foaming froth or bubbles, which may or may not be filled with noninflammable gas. The bubbles adhere to the burning fuel and reduce combustion by excluding oxygen as well as by cooling and moistening.

Firebrands	Burning debris carried aloft by the rising air from the fire and carried great distances downwind – when burning firebrands land on flammable material, such as wood shake roofs, they trigger new ignitions
Firestorm	Phenomenon typically associated with only largest conflagrations, in which rising hot air causes inrush of air at ground level. Wind speeds at ground level can exceed 100 mph. A firestorm is a generally radially symmetric phenomenon. Winds in a firestorm are a result of the fire, and further enhance the fire, and should not be confused with a Foehn.
Foehn	A hot dry wind coming down the lee slopes of a mountain range, rising in temperature as it descends (adiabatic compression). The term derives from Foehn winds off the northern slopes of the Alps, and the phenomena is known in various regions as Chinooks (Pacific Northwest), Santa Ana (southern California), Diablo or Mono winds (northern California), etc. A Foehn is an asymmetric phenomenon – it will greatly enhance a fire. However, it is not caused by or generally enhanced by the fire, and should not be confused with a firestorm.
Hypocenter	The location of initial radiation of seismic waves (i.e., the first location of dynamic rupture)
Intensity	A metric of the effect, or the strength, of an earthquake hazard at a specific location, commonly measured on qualitative scales such as MMI, MSK and JMA.
Lateral force resisting system	A structural system for resisting horizontal forces, due for example to earthquake or wind [as opposed to the vertical force resisting system, which provides support against gravity]
LDH	Large diameter hose (5 to 6 inch)
Liquefaction	A process resulting in a soil's loss of shear strength, due to a transient excess of pore water pressure.
Magnitude	A unique measure of an individual earthquake's release of strain energy, measured on a variety of scales, of which the moment magnitude M_w (derived from seismic moment) is preferred
Meizoseismal	The area of strong shaking and damage.
MMI	Modified Mercalli Intensity[30], a measure of the strength of shaking and other earthquake effects at a specific location, given an earthquake. MMI varies from I to XII, is based on observed effects such as building damage and people's reactions, and is denoted using Roman numerals to indicate its qualitative non-instrumental character.
Mop-up time	Elapsed time from completion of the controlling process until enough mop-up has been done to insure that the fire will not break out and that structures are safe to re-occupy.

---

[30] Wood, H.O. and Neumann, Fr. (1931) Modified Mercalli intensity scale of 1931, *Bulletin Seis. Soc. Am.*, v. 21, pp. 277-283.

Mutual aid	An arrangement whereby one jurisdiction aids another jurisdiction on request, in an incident, which is overwhelming the first jurisdiction's resources.
Near-field	Within one source dimension of the epicenter, where source dimension refers to the length or width of faulting, whichever is less.
NFPA	National Fire Protection Association
Normal fault	A fault that exhibits dip-slip motion, where the two sides are in tension and move away from each other
Peak ground acceleration (PGA)	The maximum amplitude of recorded acceleration (also termed the ZPA, or zero period acceleration)
PG&E	Pacific Gas & Electric Company
Pounding	The collision of adjacent buildings during an earthquake due to insufficient lateral clearance.
PWSS	Portable Water Supply System
Report time	Elapsed time from discovery of a fire until it is reported to a fire agency that will respond with personnel, supplies and equipment to the fire.
Response spectrum	A plot of maximum amplitudes (acceleration, velocity or displacement) of a single degree of freedom oscillator (SDOF), as the natural period of the SDOF is varied across a spectrum of engineering interest (typically, for natural periods from .03 to 3 or more sec., or frequencies of 0.3 to 30+ hz).
Sand boils or mud volcanoes	Ejecta of solids (i.e., sand, silt) carried to the surface by water, due to liquefaction
Santa Ana	See Foehn
SCBA	Self-contained breathing apparatus
Seismic hazards	The phenomena and/or expectation of an earthquake-related agent of damage, such as fault rupture, vibratory ground motion (i.e., shaking), inundation (e.g., tsunami, seiche, dam failure), various kinds of permanent ground failure (e.g. liquefaction), fire or hazardous materials release.
Seismic moment	The moment generated by the forces generated on an earthquake fault during slip
Seismic risk	The product of the hazard and the vulnerability (i.e., the expected damage or loss, or the full probability distribution.
Simulation	A computer model consisting of a series of equations and data, intended to simulate or replicate mathematically a complex series of interlinked phenomena, to determine the overall result. Simulation models are routinely used to model earthquake damage, chemical processes, and stock market performance. Simulation models may be deterministic (no consideration of uncertainty) or stochastic (uncertainty of key variables is considered). HAZUS is an example of a simulation model.

Soft story	A story of a building significantly less stiff than adjacent stories (that is, the lateral stiffness is 70% or less than that in the story above, or less than 80% of the average stiffness of the three stories above [BSSC, 1994[31]]).
Still alarm	See Citizen Alarm
Stochastic	Adj. meaning that probability is involved, as in a stochastic simulation.
Strike team	A group of fire companies, typically responding to a larger incident, often via mutual aid. A typical strike team will be five engine companies lead by a Battalion Chief, although specialized strike teams may differ.
Subduction	Refers to the plunging of a tectonic plate (e.g., the Pacific) beneath another (e.g., the North American) down into the mantle, due to convergent motion
TCLEE	Technical Council on Lifeline Earthquake Engineering
Thrust fault	Low-angle reverse faulting (blind thrust faults are faults at depth occurring under anticline folds - they have only subtle surface expression)
TIM	Traffic incident management
Transform or strike slip fault	A fault where relative fault motion occurs in the horizontal plane, parallel to the strike of the fault
ULDH	Ultra large diameter hose, usually 8" to 12" diameter.
Ups	Uninterruptible power supply – A power supply designed to back up the normal power supply, and provide power without interruption when the normal power supply is interrupted. Usually consists of banks of batteries and real-time fault detection circuitry and switching.

---

[31] BSSC (1994) *NEHRP Recommended Provisions for Seismic Regulations for New Buildings*, Part 1: Provisions, prepared by the Building Seismic Safety Council for the Federal Emergency Management Agency, Building Seismic Safety Council, Washington.

Index

alluvial, 177, 276, 281, 282
alluvium, 282
anchor, 283, 284, 285, 304
artificial fill, 176, 177
AWSS, xii, xvi, xix, 25, 27, 28, 29, 31,
 176, 177, 263, 268, 269, 270, 271
bay mud, 281, 282
bell and spigot, 174, 176
benefit cost, 194, 304, 305, 306, 307,
 308, 310, 312, 315, 317, 318,
 319, 320
Berkeley, ix, xii, xvi, xvii, xix, 5, 9, 49,
 50, 53, 61, 62, 63, 64, 65, 70, 71,
 73, 75, 76, 77, 95, 97, 98, 106,
 136, 138, 150, 168, 264, 268,
 275, 276, 277, 278, 279, 280,
 281, 282, 285, 286
break, 175, 176, 177, 276, 277, 286, 319
burial depth, 286
burnt area, xvi, 60, 118
butt weld, 283
canal, 316
cast iron, 176, 188, 189, 295
casualty, 307, 311, 312
cistern, xix, 271
communication, v, xi, xii, 50, 171, 217,
 224, 225, 229, 232, 233, 234,
 235, 238, 248
concrete, 174, 176, 177, 188, 189
conflagration, x, 1, 6, 7, 15, 48, 97, 150,
 152, 160, 326
corrosion, 283
customer service, 304, 305, 307, 309,
 310, 311, 314, 316, 318
Diablo wind, 61, 70, 326
discount rate, 317
ductile iron, 185, 188, 189
East Bay Municipal Utility District, 3,
 65, 73, 75, 135, 305, 311
 EBMUD, 311
economic, xiii, 6, 151, 293, 304, 305,
 307, 310, 311, 313, 314, 315,
 316, 317, 318, 324, 325

electric power, 4, 43
emergency planning, 304
emergency power, 92
Federal Emergency Management
 Agency, 305
FEMA, 4, 83, 90, 97, 98, 127, 129, 150,
 151, 168, 197, 208, 260, 305,
 306, 317, 324, 329
FFE, iii, xi, xii, 3, 4, 237, 239, 246
fills, 177
fire, iii, iv, viii, ix, x, xi, xii, xiv, xv, xvi,
 xvii, xviii, xix, 1, 6, 7, 11, 12, 13,
 15, 21, 25, 27, 29, 30, 31, 32, 34,
 35, 36, 37, 38, 39, 40, 42, 46, 47,
 48, 49, 50, 51, 55, 57, 59, 60, 61,
 63, 65, 69, 70, 71, 73, 74, 75, 79,
 81, 82, 83, 87, 88, 89, 91, 92, 93,
 94, 95, 97, 98, 99, 100, 101, 102,
 103, 109, 112, 113, 115, 116,
 118, 119, 120, 121, 124, 125,
 126, 129, 135, 136, 139, 140,
 144, 146, 147, 148, 149, 150,
 151, 152, 153, 154, 159, 160,
 164, 168, 175, 176, 177, 197,
 199, 210, 213, 215, 221, 222,
 224, 232, 233, 237, 245, 249,
 250, 257, 259, 260, 261, 263,
 264, 265, 267, 268, 269, 270,
 272, 273, 275, 276, 277, 278,
 280, 285, 286, 292, 293, 294,
 300, 302, 304, 305, 307, 314,
 315, 316, 318, 321, 324, 326, 328
fire fighting, 176, 275
fire flow, 326
fire following earthquake, xvii, xviii, 99,
 100, 129, 144, 250, 310, 311,
 314, 316
fireboat, 27, 28, 29, 30, 41, 61, 65, 177,
 270
firestorm, ix, 75, 327
Foehn, 61, 327, 328

fragility, xiv, xv, xviii, 183, 186, 188, 189, 191, 192, 195, 198, 308, 319, 324
gross regional product, 310
Ground shaking, xiv, xv, 177, 280, 306, 319
groundwater, 177
Hayward, 275, 280, 281, 282, 286, 287, 288, 311
hazard curve, 306, 307, 308
HAZUS, ix, xiv, xvii, 4, 99, 105, 116, 127, 128, 129, 150, 151, 185, 208, 213, 328
high groundwater, 177
highrise, 86, 88, 89, 90, 91, 92, 120, 146, 263
hose, 177, 276, 277, 278, 305, 316, 320
Hydrant, xix, 176, 277, 284, 286, 300
ignition, ix, x, xiv, xvii, xviii, 35, 42, 99, 104, 105, 109, 110, 111, 113, 114, 130, 203, 209, 244, 257, 275, 311
Ignition rate, xiv, xvii, 113, 130
Insurance, ix, 47, 58, 144, 151, 152
inundation, 312
Kobe, v, vii, viii, xvi, xviii, 2, 3, 6, 39, 40, 41, 42, 67, 94, 107, 115, 124, 126, 128, 144, 152, 159, 160, 169, 177, 178, 180, 181, 185, 198, 210, 223, 231, 241, 247, 268, 289
landslide, 240, 282, 306
lap weld, 283, 286, 287, 288
large diameter hose, 327
lateral spread, 281, 283
leak, 174, 176, 283, 286
life safety, 304, 311, 320
liquefaction, 48, 177, 187, 243, 276, 281, 282, 283, 284, 286, 287, 288, 306, 307, 327
Loma Prieta, viii, xv, xvi, xviii, 2, 3, 9, 24, 25, 26, 27, 30, 31, 48, 49, 74, 96, 113, 126, 129, 136, 137, 159, 169, 176, 179, 184, 185, 188, 190, 198, 210, 219, 221, 230, 240, 241, 242, 243, 245, 248, 254, 255, 271, 280, 281, 287, 288
MMI, xiv, xvi, xvii, 9, 10, 27, 31, 39, 42, 105, 106, 107, 108, 134, 145, 146, 165, 307, 327
Modified Mercalli Intensity, 307
Monte Carlo, 131, 285, 286
mutual aid, 155, 304, 309, 328
natural gas, 199, 200, 203, 209
net present value, 305, 306, 315, 316, 317, 318
Northridge, iii, v, viii, xiv, xv, xvi, xvii, xviii, 2, 3, 9, 31, 32, 33, 34, 36, 38, 42, 48, 49, 96, 110, 113, 115, 116, 126, 128, 129, 151, 152, 159, 162, 169, 174, 178, 180, 184, 188, 190, 203, 210, 219, 231, 237, 240, 242, 243, 244, 247, 261
offset, 306
outage, 304, 305, 307, 310, 311, 314
peak ground acceleration, 43, 328
  PGA, 280, 306
peat, 286, 287
performance goals, 275
permanent ground deformation, xiv
permanent ground deformations
  PGD, 283, 306
PGA, xvii, 43, 105, 106, 107, 108, 140, 190, 191, 192, 281, 286, 306, 307, 308, 309, 310, 312, 314, 315, 320, 328
PGD, 183, 184, 185, 187, 283, 284, 286
PGV, 183, 184, 187, 286
  peak ground velocity, 286
pipe, xii, xvi, xviii, 23, 26, 27, 183, 185, 186, 187, 188, 189, 284, 287, 295, 297, 302
pipe diameter, 284
Portable Water Supply System, 177
pressure, x, xii, 28, 219, 268
probabilistic, 304, 305, 307
PVC, 172, 178, 180, 188, 189, 198, 296
PWSS, viii, xii, xvi, xix, 6, 25, 26, 28, 29, 31, 74, 177, 264, 265, 266, 300, 328

redundancy, 285, 286, 304, 310
refinery, viii, 43, 44, 93, 95
reliability, 283, 285, 286, 287, 288
repair, 46, 174, 177, 286, 304, 309, 310
reservoir, xix, 10, 17, 19, 80, 176, 270
retrofit, xii, 48, 197, 295, 304, 308, 323
risk, 304, 311, 319, 320
San Andreas, 280, 281, 287, 288
San Francisco, viii, xii, xvi, xvii, xviii, xix, 1, 2, 3, 5, 6, 7, 9, 10, 11, 12, 13, 14, 15, 16, 17, 18, 19, 24, 25, 26, 27, 28, 29, 31, 47, 48, 49, 50, 52, 53, 61, 68, 73, 74, 80, 84, 91, 97, 102, 105, 106, 124, 126, 127, 135, 136, 140, 141, 142, 143, 150, 151, 152, 159, 160, 169, 176, 177, 179, 184, 198, 220, 221, 230, 237, 238, 239, 240, 248, 258, 259, 261, 263, 264, 265, 267, 268, 269, 270, 271, 272, 276, 281, 293, 294
Santa Ana, 61, 85, 327, 328
scenario, 280, 285, 286, 287, 304, 307, 308, 309, 312, 314, 315, 316, 318, 319, 320
service line, 234
SFS, xii, 275, 276, 277, 278, 280, 281, 282, 283, 284, 285, 286
soft soil, 307
sprinkler, 82, 86, 88, 89, 90, 91, 92, 95, 257, 263
steel, 174, 175, 188, 189, 276, 277, 278, 282, 283, 285, 286, 288
strain, 283, 284, 285
Suppression, ix, xii, 39, 99, 124, 150, 197, 257, 259
tank, 19, 29, 190, 191
TCLEE, iii, iv, v, vi, vii, viii, 6, 150, 151, 197, 198, 329

Tokyo, viii, xvi, 1, 2, 3, 6, 10, 19, 21, 22, 23, 24, 47, 48, 50, 66, 67, 68, 112, 121, 151, 153, 169, 263, 289
transmission pipeline, 308
transportation, iv, 195, 206, 237, 247, 248, 249, 313
tunnel, 174, 177
ultra large diameter hose, 329
valves, 176, 283, 284, 285, 286, 287, 288
Vancouver, ix, xii, xviii, xix, 5, 27, 144, 145, 147, 148, 151, 255, 256, 261, 263, 264, 267, 268, 272, 273, 274, 293, 294
vegetation, 83
vulnerability, 283, 305, 306, 307, 308, 309, 310
water, iv, v, vi, viii, ix, x, xii, xiv, xvii, xviii, xix, 16, 17, 19, 23, 25, 32, 36, 38, 41, 42, 43, 48, 49, 63, 65, 74, 75, 78, 79, 80, 81, 92, 97, 103, 131, 135, 147, 150, 151, 152, 155, 162, 163, 165, 169, 170, 171, 172, 173, 174, 176, 177, 178, 180, 181, 189, 192, 194, 197, 198, 221, 238, 243, 249, 263, 264, 265, 267, 268, 269, 277, 280, 281, 293, 295, 303, 305, 308, 313, 321, 322, 323, 324, 328
water treatment plant, 174, 308
weld, 174, 175, 188, 189, 276, 283, 286, 287, 288
wildland, 55, 168
wind, 31, 58, 61, 66, 67, 68, 70, 94, 104, 116, 117, 120, 121, 122, 123, 124, 126, 129, 133, 137, 138, 140, 164, 202, 208, 218, 231, 311, 322, 324, 327